Foodborne Pathogens and Food Safety

Foodborne Pathogens and Food Safety

Editors

Antonio Afonso Lourenco
Catherine Burgess
Timothy Ells

MDPI • Basel • Beijing • Wuhan • Barcelona • Belgrade • Manchester • Tokyo • Cluj • Tianjin

Editors

Antonio Afonso Lourenco
Food Biosciences Department
Teagasc — Agriculture and Food Development Authority
Fermoy, co. Cork
Ireland

Catherine Burgess
Food Safety Department
Teagasc — Agriculture and Food Development Authority
Dublin
Ireland

Timothy Ells
Agriculture and Agri-Food Canada
Kentville Research and Development Centre
Kentville, NS
Canada

Editorial Office
MDPI
St. Alban-Anlage 66
4052 Basel, Switzerland

This is a reprint of articles from the Special Issue published online in the open access journal *Foods* (ISSN 2304-8158) (available at: www.mdpi.com/journal/foods/special_issues/foodborne_pathogens_safety).

For citation purposes, cite each article independently as indicated on the article page online and as indicated below:

LastName, A.A.; LastName, B.B.; LastName, C.C. Article Title. *Journal Name* **Year**, *Volume Number*, Page Range.

ISBN 978-3-0365-4432-8 (Hbk)
ISBN 978-3-0365-4431-1 (PDF)

Contents

Preface to "Foodborne Pathogens and Food Safety" . . . . **vii**

Edel Stone, Vincenzo Pennone, Kerri Reilly, Irene R. Grant, Katrina Campbell and Eric Altermann et al.
Inhibition of *Listeria monocytogenes* by Phage Lytic Enzymes Displayed on Tailored Bionanoparticles
Reprinted from: *Foods* **2022**, *11*, 854, doi:10.3390/foods11060854 . . . . **1**

So Young Yang and Ki Sun Yoon
Quantitative Microbial Risk Assessment of *Listeria monocytogenes* and Enterohemorrhagic *Escherichia coli* in Yogurt
Reprinted from: *Foods* **2022**, *11*, 971, doi:10.3390/foods11070971 . . . . **13**

Juan F. De Villena, David A. Vargas, Rossy Bueno López, Daniela R. Chávez-Velado, Diego E. Casas and Reagan L. Jiménez et al.
Bio-Mapping Indicators and Pathogen Loads in a Commercial Broiler Processing Facility Operating with High and Low Antimicrobial Intervention Levels
Reprinted from: *Foods* **2022**, *11*, 775, doi:10.3390/foods11060775 . . . . **31**

Thida Kong-Ngoen, Sirijan Santajit, Witawat Tunyong, Pornpan Pumirat, Nitat Sookrung and Wanpen Chaicumpa et al.
Antimicrobial Resistance and Virulence of Non-Typhoidal *Salmonella* from Retail Foods Marketed in Bangkok, Thailand
Reprinted from: *Foods* **2022**, *11*, 661, doi:10.3390/foods11050661 . . . . **51**

Eva Fuchs, Christina Raab, Katharina Brugger, Monika Ehling-Schulz, Martin Wagner and Beatrix Stessl
Performance Testing of *Bacillus cereus* Chromogenic Agar Media for Improved Detection in Milk and Other Food Samples
Reprinted from: *Foods* **2022**, *11*, 288, doi:10.3390/foods11030288 . . . . **63**

Mayra Aguirre Garcia, Killian Hillion, Jean-Michel Cappelier, Michel Neunlist, Maxime M. Mahe and Nabila Haddad
Intestinal Organoids: New Tools to Comprehend the Virulence of Bacterial Foodborne Pathogens
Reprinted from: *Foods* **2022**, *11*, 108, doi:10.3390/foods11010108 . . . . **79**

Xiaojie Qin, Yanhong Liu and Xianming Shi
Resistance-Nodulation-Cell Division (RND) Transporter AcrD Confers Resistance to Egg White in *Salmonella enterica* Serovar Enteritidis
Reprinted from: *Foods* **2021**, *11*, 90, doi:10.3390/foods11010090 . . . . **97**

Winnie Mukuna, Abdullah Ibn Mafiz, Bharat Pokharel, Aniume Tobenna and Agnes Kilonzo-Nthenge
Antibiotic Resistant *Enterobacteriaceae* in Milk Alternatives
Reprinted from: *Foods* **2021**, *10*, 3070, doi:10.3390/foods10123070 . . . . **109**

Tianmei Sun, Yangtai Liu, Xiaojie Qin, Zafeiro Aspridou, Jiaming Zheng and Xiang Wang et al.
The Prevalence and Epidemiology of *Salmonella* in Retail Raw Poultry Meat in China: A Systematic Review and Meta-Analysis
Reprinted from: *Foods* **2021**, *10*, 2757, doi:10.3390/foods10112757 . . . . **121**

Eleni-Anna Kokkoni, Nikolaos Andritsos, Christina Sakarikou, Sofia Michailidou, Anagnostis Argiriou and Efstathios Giaouris
Investigating Transcriptomic Induction of Resistance and/or Virulence in *Listeria monocytogenes* Cells Surviving Sublethal Antimicrobial Exposure
Reprinted from: *Foods* **2021**, *10*, 2382, doi:10.3390/foods10102382 . . . . **133**

Peter Zangerl, Dagmar Schoder, Frieda Eliskases-Lechner, Abdoulla Zangana, Elisabeth Frohner and Beatrix Stessl et al.
Monitoring by a Sensitive Liquid-Based Sampling Strategy Reveals a Considerable Reduction of *Listeria monocytogenes* in Smeared Cheese Production over 10 Years of Testing in Austria
Reprinted from: *Foods* **2021**, *10*, 1977, doi:10.3390/foods10091977 . . . . **147**

Lianger Dong and Yong Li
Fate of *Salmonella* Typhimurium and *Listeria monocytogenes* on Whole Papaya during Storage and Antimicrobial Efficiency of Aqueous Chlorine Dioxide Generated with HCl, Malic Acid or Lactic Acid on Whole Papaya
Reprinted from: *Foods* **2021**, *10*, 1871, doi:10.3390/foods10081871 . . . . **159**

Olugbenga Ehuwa, Amit K. Jaiswal and Swarna Jaiswal
Salmonella, Food Safety and Food Handling Practices
Reprinted from: *Foods* **2021**, *10*, 907, doi:10.3390/foods10050907 . . . . **175**

Joshua Hadi, Shuyan Wu and Gale Brightwell
Antimicrobial Blue Light versus Pathogenic Bacteria: Mechanism, Application in the Food Industry, Hurdle Technologies and Potential Resistance
Reprinted from: *Foods* **2020**, *9*, 1895, doi:10.3390/foods9121895 . . . . **191**

Paul Culliney and Achim Schmalenberger
Growth Potential of *Listeria monocytogenes* on Refrigerated Spinach and Rocket Leaves in Modified Atmosphere Packaging
Reprinted from: *Foods* **2020**, *9*, 1211, doi:10.3390/foods9091211 . . . . **231**

Preface to "Foodborne Pathogens and Food Safety"

Foodborne pathogens represent a major burden on society as they are the cause of high numbers of illnesses, hospitalizations, and deaths each year. In addition to their detrimental impact on human health, these microorganisms, which include pathogenic bacteria, viruses, fungi, and a range of parasites, also represent a significant economic cost to food companies in the implementation and constant oversight of food hygiene and safety programs, product recalls, and potential litigation if outbreaks occur. Advancing our current knowledge of the food processing chain and its vulnerabilities to the many factors related to foodborne pathogens (e.g., their stress response, survival and persistence in processing environments, acquisition of virulence factors and antimicrobial drug resistance) is paramount to the development of effective strategies for early detection and control of pathogens, thereby improving food safety.

This Special Issue compiled original research articles contributing to a better understanding of the impact of all aspects of foodborne pathogens on food safety.

Antonio Afonso Lourenco, Catherine Burgess, and Timothy Ells
Editors

foods

Article

Inhibition of *Listeria monocytogenes* by Phage Lytic Enzymes Displayed on Tailored Bionanoparticles

Edel Stone [1,2], Vincenzo Pennone [1], Kerri Reilly [3], Irene R. Grant [2], Katrina Campbell [2], Eric Altermann [3,4] and Olivia McAuliffe [1,*]

1 Teagasc Food Research Centre, Moorepark, Fermoy, P61 C996 Cork, Ireland; edelmcstone@gmail.com (E.S.); vincenzopennone@hotmail.it (V.P.)
2 Institute for Global Food Security, School of Biological Sciences, Queens University, 19 Chlorine Gardens, BT9 5DL Belfast, Ireland; i.grant@qub.ac.uk (I.R.G.); katrina.campbell@qub.ac.uk (K.C.)
3 AgResearch Ltd., Palmerston North 4410, New Zealand; kerri.reilly@agresearch.co.nz (K.R.); eric.altermann@agresearch.co.nz (E.A.)
4 Riddet Institute, Massey University, Palmerston North 4442, New Zealand
* Correspondence: olivia.mcauliffe@teagasc.ie; Tel.: +353-(0)25-42609

Abstract: The high mortality rate associated with *Listeria monocytogenes* and its ability to adapt to the harsh conditions employed in food processing has ensured that this pathogen remains a serious problem in the ready-to-eat food sector. Bacteriophage-derived enzymes can be applied as biocontrol agents to target specific foodborne pathogens. We investigated the ability of a listeriophage endolysin and derivatives thereof, fused to polyhydroxyalkanoate bionanoparticles (PHA_BNPs), to lyse and inhibit the growth of *L. monocytogenes*. Turbidity reduction assays confirmed the lysis of *L. monocytogenes* cells at 37 °C upon addition of the tailored BNPs. The application of BNPs also resulted in the growth inhibition of *L. monocytogenes*. BNPs displaying only the amidase domain of the phage endolysin were more effective at inhibiting growth under laboratory conditions (37 °C, 3×10^7 CFU/mL) than BNPs displaying the full-length endolysin (89% vs. 83% inhibition). Under conditions that better represent those found in food processing environments (22 °C, 1×10^3 CFU/mL), BNPs displaying the full-length endolysin demonstrated a greater inhibitory effect compared to BNPs displaying only the amidase domain (61% vs. 54% inhibition). Our results demonstrate proof-of-concept that tailored BNPs displaying recombinant listeriophage enzymes are active inhibitors of *L. monocytogenes*.

Keywords: *Listeria monocytogenes*; bacteriophage; endolysin; amidase; bionanoparticles; BNPs

Citation: Stone, E.; Pennone, V.; Reilly, K.; Grant, I.R.; Campbell, K.; Altermann, E.; McAuliffe, O. Inhibition of *Listeria monocytogenes* by Phage Lytic Enzymes Displayed on Tailored Bionanoparticles. *Foods* **2022**, *11*, 854. https://doi.org/10.3390/foods11060854

Academic Editor: Antonio Bevilacqua

Received: 10 February 2022
Accepted: 15 March 2022
Published: 17 March 2022

1. Introduction

Listeria monocytogenes is a foodborne pathogen that is often associated with ready-to-eat food products such as deli meats, mixed salads, fresh dairy products and leafy greens [1,2]. If consumed in a contaminated food product, the organism can cause listeriosis; this is a rare but serious illness, particularly for at-risk groups including the young, the elderly and the immunocompromised [3]. The high mortality rate (20–30%) associated with the illness has resulted in stringent detection and control measures for *L. monocytogenes* in food processing environments. Despite these controls, the physiological resistance of the organism against low temperatures and high salt concentrations, and its ability to form biofilms, make this pathogen difficult to manage [4].

The use of bacteriophages (phages) as natural biocontrol agents against foodborne pathogens including *L. monocytogenes* has been investigated elsewhere [5,6]. As reported in these and other studies, the application of whole phages has been shown to significantly inhibit the growth of *L. monocytogenes* on different food matrices. Recombinant production of phage proteins, such as endolysins, is a useful alternative to the use of whole phages. Endolysins (lysins) are phage-encoded peptidoglycan hydrolases produced in phage-infected bacterial cells toward the end of the replication cycle [7]. Holins form membrane lesions so that lysins can reach the peptidoglycan and cleave the bacterial membrane, subsequently leading to host cell death and the release of newly formed phages into the environment [8]. Lysins acting against Gram-positive bacteria typically show a modular design, in which catalytic function and specific cell-wall recognition areseparated into two or more functional domains. Simplistically, lysins contain one N-terminal enzymatically active domain (EAD) and one C-terminal cell-wall-binding domain (CBD) [9]. The use of recombinant lysins allows the exploitation of phages that have a lysogenic life cycle and reduces the risk of the emergence of bacteriophage-insensitive mutants [9]. Lysins are also considered to be less host-specific and do not necessarily require actively growing host cells to bring about inhibition [10–12]. Previous work by our group demonstrated the inhibitory effect of the catalytic domain of the *L. monocytogenes* phage vB_LmoS_293 lysin on the formation of *L. monocytogenes* biofilms [13].

Polyhydroxyalkanoate bionanoparticles, or PHA_BNPs, have gained significant interest in a variety of applications in the biotechnology sector as an economically efficient, non-toxic, biodegradable method for the delivery of functional proteins and enzymes [14,15]. Polyhydroxyalkanoates (PHAs) are biopolyesters synthesized by cells in which they function as carbon reservoirs [16]. The enzyme PhaC permits protein fusions to both its C- and N-termini. As a result, the tailored BNPs can display proteins and enzymes on the surface in an orientated fashion without the enzymatic activity of the enzyme being lost [14]. PHA_BNPs offer distinct advantages over other possible expression methods. These include the covalent binding and stabilization of the protein in a uniform direction to the surface of the nanobead. The stabilizing matrix on the nanobeads enables ready deployment of proteins and enzymes in liquids or on surfaces, the expression of proteins in a one-step process, and the resulting high yield of product [17]. Effective uses of these BNPs have previously been demonstrated by Altermann et al. [14] wherein tailored BNPs lysed a range of rumen methanogen strains and reduced methane production by 97%. Similarly, Davies et al. [17] reported that tailored BNPs could act as a successful protective layer in PPE against Mycobacteria after a one log (91%) reduction was reported.

In this study, the hypothesis that PHA_BNPs can be successfully deployed as a potential production and delivery system for *L. monocytogenes*-specific phage-derived endolysins and their catalytic domains was validated. The objectives of this study were (1) to determine if PHA_BNPs displaying either lysin293 or amidase 293 can be produced in *E. coli* and subsequently purified; (2) to determine if assays can be developed to successfully measure the lytic activity of these proteins displayed on PHA_BNPs; (3) to determine if amidase293 will have equal or greater efficacy compared to lysin293 when displayed on PHA_BNPs; (4) to determine the effect of temperature on the activity of the PHA_BNPs; and (5) to determine if the concentration of bacterial cells (CFU/mL) has an effect on the activity of the PHA_BNPs. By meeting each of these objectives, this study would act as a proof-of-concept that these tailored BNPs could be exploited in the future as natural antimicrobials or sanitizing agents. Ultimately, two separate varieties of tailored BNPs were generated: the first variety, PHA_lysin293_BNPs, displayed the full-length lysin, lysin 293, of *L. monocytogenes* phage vB_LmoS_293; the second variety, PHA_amidase293_BNPs, displayed a truncated lysin harboring only the amidase domain of lysin 293, or amidase 293. The efficacy of these lysin-displaying BNPs against *L. monocytogenes* in both turbidity reduction assays and in growth inhibition experiments was tested to determine the potential of tailored BNPs as delivery mechanisms for phage-based biocontrol agents.

2. Materials and Methods

2.1. Bacterial Strains, Plasmids and Culture Conditions

L. monocytogenes strain 473 (serotype 4e) was streaked from −80 °C stocks onto Tryptic Soy Agar (TSA; Becton Dickinson and Company, Le Pont-de-Claix, France) and incubated at 37 °C for 48 h. Actively growing *L. monocytogenes* cells were produced by selecting a single colony from these plates and inoculating this into 10 mL of Tryptic Soy Broth (TSB) and incubating for 18 h at 37 °C. *E. coli* BL21 (DE3) cells (Thermo Fisher Scientific,

Dublin, Ireland) were grown in Lysogeny Broth (LB) liquid media (Neogen, Lancashire, UK) containing 50 μg/mL ampicillin (Amp; Merck Life Science Ltd., Wicklow, Ireland) and 64 μg/mL chloramphenicol (Cm; Merck Life Science Ltd., Wicklow, Ireland) at 37 °C with shaking. Table 1 lists the bacterial cells, plasmids and conditions used in this study.

Table 1. Plasmids used in this study, detailing insert, features, host and products.

Plasmid Name	Insert	Resistance	Host Bacterium	Product	Reference
pET14b-PHA_lysin293_BNPs	Gene fusion of lysin293 and PhaC	Amp^R	*E. coli* BL21 (DE3)	PHA_lysin293_BNPs	This study
pET14b-PHA_amidase293_BNPs	Gene fusion of amidase293 and PhaC	Amp^R	*E. coli* BL21 (DE3)	PHA_amidase293_BNPs	This study
pET14b-PHA_BNPs	PhaC sequence	Amp^R	*E. coli* BL21 (DE3)	PHA_BNPs	This study
pMCS69 (helper plasmid)	N/A	Cm^R	*E. coli* BL21 (DE3)	N/A	[18]

2.2. *Bioinformatic Analysis of Phage vB_LmoS_293*

The genome of phage vB_LmoS_293 has been previously sequenced and annotated, and is available in the GeneBank database with the Accession Number KP399678.1 [19]. The Basic Local Alignment Search Tool (BLAST) was used to analyze Open Reading Frame (ORF) 25 coding for lysin293, and the NCBI Conserved Domains Database [20] was used to identify the amidase domain [13].

2.3. *Plasmid Construction for PHA BNP Generation*

The constructs used in this study were created according to Altermann et al. [14]. The PHA–BNP constructs used in this study were synthesized by GeneArt (Thermo Fisher Scientific, GENEART GmbH, Regensburg, Germany). The gene sequences used in this study can be found in Table S1. Briefly, the gene fusions of lysin293 and PhaC, and amidase293 and PhaC, were optimized for expression in *E. coli*. The synthetic gene was then incorporated into the pET14b vector under the control of the LacZ promoter. pET14b containing the PHA sequence only was also synthesized as a control (Table 1). Following synthesis, the pET14b plasmids were transformed into chemically competent *E. coli* BL21 (DE3) cells (Thermo Fisher Scientific, Dublin, Ireland) that contained the helper plasmid pMCS69, harboring the *phaA* and *phaB* genes required to synthesize PHA precursors [21]. pMCS69 was transformed into chemically competent *E. coli* DE3 cells. Briefly, 100 ng of DNA (pMCS69) was transformed into 50 μL of *E. coli* competent cells and incubated on ice for 30 min. The cells were heat-shocked at 42 °C for 60 s and placed on ice for 3 min. An amount of 500 μL of LB medium was added to the cells and incubated at 37 °C for 40 min with shaking. After incubation, 200 μL of the transformation mix was plated onto LB agar plates containing 50 μg/mL Cm. The plates were incubated at 37 °C overnight. Subsequently, the pET14b plasmids containing the gene fusions of PHA_lysin293, PHA_amidase293 or the PHA sequence only were transformed into competent *E. coli* BL21 (DE3) cells containing the helper plasmid pMCS69, following the method outlined above. Double transformants containing the pET14b plasmids and pMCS69 were plated onto LB agar plates containing 50 μg/mL Amp and 64 μg/mL Cm and incubated overnight at 37 °C.

2.4. *Generation of PHA-BNPs*

PHA_BNPs were produced according to Altermann et al. [14]. Briefly, the transformants of interest were grown in 1 L of LB broth supplemented with 1% (*w*/*v*) glucose and with appropriate antibiotics (Amp (50 μg/mL), and Cm (64 μg/mL)) and at 37 °C with shaking (150 rpm). At an OD600 of 0.5, production of BNPs (PHA_lysin293_BNPs, PHA_amidase293_BNPs and PHA_BNPs) was induced by the addition of 1 mM Isopropyl β-D-1-thiogalactopyranoside (IPTG; Merck Life Science Ltd., Wicklow, Ireland). Following growth at 25 °C with agitation for 48 h, cells were harvested by centrifugation (6000× *g*, 5 min at 4 °C). Cell pellets were resuspended in 50 mM phosphate buffer with a pH of 7.5

and lysed via sonication (Vibracell Sonicator, Sonics and Materials, Newtown, CT, USA) on ice, with 20 s bursts at a medium intensity and 30 s rest intervals over a 10 min time interval. Recovery of BNPs was performed using ultracentrifugation at 21,000× g for 2 h at 4 °C in a Sorvall TH641 swing-out rotor (Thermofisher Scientific, Auckland, New Zealand) over a glycerol gradient, as described in [22]. After ultracentrifugation, the white band containing the PHA_BNPs at the glycerol gradient interface was extracted and brought to a volume of 45 mL using phosphate-buffered saline (PBS) (Life Technologies Ltd., Paisley, UK). The solution was centrifuged at 8000× g for 20 min to separate the purified PHA_BNPs from any remaining glycerol. After centrifugation, the supernatant was discarded and PHA BNP pellets were resuspended in phage buffer (10 mM Tris (pH 7.5), 10 mM MgSO4, 68 mM NaCl) at a concentration of 20 mg/mL with 20 µL/mL Tween 80 (Merck Life Science Ltd., Wicklow, Ireland). The purified PHA_BNPs were stored at −80 °C. When in use, the PHA_BNPs were stored at 4 °C and not continuously frozen and refrozen.

2.5. Lysis and Growth Inhibition Assays

2.5.1. Preparation of Bacterial Culture and Protein

L. monocytogenes strain 473 (serotype 4e) was prepared following 18 h of incubation in TSB (Becton Dickinson and Company, Le Pont-de-Claix, France) at 37 °C under aerobic conditions. The concentrations of each of the PHA_BNPs, PHA_lysin293_BNPs and PHA_amidase293_BNPs, were adjusted to 0.25 mg/mL in PBS (Life Technologies Ltd., Paisley, UK). Protein concentration was confirmed with a Qubit protein quantification assay using the Qubit 4 Fluorometer (Invitrogen, Thermo Fisher, Singapore) following the manufacturer's guidelines. Supplementary Figure S1 depicts the experimental design for the following assays.

2.5.2. Application of PHA_BNPs for Lysis of *L. monocytogenes*

An amount of 100 µL TSB (Becton Dickinson and Company) was inoculated with approximately 1×10^7 CFU/mL of *L. monocytogenes* strain 473, to which 0.25 mg/mL of PHA_lysin293_BNPs, PHA_amidase293_BNPs or control PHA_BNPs was added to give total reaction volumes of 200 µL in a 96-well plate. Samples were incubated at 37 °C, and the turbidity of the samples was measured at 30 min intervals for up to 3 h, by reading the absorbance of samples using a Synergy 2 BioTek 96-well-plate reader (BioTek Instruments, Inc., Winooski, VT, USA) at an OD of 600 nm. Optical densities were corrected according to Altermann et al. [14] using Equation (1).

Equation (1): Where n: sample taken at predefined time point; OD600 (n): corrected optical density at point n; OD600 (n)(a): measured optical density at point n; OD600 (0): measured optical density at time point 0; OD600 (n − 1): measured optical density at point n − 1; bc: test BNPs used; Lmc: *L. monocytogenes* control plus cells; bead: PHA_BNPs or PHA_lysin293_BNPs or PHA_amidase293_BNPs in the absence of *L. monocytogenes* cells.

$$OD_{600(n)} = OD_{600(n)(a)} - \left(OD_{600(0)(bc)} - OD_{600(0)(Lmc)}\right) + \left(OD_{600(n-1)(bead)} - OD_{600(n)(bead)}\right) \quad (1)$$

2.5.3. Application of PHA_BNPs for Growth Inhibition of *L. monocytogenes*

TSB was inoculated with approximately 1×10^7 CFU/mL of *L. monocytogenes* strain 473, and 0.25 mg/mL of either PHA_lysin293_BNPs, PHA_amidase293_BNPs or the control PHA_BNPs was added for a total reaction volume of 200 µL. Samples were incubated at 37 °C and plated at 30 min intervals for up to 3 h on Listeria Chromogenic agar (Harlequin, Lancashire, UK). A total volume of 100 µL was taken and serially diluted, using Maximum Recovery Diluent (Oxoid Ltd., Basingstoke, UK), to a dilution of 10^{-8}.The plates were incubated at 37 °C for 48 h. To assess the inhibitory nature of the beads at a lower starting cell number, TSB was inoculated with approximately 1×10^3 CFU/mL of *L. monocytogenes* strain 473, and 0.25 mg/mL of either PHA_lysin293_BNPs, PHA_amidase293_BNPs or the control PHA_BNPs was added for a total reaction volume of 200 µL. Samples were incubated at 22 °C and plated at 30 min intervals over a 3 h period onto Listeria Chromogenic agar (Neogen, Lancashire, UK). The plates were incubated at 37 °C for 48 h. The percentage inhibition was calculated using CFU/mL data.

2.6. Statistical Analysis

Statistical analysis was performed using Prism Software GraphPad 9. A paired *t*-test was used for comparison between two groups. The data are presented as the standard error of mean (SEM) values. A *p*-value of 0.05 was considered statistically significant. The mean OD600 nm and standard deviations were calculated from two independent experiments with duplicates in each experiment.

3. Results

3.1. PHA_BNPs Displaying Lysin293 and Amidase293 Cause Lysis of *L. monocytogenes*

To determine if the application of PHA_lysin293_BNPs and PHA_amidase293_BNPs result in the lysis of *L. monocytogenes* strain 473 (serotype 4e), turbidity reduction assays were conducted. The controls in these experiments consisted of cells of *L. monocytogenes* strain 473 in the absence of any PHA_BNPs (L. mono-PHA_BNPs) and cells of *L. monocytogenes* strain 473 in the presence of PHA_BNPs displaying no form of lysin (L. mono + PHA_BNPs).

When applied at 37 °C to 1×10^7 CFU/mL (OD 600 nm 0.2) of *L. monocytogenes* strain 473 (Experiment 1A), the addition of PHA_lysin293_BNPs resulted in a reduction in turbidity of 80% ($p = 0.0126$) and 76.71% ($p = 0.0002$) after 30 min, compared to the control without BNPs (L. mono-PHA_BNPs) and with BNPs without lysin (L. mono + PHA_BNPs), respectively (Figure 1). Under the same conditions, the application of PHA_amidase293_BNPs resulted in a reduction in turbidity of 81.5% ($p = 0.0244$) and 76.85% ($p = 0.0012$), compared to the control without BNPs (L. mono-PHA_BNPs) and with BNPs without lysin (L. mono + PHA_BNPs), respectively (Figure 1). In both cases, the reduction in optical density persisted throughout the duration of the assays, and the growth of *L. monocytogenes* strain 473 was inhibited for 3 h.

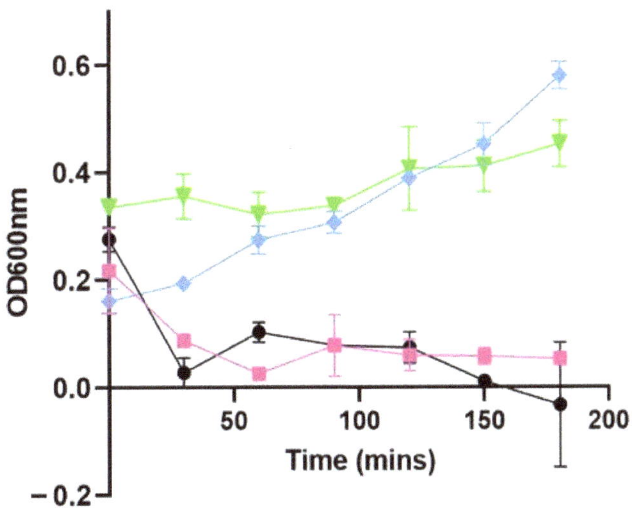

Figure 1. Experiment 1A: turbidity reduction assays performed at 37 °C using 1×10^7 CFU/mL *L. monocytogenes* 473 (serotype 4e). The data have been adjusted according to Equation (1). *L. monocytogenes* strain 473 was inoculated into TSB containing PHA_lysin293_BNPs (pink symbols) ($n = 4$), PHA_amidase293_BNPs (black symbols) ($n = 4$), L. mono + PHA_BNP control (green symbols) ($n = 4$), and L. mono-PHA_BNPs (blue symbols) ($n = 4$). Absorbance at OD 600 nm was measured at 0, 30, 60, 90, 120, 150 and 180 min.

3.2. PHA_BNPs Displaying Lysin293 and Amidase293 Cause Growth Inhibition of *L. monocytogenes*

To investigate the effects of PHA_lysin293_BNPs and of PHA_amidase293_BNPs on the growth of *L. monocytogenes* strain 473, cell counts (CFU/mL) were also determined. Two experiments were designed, one at 37 °C with a high starting inoculum (1×10^7 CFU/mL; Experiment 1B), and one at 22 °C, with a starting inoculum that represents the concentration of *L. monocytogenes* commonly isolated from contaminated plants (1×10^3 CFU/mL) (Experiment 2B) [23]. The controls in this group were similar to those used for the turbidity reduction assays. In experiment 1B (37 °C, 1×10^7 CFU/mL), when compared to the cells-only control, the addition of PHA_lysin293_BNPs and PHA_amidase293_BNPs lowered the population numbers of *L. monocytogenes* by 84.4% ($p = 0.008$) and 89.5% ($p = 0.0006$), respectively, following 3 h of incubation (Figure 2). When compared to the L. mono + PHA_BNP control, the highest inhibition was seen at 3 h for PHA_amidase293_BNPs, which reduced the rate of growth by 75% ($p = 0.0141$) and 2 h for PHA_lysin293_BNPs (83% $p = 0.0046$). This experiment shows that these PHA_BNPs have no killing effect but have a slight inhibitory effect on the growth of *L. monocytogenes*. Compared to the L. mono + PHA_BNP control, the average inhibition over the course of 3 h was 66.5% ($p = 0.0001$) and 61.3% ($p = 0.0002$) when applying the PHA_amidase293_BNPs and PHA_lysin293_BNPs, respectively. When compared to the cells-only control the average inhibition over the course of 3 h was 83.1% ($p = 0.0007$) and 81.5% ($p = 0.0008$) when applying the PHA_amidase293_BNPs and PHA_lysin293_BNPs, respectively. Although there is slight inhibition shown for the duration of this experiment, there is significance shown between the controls and the test.

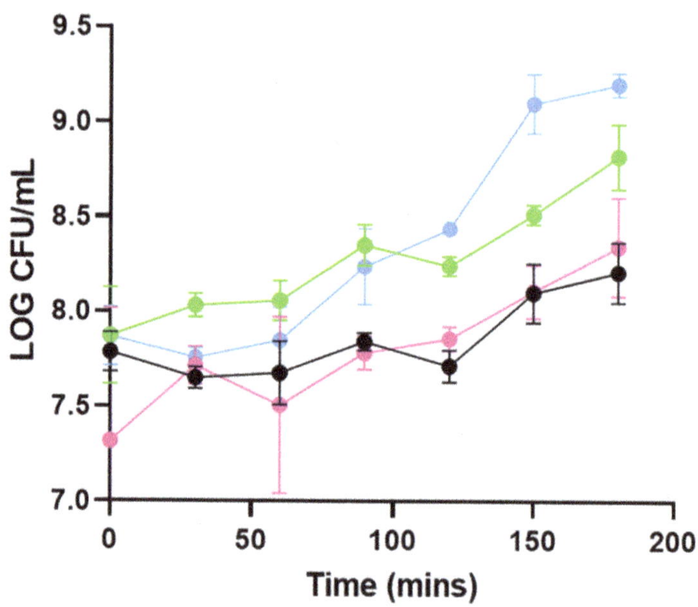

Figure 2. Experiment 1B: growth inhibition assays at 37 °C using 1×10^7 CFU/mL *L. monocytogenes* 473 (serotype 4e). *L. monocytogenes* strain 473 was inoculated into TSB containing PHA_lysin293_BNPs (pink symbols) ($n = 4$), PHA_amidase293_BNPs (black symbols) ($n = 4$), L. mono + PHA_BNP control (green symbols) ($n = 4$), and L. mono-PHA_BNPs (blue symbols) ($n = 4$). Cells were incubated at 37 °C and samples taken for plating on Listeria Chromogenic Agar at 0, 30, 60, 90, 120, 150 and 180 min. The figure depicts total counts of *L. monocytogenes*.

In experiment 2B (22 °C, 1 × 10^3 CFU/mL), the addition of the PHA_lysin293_BNPs and the PHA_amidase293_BNPs resulted in the inhibition of *L. monocytogenes* strain 473 by 61.5% ($p = 0.0246$) and 54.6% ($p = 0.0111$), respectively, compared to the L. mono-PHA_BNP control (Figure 3). The average inhibition exhibited upon addition of PHA_amidase293_BNPs over the 3 h period was 47.5% ($p = 0.0025$), and upon addition of PHA_lysin293_BNPs, was 46.7% ($p = 0.0022$). Like in experiment 1B, there is slight inhibition of *L. monocytogenes*.

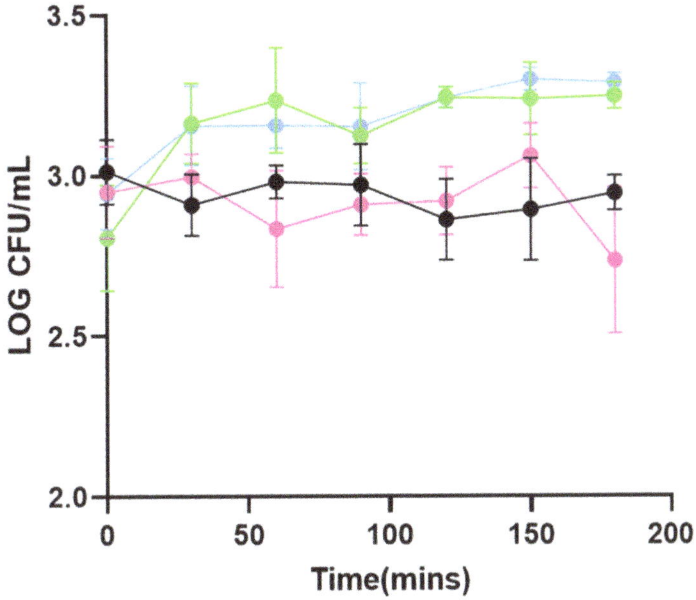

Figure 3. Experiment 2B: growth inhibition assays at 22 °C using 1 × 10^3 CFU/mL *L. monocytogenes* 473 (serotype 4e). *L. monocytogenes* strain 473 was inoculated into TSB containing PHA_lysin293_BNPs (pink symbols) ($n = 4$), PHA_amidase293_BNPs (black symbols) ($n = 4$), L. mono + PHA_BNP control (green symbols) ($n = 4$), and L. mono-PHA_BNPs (blue symbols) ($n = 4$). Cells were incubated at 22 °C and samples taken for plating on Listeria Chromogenic Agar at 0, 30, 60, 90, 120, 150 and 180 min. The figure depicts total counts of *L. monocytogenes*.

4. Discussion

This study investigated the potential for tailored PHA_BNPs (expressing a fusion of lysin293 or the amidase domain of this lysin) to lyse and inhibit the growth of *L. monocytogenes* cells in pure culture. Phage vB_LmoS_293, belonging to the family *Siphoviridae*, was previously isolated by our group from mushroom compost and was found to be specific for *L. monocytogenes* serotypes 4e and 4b [19,24]. An analysis of the genome of phage vB_LmoS_293 revealed that ORF 25 (nucleotide 19966–20916) encoded a 316-amino-acid endolysin (lysin293), belonging to the N-acetylmuramoyl-L-alanine amidase family (COG5632). BLASTp analysis revealed that the protein contained a PGRP element that functions in peptidoglycan recognition in the bacterial cell wall, as well as a catalytic domain (amidase293), belonging to the amidase 2 family (pfam015100) [13]. We have previously demonstrated the lytic capability of amidase293 on autoclaved cells of *L. monocytogenes* and its ability to inhibit the formation of an *L. monocytogenes* biofilm on stainless steel [13].

Both lysin293 and the amidase293 were successfully fused C-terminally to PhaC, which allowed the generation of PHA_BNPs. Two separate varieties of tailored BNPs were-successfully produced in *E. coli* and subsequently purified: the first displayed the lysin293 (PHA_lysin293_BNPs) and the second displayed the amidase293 (PHA_amidase293_BNPs).

A series of assays were developed and optimized to determine the efficacy of these BNPs as lytic agents and/or growth inhibitors of *L. monocytogenes* (Supplementary Figure S2).

The lytic ability of the BNPs were tested against *L. monocytogenes* strain 473, the host strain of phage vB_LmoS_293, in a series of turbidity reduction assays. At 37 °C, the application of both PHA_lysin293_BNPs and PHA_amidase293_BNPs resulted in a reduction in the turbidity of these test solutions. This reduction in turbidity is an indication that the application of these BNPs harboring the phage-derived enzymes results in the lysis of *L. monocytogenes* strain 473 cells. Interestingly, under these experimental conditions, it can be seen that amidase293 maintains the lytic ability of lysin293 when compared to the L. mono + PHA_BNP control. These turbidity-reduction assays also indicate that there is no significant difference between the rate of lysis when using amidase293 versus lysin293. With the lytic ability being maintained, and the rate of lysis not being hindered by truncating the lysin, it suggests that there is a level of substrate specificity in the N-terminal domain. Our group and others have made similar observations previously. $CHAP_K$, the catalytic domain of the LysK endolysin from the *Staphylococcus aureus* phage, phage K, was as active, if not more active, than the full-length LysK [25]. We also reported that the host range of $CHAP_K$ was broader than that of LysK [22]. More recently, Mayer et al. found that a truncated N-acetylmuramoyl-L-alanine amidase of a *Clostridium difficile* endolysin lysed cells of *C. difficile* faster than the full length lysin [23]. However, in this case, no increase in host range was observed with the truncated lysin. A host range comparison of lysin293 and amidase293 is an area that needs be further investigated.

Two experimental variables were altered in a subsequent experiment to better reflect the conditions in which *L. monocytogenes* would be found in the food processing environment. These conditions are a lower temperature (i.e., room temperature) and a lower concentration of cells (CFU/mL) that represents the levels of contamination that would generally be found in food-processing plants. When analyzing the growth kinetics of *L. monocytogenes* strain 473 in experiment 1B, the addition of PHA_lysin293_BNPs reduced the rate of growth of strain 473 by an additional 12.1% in comparison to PHA_amidase293_BNPs, although no significance was observed for this result ($p = 0.986$). This indicates that, under these experimental conditions, amidase293 retains the same lytic ability as lysin293 when displayed on PHA_BNPs. Interestingly, at 22 °C, the application of PHA_lysin293_BNPs and PHA_amidase293_BNPs resulted in the inhibition of *L. monocytogenes* strain 473, maintaining *L. monocytogenes* levels at approximately the same concentration as the starting inoculum over the course of incubation. As experiment 2B better represents the conditions of food-processing plants, it can be suggested that the application of PHA_lysin293_BNPs and PHA_amidase293_BNPs may result in an inhibition of *L. monocytogenes* in food-processing plants.

Although the inhibition of *L. monocytogenes* in experiment 2B is markedly less than in the experiment 1B, there is an immediate decrease in the CFU/mL when the PHA_amidase293_BNPs are added in experiment 2B suggesting that under conditions where there is a lower starting inoculum and a lower temperature, the PHA_amidase293_BNPs not only inhibit the growth of *L. monocytogenes*, but reduce it (reduction in the concentration of the starting inoculum of *L. monocytogenes*) by up to 28.75% (120 min) ($p = 0.002$); however, a marginal (17.5% average) reduction in *L. monocytogenes* is seen throughout the entire 3 h timeline. A hypothesis as to why an inhibitory effect and no reduction are seen in experiment 1B may be due to the PHA_lysin293_BNP: *L. monocytogenes* ratio. In experiment B, the concentration of PHA_amidase/lysin293_BNPs per cell of *L. monocytogenes* is approximately 0.25 µg/mL (0.25 mg/mL/1 $\times$ 10^3 CFU/mL); in experiment 1B, the concentration of PHA_amidase/lysin293_BNPs per cell of *L. monocytogenes* is 0.025 ng/mL (0.25 mg/mL/1 $\times$ 10^7 CFU/mL). Thus, the ratio of PHA_amidase/lysin293_BNPs: cell of *L. monocytogenes* is 10,000 times greater in experiment 2B vs. experiment 1B. To achieve the same PHA_amidase/lysin293_BNP: cell of *L. monocytogenes* ratio in experiment 1B as in experiment 2B, a concentration of 2.5 mg/mL of proteins would be required. Additionally, a achieving the reduction of 17.5% seen in experiment 2B would mathematically mean adding 1.4 mg/mL of protein to achieve a result of 99.9% (3-log reduction). However, preliminary studies performed using varying concentrations of protein revealed that concentrations above 0.25 mg/mL increased the growth of *L. monocytogenes*. Changes in the storage buffer of the tailored PHA_BNPs may allow the use of higher concentrations of tailored PHA_BNPs and, inversely, lead to a greater decrease in *L. monocytogenes*.

Other studies have been conducted using phage lysins linked to nanoparticles for the reduction of *L. monocytogenes*, but using autoclaved cells. Pennone described experiments similar to those outlined is this work, wherein PHA_amidase293_BNPs were applied to *L. monocytogenes* strain 473 that had been subjected to autoclaving (121 °C/15 min) [26]. Turbidity reduction assays showed a reduction of 33.9% and 38% when using 1 mg and 5 mg of PHA_amidase_BNPs, respectively [26]. In another report, Solanki et al. conjugated lysin Ply500 to silica nanoparticles and, when applied to iceberg lettuce, a 4-log reduction in *Listeria innocua* was observed [27].

These lysin PHA_BNPs are natural and decomposable, which is an advantage to chemical-based antimicrobials that may be applied in the food processing environment. The key findings from this research are that PHA_BNPs may act as a suitable delivery system of phage vB_LmoS_293 endolysin and amidase domains, maintaining the enzymes in a stable form and preserving their lytic ability without the use of any chaperone proteins for lysis.

The results show an initial proof-of-concept for the application of these tailored PHA_BNPs in the inhibition of *L. monocytogenes*. Future experiments will determine if the tailored PHA_BNPs can be applied to inhibit *L. monocytogenes* present on surfaces in food-processing plants in an approach similar to that used by Davies et al. 2021 [17], where Mycobacteriophage endolysins fused to biodegradable nanobeads were applied to solid surfaces (filter paper). As the PHA_BNPs are active in liquid suspensions, as indicated in this study, a potential option for their application includes spraying onto food-contact surfaces, as with traditional sanitizers. It is unlikely that these tailored BNPs will replace traditional sanitizers, but may act as an additional hurdle to controlling *L. monocytogenes* where this organism is particularly problematic. It should also be noted that the conditions tested in these sets of experiments are not reflective of the conditions found in food-processing plants. Although this study showed that a reduction in temperature (37 °C to 22 °C) and CFU/mL maintained the activity of these BNPs, future experiments will focus on the application of these BNPs at refrigeration temperatures, given the ability of *L. monocytogenes* to grow at 4 °C. Preliminary findings also indicate that the application of these tailored BNPs is time-limited, as they were shown to be ineffective when applied for more than 3 h. These findings suggest that the tailored BNPs may be ineffective when applied as an antimicrobial for long durations; however, they may be applied for sanitization purposes over shorter periods of time. Future experiments may focus on the optimization of cells to tailored PHA_BNP ratios, to determine if this inhibitory effect can be further increased. The effect of these BNPs on biofilms would also be an area of interest in the future, as Pennone et al. have reported that the amidase domain from this lysin inhibits *L. monocytogenes* biofilm formation on stainless steel surfaces.

5. Conclusions

To summarize, the findings of this study show that when displayed on PHA_BNPs, the amidase domain of lysin293 exhibits the same lytic ability as the full-length lysin293 at both 22 °C and 37 °C. Preliminary results also indicate that the application of these tailored BNPs is time-limited, as they were shown to be ineffective when applied for longer than 3 h.

The results are promising and show an initial proof-of-concept for the use of PHA_BNPs displaying listeriophage lysins as a potential biocontrol agent against *L. monocytogenes*. The production of these bionanoparticles does not entail any complex or expensive post-production processes. In this study, bacterial cells (*E. coli* DE3) produced PHA_BNPs in a one-step process that only requires simple disruption of the bacterial cells to free the

PHA_BNPs. This holds promise for rendering future large-scale production of PHA_BNPs cost-effective. The application of these tailored BNPs was shown to be successful at both 37 °C and 22 °C, and at *L. monocytogenes* concentrations of approximately 1×10^7 CFU/mL and 1×10^3 CFU/mL. An advantage of using this technology over chemical-based sanitizers or chemical-inhibition techniques is that these BNPs are biodegradable and, therefore, could be released in the food processing plant and naturally degraded over time, thus posing no threat to human health. Further studies are required on an extensive strain set, at a larger scale and, ultimately, in food production environments to demonstrate the efficacy of tailored BNPs in food-production environments. The results obtained to date are encouraging, considering the potential future applications in food-processing plants where cross contamination of *L. monocytogenes* poses a major concern.

Supplementary Materials: The following supporting information can be downloaded at: https://www.mdpi.com/article/10.3390/foods11060854/s1, Figure S1: pET-14b vector used in this experiment, Figure S2: Flow chart depicting the experimental design of the assays, Table S1: Gene sequences of lysin293 and the amidase domain, amidase293.

Author Contributions: E.S.: original research and writing; V.P.: original research; K.R.: technical assistance; I.R.G. and K.C.: editing; E.A.: conceptualization and editing; O.M.: conceptualization, resourcing, writing, and editing. All authors have read and agreed to the published version of the manuscript.

Funding: This research was funded by Teagasc (grant number: MDBY0027) and by the Department of Food, Agriculture and the Marine—Food Institutional Research Measure (grant number: 14/F/881). Edel Stone was funded by a Teagasc Walsh Scholarship (grant number: 2016034).

Institutional Review Board Statement: Not applicable.

Informed Consent Statement: Not applicable.

Data Availability Statement: The raw data supporting the conclusions of this article will be made available by the authors, without undue reservation, to any qualified researcher.

Conflicts of Interest: The authors declare no conflict of interest. The funders had no role in the design of the study; in the collection, analyses, or interpretation of data; in the writing of the manuscript; or in the decision to publish the results.

References

1. Zhu, Q.; Gooneratne, R.; Hussain, M.A. *Listeria monocytogenes* in Fresh Produce: Outbreaks, Prevalence and Contamination Levels. *Foods* **2017**, *6*, 21. [CrossRef]
2. Desai, A.N.; Anyoha, A.; Madoff, L.C.; Lassmann, B. Changing epidemiology of *Listeria monocytogenes* outbreaks, sporadic cases, and recalls globally: A review of ProMED reports from 1996 to 2018. *Int. J. Infect. Dis.* **2019**, *84*, 48–53. [CrossRef]
3. de Noordhout, C.M.; Devleesschauwer, B.; Angulo, F.J.; Verbeke, G.; Haagsma, J.; Kirk, M.; Havelaar, A.; Speybroeck, N. The global burden of listeriosis: A systematic review and meta-analysis. *Lancet Infect. Dis.* **2014**, *14*, 1073–1082. [CrossRef]
4. Martins, E.A.; Germano, P.M.L. *Listeria monocytogenes* in ready-to-eat, sliced, cooked ham and salami products, marketed in the city of São Paulo, Brazil: Occurrence, quantification, and serotyping. *Food Control* **2011**, *22*, 297–302. [CrossRef]
5. Oliveira, M.; Viñas, I.; Colàs, P.; Anguera, M.; Usall, J.; Abadias, M. Effectiveness of a bacteriophage in reducing *Listeria monocytogenes* on fresh-cut fruits and fruit juices. *Food Microbiol.* **2014**, *38*, 137–142. [CrossRef]
6. Stone, E.; Lhomet, A.; Neve, H.; Grant, I.; Campbell, K.; McAuliffe, O. Isolation and Characterization of *Listeria monocytogenes* phage vB_LmoH_P61, a phage with biocontrol potential on different food matrices. *Front. Sustain. Food Syst.* **2020**, *4*, 521645. [CrossRef]
7. Oechslin, F.; Menzi, C.; Moreillon, P.; Resch, G. The multi-domain architecture of a bacteriophage endolysin enables intramolecular synergism and regulation of bacterial lysis. *J. Biol. Chem.* **2021**, *296*, 100639. [CrossRef]
8. Borysowski, J.; Weber-Dąbrowska, B.; Górski, A. Bacteriophage endolysins as a novel class of antibacterial agents. *Exp. Biol. Med.* **2006**, *231*, 366–377. [CrossRef]
9. Haddad Kashani, H.; Schmelcher, M.; Sabzalipoor, H.; Seyed Hosseini, E.; Moniri, R. Recombinant endolysins as potential therapeutics against antibiotic-resistant *Staphylococcus aureus*: Current status of research and novel delivery strategies. *Clin. Microbiol. Rev.* **2018**, *31*, e00071-17. [CrossRef]

10. Radford, D.; Guild, B.; Strange, P.; Ahmed, R.; Lim, L.-T.; Balamurugan, S. Characterization of antimicrobial properties of *Salmonella* phage Felix O1 and *Listeria* phage A511 embedded in xanthan coatings on poly(lactic acid) films. *Food Microbiol.* **2017**, *66*, 117–128. [CrossRef]
11. Schmelcher, M.; Loessner, M.J. Bacteriophage endolysins: Applications for food safety. *Curr. Opin. Biotechnol.* **2016**, *37*, 76–87. [CrossRef]
12. Shannon, R.; Radford, D.R.; Balamurugan, S. Impacts of food matrix on bacteriophage and endolysin antimicrobial efficacy and performance. *Crit. Rev. Food Sci. Nutr.* **2020**, *60*, 1631–1640. [CrossRef]
13. Pennone, V.; Sanz-Gaitero, M.; O'Connor, P.; Coffey, A.; Jordan, K.; van Raaij, M.J.; McAuliffe, O. Inhibition of *L. monocytogenes* Biofilm Formation by the Amidase Domain of the Phage vB_LmoS_293 Endolysin. *Viruses* **2019**, *11*, 722. [CrossRef]
14. Altermann, E.; Schofield, L.R.; Ronimus, R.S.; Beatty, A.K.; Reilly, K. Inhibition of Rumen Methanogens by a Novel Archaeal Lytic Enzyme Displayed on Tailored Bionanoparticles. *Front. Microbiol.* **2018**, *9*, 2378. [CrossRef]
15. Pakalapati, H.; Chang, C.K.; Show, P.L.; Arumugasamy, S.K.; Lan, J.C. Development of polyhydroxyalkanoates production from waste feedstocks and applications. *J. Biosci. Bioeng.* **2018**, *126*, 282–292. [CrossRef]
16. Raza, Z.A.; Abid, S.; Banat, I.M. Polyhydroxyalkanoates: Characteristics, production, recent developments and applications. *Int. Biodeterior. Biodegrad.* **2018**, *126*, 45–56. [CrossRef]
17. Davies, C.G.; Reilly, K.; Altermann, E.; Hendrickson, H.L. PLAN-M; Mycobacteriophage Endolysins Fused to Biodegradable Nanobeads Mitigate Mycobacterial Growth in Liquid and on Surfaces. *Front. Microbiol.* **2021**, *12*, 562748. [CrossRef]
18. Amara, A.A.; Rehm, B.H. Replacement of the catalytic nucleophile cysteine-296 by serine in class II polyhydroxyalkanoate synthase from *Pseudomonas aeruginosa*-mediated synthesis of a new polyester: Identification of catalytic residues. *Biochem. J.* **2003**, *374*, 413–421. [CrossRef]
19. Casey, A.; Jordan, K.; Coffey, A.; McAuliffe, O. Complete Genome Sequences of vB_LmoS_188 and vB_LmoS_293, Two Bacteriophages with Specificity for *Listeria monocytogenes* Strains of Serotypes 4b and 4e. *Genome Announc.* **2015**, *3*, e00040-15. [CrossRef]
20. Marchler-Bauer, A.; Panchenko, A.R.; Shoemaker, B.A.; Thiessen, P.A.; Geer, L.Y.; Bryant, S.H. CDD: A database of conserved domain alignments with links to domain three-dimensional structure. *Nucleic Acids Res.* **2002**, *30*, 281–283. [CrossRef]
21. González-Miró, M.; Rodríguez-Noda, L.M.; Fariñas-Medina, M.; Cedré-Marrero, B.; Madariaga-Zarza, S.; Zayas-Vignier, C.; Hernández-Cedeño, M.; Kleffmann, T.; García-Rivera, D.; Vérez-Bencomo, V.; et al. Bioengineered polyester beads co-displaying protein and carbohydrate-based antigens induce protective immunity against bacterial infection. *Sci Rep.* **2018**, *8*, 1888. [CrossRef]
22. Fenton, M.; Keary, R.; McAuliffe, O.; Ross, R.P.; O'Mahony, J.; Coffey, A. Bacteriophage-Derived Peptidase CHAP(K) Eliminates and Prevents Staphylococcal Biofilms. *Int. J. Microbiol.* **2013**, *2013*, 625341. [CrossRef]
23. Mayer, M.J.; Garefalaki, V.; Spoerl, R.; Narbad, A.; Meijers, R. Structure-based modification of a *Clostridium difficile*-targeting endolysin affects activity and host range. *J. Bacteriol.* **2011**, *193*, 5477–5486. [CrossRef]
24. Casey Jordan, K.; Neve, H.; Coffey, A.; McAuliffe, O. A tail of two phages: Genomic and functional analysis of *Listeria monocytogenes* phages vB_LmoS_188 and vB_LmoS_293 reveal the receptor-binding proteins involved in host specificity. *Front. Microbiol.* **2015**, *6*, 1107. [CrossRef]
25. Horgan, M.; O'Flynn, G.; Garry, J.; Cooney, J.; Coffey, A.; Fitzgerald, G.F.; Ross, R.P.; McAuliffe, O. Phage lysin LysK can be truncated to its CHAP domain and retain lytic activity against live antibiotic-resistant staphylococci. *Appl. Environ. Microbiol.* **2009**, *75*, 872–874. [CrossRef]
26. Pennone, V. The Occurrence of Listeria Monocytogenes in the Mushroom Production Chain and the Use of Bacteriophage for Its Control. Ph.D. Thesis, Munster Technological University, Cork, Ireland, March 2019.
27. Solanki, K.; Grover, N.; Downs, P.; Paskaleva, E.E.; Mehta, K.K.; Lee, L.; Schadler, L.S.; Kane, R.S.; Dordick, J.S. Enzyme-based listericidal nanocomposites. *Sci. Rep.* **2013**, *3*, 1584. [CrossRef]

foods

MDPI

Article

Quantitative Microbial Risk Assessment of *Listeria monocytogenes* and Enterohemorrhagic *Escherichia coli* in Yogurt

So Young Yang and Ki Sun Yoon *

Department of Food and Nutrition, College of Human Ecology, Kyung Hee University, 26 Kyungheedae-ro, Dongdaemun-gu, Seoul 02447, Korea; sy2ang@khu.ac.kr
* Correspondence: ksyoon@khu.ac.kr; Tel.: +82-2-961-0264

Abstract: *Listeria monocytogenes* can survive in yogurt stored at a refrigeration temperature. Enterohemorrhagic *Escherichia coli* (EHEC) has a strong acid resistance that can survive in the yogurt with a low pH. We estimated the risk of *L. monocytogenes* and EHEC due to yogurt consumption with @Risk. Predictive survival models for *L. monocytogenes* and EHEC in drinking and regular yogurt were developed at 4, 10, 17, 25, and 36 °C, and the survival of both pathogens in yogurt was predicted during distribution and storage at home. The average initial contamination level in drinking and regular yogurt was calculated to be −3.941 log CFU/g and −3.608 log CFU/g, respectively, and the contamination level of both LM and EHEC decreased in yogurt from the market to home. Mean values of the possibility of illness caused by EHEC were higher (drinking: 1.44×10^{-8}; regular: 5.09×10^{-9}) than *L. monocytogenes* (drinking: 1.91×10^{-15}; regular: 2.87×10^{-16}) in the susceptible population. Both pathogens had a positive correlation with the initial contamination level and consumption. These results show that the foodborne illness risk from *L. monocytogenes* and EHEC due to yogurt consumption is very low. However, controlling the initial contamination level of EHEC during yogurt manufacture should be emphasized.

Keywords: *Listeria monocytogenes*; enterohemorrhagic *Escherichia coli*; yogurt; quantitative microbial risk assessment

Citation: Yang, S.Y.; Yoon, K.S. Quantitative Microbial Risk Assessment of *Listeria monocytogenes* and Enterohemorrhagic *Escherichia coli* in Yogurt. *Foods* **2022**, *11*, 971. https://doi.org/10.3390/foods11070971

Academic Editors: Antonio Afonso Lourenco, Catherine Burgess and Timothy Ells

Received: 7 February 2022
Accepted: 25 March 2022
Published: 27 March 2022

1. Introduction

Yogurt is a dairy product fermented by *Streptococcus thermophilus* and *Lactobacillus bulgaricus* [1]. Yogurt provides probiotics known to be beneficial bacteria that can promote health. Worldwide, the consumption of probiotics and yogurt is increasing every year [2–4].

Pathogenic *Escherichia coli* (*E. coli*) are a group of facultative anaerobes that can cause diseases in healthy individuals with a combination of certain virulence factors, including adhesins, invasins, toxins, and capsules. Pathogenic *E. coli* are classified into six pathotypes based on clinical, epidemiological, and virulence traits: enteropathogenic *E. coli* (EPEC), enteroaggregative *E. coli* (EAEC), diffusely adherent *E. coli* (DAEC), enterotoxigenic *E. coli* (ETEC), enteroinvasive *E. coli* (EIEC) and enterohemorrhagic *E. coli* (EHEC) [5]. EPEC (60.5%) is the primary cause of pathogenic *E. coli* outbreaks in Korea, followed by ETEC (31.2%), EHEC (6.8%), and EIEC (1.5%) [6]. Among them, EHEC can cause diarrhea with a mechanism of attaching-effacing (A/E) lesions with only a low infectious dose (1–100 CFU) [7]. EHEC has strong acid resistance that can make it viable in food with a low pH [8]. Morgan et al. [9] reported 16 cases of *E. coli* O157:H7 Phage Type 49 due to the consumption of a locally produced yogurt occurring in the northwest of England in 1991. In a study by Cutrim et al. [10], *E. coli* O157:H7 was shown to survive for 10 days in both traditional inoculated yogurt and pre-hydrolyzed inoculated yogurt, whereas its survival increased to 22 days in lactose-free yogurt. The populations of *E. coli* O157:H7 decreased by only about 1.4 log CFU/g after 28 days in Greek-style yogurt [11].

Listeria monocytogenes (LM) are facultatively anaerobic opportunistic pathogens that can grow between 0 and 45 °C; optimal growth occurs at 30~37 °C [12]. It can grow at pH 4–9.6 [13]. Listeriosis is caused by LM, which can cross the intestinal barrier and spread to lymph and blood to reach target organs such as the liver and spleen. Moreover, LM can be fatal to immunocompromised individuals, newborns, older adults, and pregnant women since LM can penetrate the blood–brain barrier or the fetoplacental barrier [14,15]. The approximate infective dose of LM is estimated to be 10 to 100 million CFU in healthy hosts and only 0.1 to 10 million CFU in individuals at high risk of infection [16]. In the US, a significant number of LM outbreaks are caused by raw milk, unpasteurized milk, cheeses, and ice cream [17]. Improper management of pasteurization temperature or technical imperfections can lead to the contamination of dairy products [18].

Risk assessment can estimate the probability of occurrence and severity of adverse effects in humans exposed to foodborne hazards [19]. Quantitative microbial risk assessment (QMRA) provides numerical estimates of risk exposure to identify which factors affect the exposure [20]. QMRA consists of hazard identification, hazard characterization, exposure assessment, and risk characterization [19]. Hazard identification is the step that identifies the presence of microorganisms or microbial toxins in a particular food based on the scientific literature. In the hazard characterization step, it is possible to perform qualitative and quantitative assessments of the adverse effects of consuming food contaminated by microorganisms [21]. Exposure assessment is the process that characterizes the level of hazard exposed to the population [22]. The final step of QMRA is risk characterization that provides the possibility of illness/person/day of pathogens when consuming contaminated food [21]. A risk assessment study of *Staphylococcus aureus* in milk and homemade yogurt was reported in Ethiopia [23]. Results showed the importance of traditional food preparation methods, such as fermentation, in risk mitigation; yogurt, traditional milk fermentation, reduced the risk by 93.7%. QMRA of LM and enterohemorrhagic *E. coli* in yogurt has not been reported yet. Therefore, the objective of this study was to conduct a microbial risk assessment for *L. monocytogenes* and enterohemorrhagic *E. coli* to compare their risks in drinking and regular yogurt.

2. Materials and Methods

2.1. Prevalence and Initial Contamination Level in an Offline Market

To derive prevalence (PR) data of LM and EHEC in yogurt by season and location, results of yogurt monitoring (195 drinking yogurts and 90 regular yogurts) were used [24]. LM and EHEC were identified with methods as described in the Korean Food Code [25]. The distribution of PR was fitted using Beta distribution (α, β), with α meaning "number of positive samples+1" and β meaning "number of total samples-number of positive samples +1" [26]. Initial contamination levels of LM and EHEC were estimated using the equation [Log (-ln(1-PR)/weight)] of Sanaa et al. [27].

2.2. Physicochemical and Microbiological Analyses of Yogurt

Ten products of two types of yogurt (drinking and regular) were purchased from an offline market. The pH, water activity (Aw), total aerobic bacteria, coliform, and *E. coli* were measured. Briefly, 10 g of sample was aseptically placed in a stomacher bag with 90 mL of distilled water and homogenized with a stomacher (Interscience, Paris, France). The pH was measured with a pH meter (Orion™ Star A211, ThermoFisher Scientific Co., Waltham, MA, USA). The Aw of each sample (15 g) was measured in triplicate using a water activity meter (Rotronic HP23-AW-A, Rotronic AG, Bassersdorf, Switzerland). To measure total aerobic bacteria (AC), coliform, and *E. coli* (EC), 25 g of sample was homogenized with 225 mL of 0.1% sterile peptone water (BD, Sparks, MD, USA) and serially diluted 10-fold with 0.1% peptone water. After inoculating 1 mL aliquot of each dilution onto two or more sheets of 3M Petrifilm *E. coli*/Coliform Count Plate (3M corporation, St. Paul, MN, USA), AC and EC plates were incubated at 36 ± 1 °C for 48 h and 24 h, respectively.

2.3. Strain Preparation

An LM strain isolated from the gloves of a slaughterhouse worker [28] was stored in tryptic soy broth (TSB, MB cell, Seoul, Korea) containing 0.6% yeast extract with 20% glycerol (Duksan, South Korea) at −80 °C. After thawing at ambient temperature, 10 μL of LM inoculum was added into 10 mL of TSB containing 0.6% yeast extract and then cultured at 36 ± 1 °C for 24 h in a 140 rpm rotary shaker (VS-8480, Vision Scientific, Daejeon, Korea).

E. coli (EHEC) strains (NCCP 13720, 13721) including *E. coli* O157:H7 (NCTC 12079) were obtained from the Ministry of Food and Drug Safety (MFDS) in Korea. After thawing frozen strains that were stored at −80 °C, they were cultured in the same way as described above. All strains were centrifuged at 4000 rpm for 10 min (VS-550, Vision Scientific, Daejeon, Korea) and the supernatant was removed. Pellets were harvested by centrifugation (4000 rpm for 10 min), washed with 10 mL of 0.1% peptone water, and resuspended with 0.1% peptone water to a final concentration of approximately 9.0 log CFU/mL.

2.4. Sample Preparation and Inoculation

For model development, the popularity of yogurt samples and results of physicochemical (high pH value) and microbiological analyses of yogurt were considered. Drinking and regular yogurt were purchased from an offline market (Seoul, Korea) and aseptically divided into 30 mL and 10 g, respectively, into 50 mL conical tubes (SPL Life Science Co., Daejeon, Seoul). LM and the cocktail of *E. coli* strains were independently inoculated into drinking (4~5 log CFU/g) and regular yogurts (5~6 log CFU/g). Each sample was then stored at 4, 10, 17, 25, and 36 °C until no colonies were detected for up to 21 days. At a specific time, each yogurt sample was homogenized with sterilized 0.1% peptone water for 120 s using a stomacher. Then 1 mL of the aliquot of the homogenate was serially diluted ten-fold with 0.1% peptone water and spread onto PALCAM agar (Oxoid, Basingstoke, Hampshire, UK) for LM and EMB agar (Oxoid, Basingstoke, Hampshire, UK) for EHEC, which were incubated at 36 ± 1 °C for 24 h to analyze the change in pathogen populations.

2.5. Development of Primary and Secondary Model

The Weibull model [29] (Equation (1)) and GinaFit V1.7 program [30] were used to develop the primary survival model of yogurt as a function of temperature. Delta value (time for the first decimal reduction) and *p*-value (shape of graph) were then calculated.

$$\text{Weibull equation}: \ \text{Log(N)} = \text{Log}(N_0) - \left(\frac{t}{delta}\right)^p \quad (1)$$

N_0: log initial number of cells
t: time
delta: time for the first decimal reduction
p: shape ($p > 1$: concave downward curve, $p < 1$: concave upward curve, $p = 1$: log-linear)

From results obtained through the primary predictive model, the secondary model was developed by applying the third-order polynomial model (Equation (2)) to delta values of both LM and EHEC as a function of temperature.

$$\text{Third} - \text{order polynomial model}: \ \text{Y} = b_0 + b_1 \times T + b_2 \times T^2 + b_3 \times T^3 \quad (2)$$

Y: delta (d)
b_0, b_1, b_2, b_3: constant
T: temperature

2.6. Validation

To verify the applicability of the predictive model of LM, the delta value was obtained with temperatures not used for model development in this study, which was 7 °C for drinking yogurt and 13 °C for regular yogurt (interpolation). The predictive model of EHEC was verified with enteropathogenic (EPEC) strain (extrapolation), which was detected in a dairy farm [24]. The root mean square error (RMSE; Equation (3)) [31] was used as a measure of applicability:

$$RMSE = \sqrt{\frac{1}{n} \times \sum (observed\ value - predicted\ value)^2} \quad (3)$$

n: the total number of experimental values (values obtained from independent variables) or predicted values (values obtained from the developed survival model).

2.7. Development of Scenario from Market to Home

The exposure assessment scenario for the risk assessment of yogurt was divided into three stages: "market storage", "transportation to home", and "home storage".

The storage temperature of yogurt in the market was investigated for an offline market, which was used as an input variable into an Excel (Microsoft@ Excel 2019, Microsoft Corp., USA) spreadsheet. PERT distribution was confirmed as the most suitable probability distribution model using @RISK 7.5 (Palisade Corp., Ithaca, NY, USA). The minimum, mode, and maximum values of storage temperature were 2.1, 7, and 9.7 °C, respectively. Storage time was also input based on the shelf-life of yogurt. The PERT distribution was confirmed as the most suitable model using @RISK 7.5 (Palisade Corp., Ithaca, NY. USA). The minimum, mode, and maximum values of storage time were 0, 240, and 312 h for drinking yogurt and 0, 240, and 480 h for regular yogurt, respectively.

At the stage of transporting from market to home, the pert distribution was applied to transportation time and temperature according to Jung [32]. Values of minimum (0.325 h, 10 °C), mode (0.984 h, 18 °C), and maximum (1.643 h, 25 °C) time and temperature were applied.

According to data from the MFDS [33], 69.2% of respondents answered that the most frequent storage period for milk was 2–3 days at the refrigeration temperature and the maximum storage period was 30 days or more. As a result, RiskPert (0, 60, 720 h) distribution was input in the scenario and a RiskLogLogistic (−10.407, 13.616, 8.611) distribution was used as the storage temperature [34].

2.8. Estimation of Consumption Data of Yogurt

The appropriate probability distribution model for consumption amount and intake rate of yogurt was confirmed using data from "Estimation of amount and frequency of consumption of 50 domestic livestock and processed livestock products" from the MFDS [35].

2.9. Hazard Characterization

For hazard characterization, the exponential model was used for the dose–response model of LM [36] (Equation (4)) and the Beta-Poisson model [37] was used for the dose–response model of EHEC (Equation (5)):

$$p = 1 - \exp(\mathrm{r} \times \mathrm{N}) \quad (4)$$

P: the probability of foodborne illness for the intake of LM

r: the probability that one cell can cause disease (susceptible population: 1.06×10^{-12}, general population: 2.37×10^{-14})

N: the number of cells exposed to the consumption of LM

$$\mathrm{P} = 1 - \left(1 + \frac{\mathrm{N}}{\beta}\right)^{-\alpha} \quad (5)$$

P: the probability of foodborne illness for the intake of EHEC

N: the consumption dose of EHEC

α: constant (0.49)

β: constant (1.81×10^5)

2.10. Risk Characterization

To estimate the probability of foodborne illness per person per day for the intake of drinking and regular yogurt contaminated by LM or EHEC, formulas and inputs of exposure scenarios were written in an Excel spreadsheet. The risk was then calculated through a Monte Carlo simulation of @RISK. Median Latin hypercube sampling was used for sampling type, and a random method was used for generator seed. Finally, the correlation coefficient was calculated based on sensitivity analysis results to analyze factors affecting the probability of occurrence of foodborne illness.

2.11. Statistical Analysis

All experiments were repeated at least three times. All statistical analyses were performed using SAS version 9.4 (SAS Institute Inc., Cary, NC, USA). To describe significant variations of delta values between LM and EHEC at the same temperature, a t-test was used. Differences were considered significant at $p < 0.05$.

3. Results and Discussion

3.1. Prevalence and Intial Contamination Level in an On- an Offline Market

As a first step in the exposure assessment, initial contamination levels for LM and EHEC were analyzed for drinking yogurt (n = 195) and regular yogurt (n = 90) purchased from on and offline markets in Korea. LM and EHEC were not detected in any samples [24]. The average contamination level was calculated using the equation [Log (−ln(1−PR)/weight)] by Sanaa et al. [27]. The average initial contamination level of both LM and EHEC was −3.941 log CFU/g in the drinking yogurt and −3.608 log CFU/g in the regular yogurt (Figure 1).

Drinking yogurt

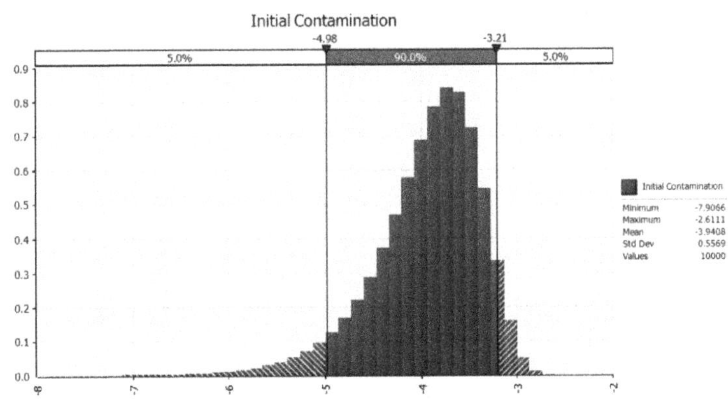

(A)

Figure 1. *Cont.*

Regular yogurt

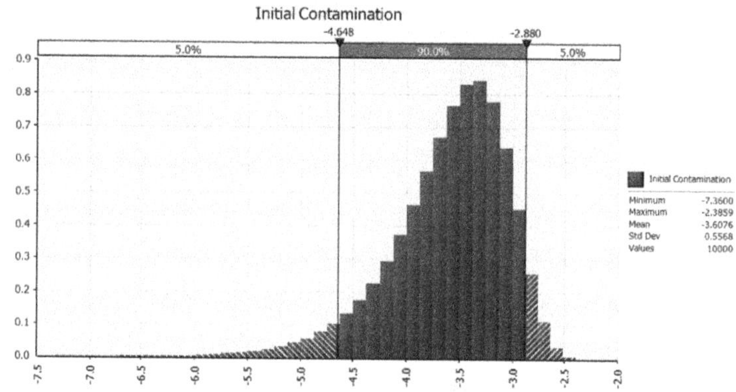

(B)

Figure 1. The probability distribution of initial contamination level of *Listeria monocytogenes* and EHEC in drinking (**A**) and regular yogurt (**B**).

3.2. Development of Primary and Secondary Predictive Model

The primary models of LM and EHEC in yogurt are shown in Figure 2. Secondary predictive models of delta values for LM and EHEC and equations are shown in Figure 3. Delta values of LM at 4, 10, 17, 25, and 36 °C were 20.31, 7.16, 2.15, 1.81, and 0.62 days in drinking yogurt and 9.04, 4.76, 1.89, 0.66, and 0.14 days in regular yogurt, respectively. Delta values of EHEC at 4, 10, 17, 25, and 36 °C were 67.61, 38.31, 13.42, 5.51, and 1.42 days in drinking yogurt and 14.93, 10.41, 8.21, 2.23, and 0.42 days in regular yogurt, respectively (Table 1). The delta value corresponds to the time for the first decimal reduction of the surviving populations of LM and EHEC. Overall, the higher the temperature, the lower the delta value, indicating that survival of LM and EHEC is better in yogurt stored at refrigeration temperature. Lactic acid bacteria (LAB) activity in yogurt increases as the temperature increases. Thus, the viability of LM and EHEC can be decreased. LAB can produce large amounts of organic acids and lower the pH value [38]. Some LAB can also produce bacteriocins and bacteriocin-like compounds to inhibit pathogens [39]. The temperature can affect the growth of LAB, and LAB isolated from Calabrian cheeses can inhibit the growth of LM in soft cheese [40]. LAB has the highest specific growth rate at 42–44 °C, the optimum growth temperature for LAB [41]. LAB starters can reduce the survival ability of EHEC in kimchi [42]. Bachrouri et al. [43] have reported that the viability of *E. coli* O157:H7 decreased as the temperature increased and *E. coli* O157:H7 is more resistant to death than nonpathogenic *E. coli* at 4 and 8 °C. The survival ability of LM is drastically decreased at 15 °C, but not significantly changed at 3~12 °C [44].

This work also noticed that LM and EHEC died faster in regular yogurt than in drinking yogurt due to the lower pH of regular yogurt (4.14 ± 0.02) than drinking yogurt (4.60 ± 0.02). This result is consistent with the study of Millet et al. [45], showing that low pH can decrease the growth of LM in raw-milk cheese. Guraya et al. [46] have also suggested that the viability of EHEC is drastically decreased in yogurt with pH below 4.1. Additionally, drinking yogurt has higher water activity (0.961 ± 0.001) than regular yogurt (0.943 ± 0.002) in this work. The Aw is the availability of the water in the product for microbes, and the higher the Aw, the better microorganism can survive. At 10 °C, the highest survival ability of EHEC was observed in drinking yogurt, followed by EHEC in regular yogurt, LM in drinking yogurt, and LM in regular yogurt (Figure 2). Overall, EHEC survived better than LM at especially low temperatures, regardless of the kind of yogurt in this work (Figure 3).

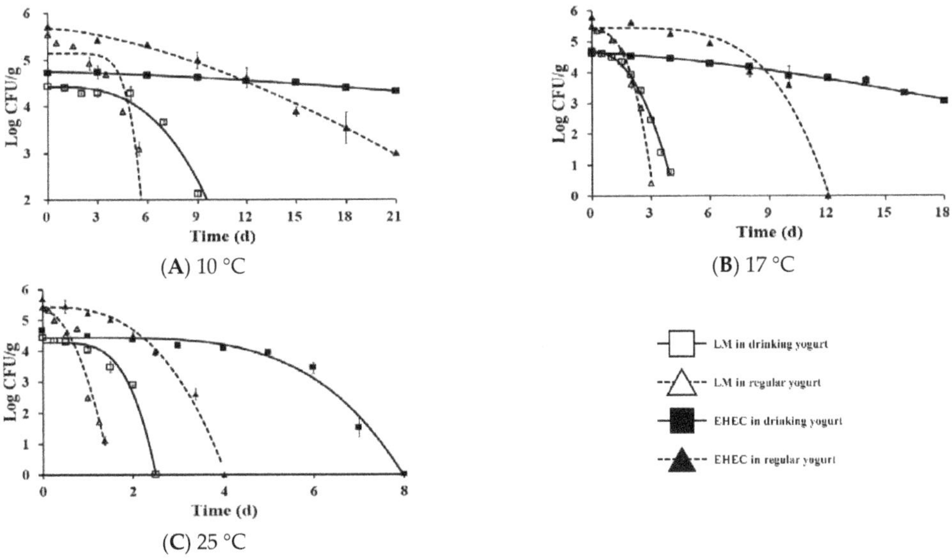

Figure 2. Primary survival models of *Listeria monocytogenes* (LM) and EHEC in yogurt as a function of temperature. LM in drinking yogurt: □, LM in regular yogurt: △, EHEC in drinking yogurt: ■, EHEC in regular yogurt: ▲.

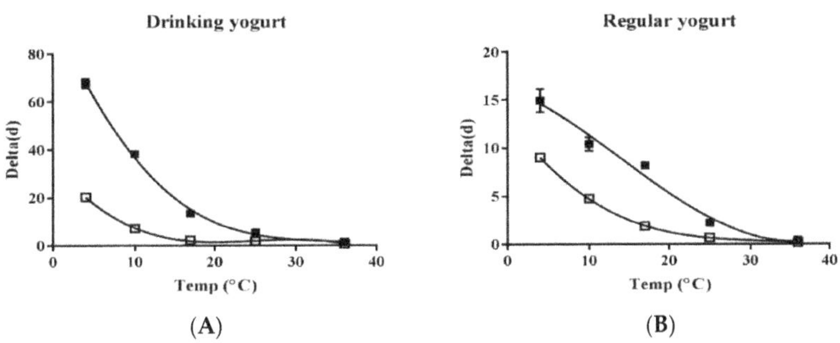

Figure 3. Secondary models for delta values of *Listeria monocytogenes* (□); and EHEC (■) in drinking (**A**) and regular yogurt (**B**).

Table 1. Survival kinetic parameters of *Listeria monocytogenes* (LM) and EHEC in yogurt [1].

Temperature (°C)	Pathogens	Drinking		Regular	
		Delta (Day) [2]	p [3]	Delta (Day)	p
4	LM [4]	20.31 ± 0.20 *	0.73	9.04 ± 0.13 *	1.07 ± 0.07
	EHEC [5]	67.61 ± 1.92 *	1.25 ± 0.01	14.93 ± 1.20 *	1.12 ± 0.06
10	LM	7.16 *	3.1 ± 0.08	4.76 ± 0.08 *	6.88 ± 0.44
	EHEC	38.31 ± 0.37 *	1.45 ± 0.01	10.41 ± 0.71 *	1.45 ± 0.12
17	LM	2.15 ± 0.01 *	2.27 ± 0.03	1.89 ± 0.06 *	3.32 ± 0.08
	EHEC	13.42 *	1.35 ± 0.08	8.21 ± 0.11 *	4.17 ± 0.16
25	LM	1.81 ± 0.03 *	4.43 ± 0.21	0.66 *	1.98 ± 0.02
	EHEC	5.51 ± 0.12 *	4.09 ± 0.21	2.23 ± 0.01 *	2.84 ± 0.03
36	LM	0.62 *	2.83 ± 0.03	0.14 *	1.17 ± 0.06
	EHEC	1.42 *	3.90 ± 0.05	0.42 ± 0.04 *	2.57 ± 0.33

[1] Values are expressed as mean ± SD (n = 3). [2] Delta: Time for 1 log reduction. [3] p: Shape of graph. [4] LM: *Listeria monocytogenes*. [5] EHEC: Enterohemorrhagic *Escherichia coli*. * Significant difference of delta values was observed between LM and EHEC at the same temperature by t-test at $p < 0.05$.

3.3. Validation

RMSE value is one of the parameters that can estimate the accuracy of the predictive model, and it was used to calculate the suitability of the model. The predictive model can be considered perfect if RMSE values are close to zero [47]. According to the study of model development using the Weibull model in heat-stressed *E. coli* O157:H7 and *L. monocytogenes* in kefir, RMSE values ranged from 0.13 to 0.52 in *E. coli* O157:H7 and 0.06 to 0.82 in *L. monocytogenes* [48]. The RMSE value calculated from the estimated data of LM was 0.185 in drinking yogurt and 0.115 in regular yogurt for interpolation. The RMSE value of EPEC was 1.079 in drinking yogurt and 1.001 in regular yogurt for extrapolation. As a result, the developed models in this study were judged to be appropriate to predict the survival of LM, EHEC, and EPEC in drinking and regular yogurt.

3.4. Change in Contamination Level of Listeria Monocytogenes and EHEC from Market to Home

The average contamination level of LM decreased −4.396 log CFU/g in drinking yogurt and −7.965 log CFU/g in regular yogurt at the market. The average contamination level of drinking yogurt during transportation from market to home was slightly decreased to −4.396 log CFU/g, and there was no change in regular yogurt. It was further decreased −5.00 log CFU/g for drinking yogurt and −10.25 log CFU/g for regular yogurt during storage at home before consumption.

The initial contamination level of EHEC was the same as that of LM. The contamination level of EHEC was −3.957 log CFU/g in drinking yogurt and −4.244 log CFU/g in regular yogurt at the market, which was maintained when yogurt was transported from market to home. The contamination level decreased −3.969 log CFU/g in drinking yogurt and −4.71 log CFU/g in regular yogurt before consumption at home. The contamination level of both LM and EHEC decreased in yogurt from the market to home because both pathogens cannot grow in yogurt, regardless of the type of yogurt. In this work, a more rapid decrease of contamination level of LM was observed than EHEC in regular yogurt.

Hu et al. [49] observed that organic acid produced from *Lactobacillus plantarum* isolated from traditional dairy products (kumis, milk thistle, yogurt) exhibits antimicrobial activity against pathogenic bacteria. They found that different proportions of organic acid (primarily lactic and acetic acid) show different antimicrobial activity against pathogenic bacteria. The difference in the proportion of organic acid between drinking and regular yogurt may affect the behavior of pathogens in yogurt.

3.5. Consumption Data of Yogurt

The consumption amount and intake rate of yogurt are shown in Figure 4. As a result of fitting the distribution with @Risk, the RiskLaplace model was found to be the most suitable. Daily average consumption amounts of drinking yogurt and regular yogurt were 140 g and 97.046 g, respectively. Intake rates for drinking yogurt and regular yogurt were calculated to be 0.184 and 0.146, respectively. It could be concluded that the consumption of drinking yogurt was higher than that of regular yogurt.

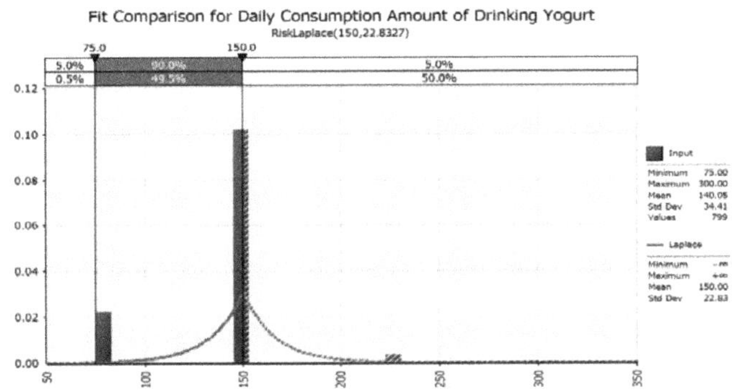

(**A**) Drinking yogurt

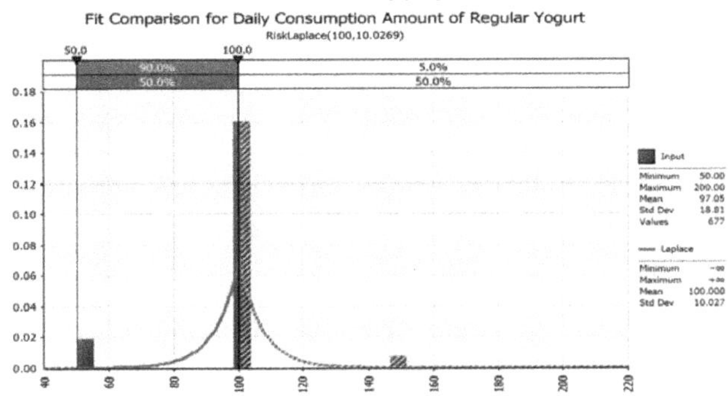

(**B**) Regular yogurt

Figure 4. Probabilistic distribution for daily consumption amount of yogurt with @Risk.

3.6. Hazard Characterization and Risk Characterization

Final risks of LM and EHEC in yogurt were analyzed by separating susceptible population and general population using contamination level, consumption data, and dose–response model derived according to the scenario of the market to home (Tables 2 and 3). As a result, no risk was estimated for the general group due to LM. However, the probability risk of foodborne illness due to LM was 1.91×10^{-15} in drinking yogurt and 2.87×10^{-16} in regular yogurt for susceptible populations per day. It is concluded that the risk of listeriosis is very low with yogurt consumption. The risk assessment result on LM in milk [36] demonstrates that the risk of milk consumption is also low (5.0×10^{-9} cases per serving).

Table 2. Simulation model and formulas in the Excel spreadsheet used to calculate the risk of *Listeria monocytogenes* (LM) in drinking and regular yogurt with @RISK.

Symbol	Unit	Definition	Formula	Reference
		Product		
PR		Prevalence of LM in drinking yogurt	=RiskBeta(1, 196)	MFDS [24]
		Prevalence of LM in regular yogurt	=RiskBeta(1, 91)	
CL	CFU/g	Contamination level of LM	=−LN(1 − PR)/25	Sanna et al. [27]
IC	log CFU/g	Initial contamination level	=Log(CL)	

Table 2. *Cont.*

Symbol	Unit	Definition	Formula	Reference
		Market		
M_{Time}	h	Storage time in market of drinking yogurt	=RiskPert(0, 240, 312)	MFDS [24]
		Storage time in market of regular yogurt	=RiskPert(0, 240, 480)	
M_{Temp}	°C	Storage temperature in market	=RiskPert(2.1, 7, 9.7)	
		Death		
Delta	h	Drinking yogurt	$=823.8 + (-100.8) \times M_{Temp} + 4.177 \times {M_{Temp}}^2 + (-0.0556) \times {M_{Temp}}^3$	This research
		Regular yogurt	$=315.1 + (-27.57) \times M_{Temp} + 0.8396 \times {M_{Temp}}^2 + (-0.0087) \times {M_{Temp}}^3$	
p		Drinking yogurt	=2.67 (Fixed)	
		Regular yogurt	=2.882 (Fixed)	
LM survival model	log CFU/g	C1	$=IC - (M_{Time}/delta)^p$	
		Transportation to home		
T_{Time}	h	Storage time during transportation	=RiskPert(0.325, 0.984, 1.643)	Jung [32]
T_{Temp}	°C	Storage temperature during transportation	=RiskPert(10, 18, 25)	
		Death		
Delta	h	Drinking yogurt	$=823.8 + (-100.8) \times T_{Temp} + 4.177 \times {T_{Temp}}^2 + (-0.0556) \times {T_{Temp}}^3$	This research
		Regular yogurt	$=315.1 + (-27.57) \times T_{Temp} + 0.8396 \times {T_{Temp}}^2 + (-0.0087) \times {T_{Temp}}^3$	
p		Drinking yogurt	=2.67 (Fixed)	
		Regular yogurt	=2.882 (Fixed)	
LM survival model	log CFU/g	C2	$=C1-(T_{Time}/delta)^p$	
		Home		
H_{Time}	h	Storage time until consumption	=RiskPert(0, 60, 720)	MFDS [33]
H_{Temp}	°C	Storage temperature until consumption	=RiskLogLogistic(−10.407, 13.616, 8.611)	Bahk [34]
		Death		
Delta	h	Drinking yogurt	$=823.8 + (-100.8) \times H_{Temp} + 4.177 \times {H_{Temp}}^2 + (-0.0556) \times {H_{Temp}}^3$	This research
		Regular yogurt	$=315.1 + (-27.57) \times H_{Temp} + 0.8396 \times {H_{Temp}}^2 + (-0.0087) \times {H_{Temp}}^3$	
p		Drinking yogurt	=2.67 (Fixed)	
		Regular yogurt	=2.882 (Fixed)	
LM survival model	log CFU/g	C3	$=C2 - (H_{Time}/delta)^p$	

Table 2. *Cont.*

Symbol	Unit	Definition	Formula	Reference
		Consumption		
Consume (Daily consumption average amount)		Drinking yogurt Regular yogurt	=RiskLaplace(150, 22.833) =RiskLaplace(100, 10.027)	Park et al. [35]
Intake rate(Distribution for consumption frequency)		Drinking yogurt Regular yogurt	=0.184(Fixed) =0.146(Fixed)	
Amount		Daily consumption average amount considered frequency	=Consume × Intake rate	
		Dose-Response model		
Dose(D)		LM amount	$=10^{C3}$ × Amount	
1-EXP(-r × D)		Parameter of r	$=1.06 \times 10^{-12}$ (Susceptible population) $=2.37 \times 10^{-14}$ (General population)	FDA/WHO [36]
		Risk Characterization		
Risk		Probability of illness/person/day	$=1 - \exp(-r \times D)$	FDA/WHO [36]

Table 3. Simulation model and formulas in the Excel spreadsheet used to calculate the risk of EHEC in drinking and regular yogurt with @RISK.

Symbol	Unit	Definition	Formula	Reference
		Product		
PR		Prevalence of EHEC in drinking yogurt Prevalence of EHEC in regular yogurt	=RiskBeta(1, 196) =RiskBeta(1, 91)	MFDS [24]
CL	CFU/g	Contamination level of EHEC	=−LN(1 − PR)/25	Sanna et al. [27]
IC	log CFU/g	Initial contamination level	=Log(CL)	
		Market		
M_{Time}	h	Storage time in market of drinking yogurt Storage time in market of regular yogurt	=RiskPert(0, 240, 312) =RiskPert(0, 240, 480)	MFDS [24]
M_{Temp}	°C	Storage temperature in market	=RiskPert(2.1, 7, 9.7)	
		Death		
Delta	h	Drinking yogurt Regular yogurt	$=2347 + (-201.9) \times M_{Temp} + 6.044 \times {M_{Temp}}^2 + (-0.0616) \times {M_{Temp}}^3$ $=391.7 + (-8.478) \times M_{Temp} + (-0.4534) \times {M_{Temp}}^2 + (-0.0109) \times {M_{Temp}}^3$	This research
p		Drinking yogurt Regular yogurt	=2.406 (Fixed) =2.429 (Fixed)	
EHEC survival model	log CFU/g	C1	$=IC - (M_{Time}/delta)^p$	
		Transportation to home		
T_{Time}	h	Storage time during transportation	=RiskPert(0.325, 0.984, 1.643)	Jung [32]
T_{Temp}	°C	Storage temperature during transportation	=RiskPert(10, 18, 25)	

Table 3. *Cont.*

Symbol	Unit	Definition	Formula	Reference
		Death		
Delta	h	Drinking yogurt	$=2347 + (-201.9) \times T_{Temp} + 6.044 \times T_{Temp}^2 + (-0.0616) \times T_{Temp}^3$	This research
		Regular yogurt	$=391.7 + (-8.478) \times T_{Temp} + (-0.4534) \times T_{Temp}^2 + (-0.0109) \times T_{Temp}^3$	
p		Drinking yogurt Regular yogurt	=2.406 (Fixed) =2.429 (Fixed)	
EHEC survival model	log CFU/g	C2	$=C1 - (T_{Time}/\text{delta})^p$	
		Home		
H_{Time}	h	Storage time until consumption	=RiskPert(0, 60, 720)	MFDS [33]
H_{Temp}	°C	Storage temperature until consumption	=RiskLogLogistic(−10.407, 13.616, 8.611)	Bahk [34]
		Death		
Delta	h	Drinking yogurt	$=2347 + (-201.9) \times H_{Temp} + 6.044 \times H_{Temp}^2 + (-0.0616) \times H_{Temp}^3$	This research
		Regular yogurt	$=391.7 + (-8.478) \times H_{Temp} + (-0.4534) \times H_{Temp}^2 + (-0.0109) \times H_{Temp}^3$	
p		Drinking yogurt Regular yogurt	=2.406 (Fixed) =2.429 (Fixed)	
EHEC survival model	log CFU/g	C3	$=C2 - (H_{Time}/\text{delta})^p$	
		Consumption		
Consume (Daily consumption average amount)		Drinking yogurt Regular yogurt	=RiskLaplace(150, 22.833) =RiskLaplace(100, 10.027)	Park et al. [35]
Intake rate(Distribution for consumption frequency)		Drinking yogurt Regular yogurt	=0.184(Fixed) =0.146(Fixed)	
Amount		Daily consumption average amount considered frequency	=Consume × Intake rate	
		Dose-Response model		
Dose(D)		EHEC amount	$=10^{C3} \times$ Amount	Park et al. [37]
Model		Parameter of α Parameter of β	=0.49 $=1.81 \times 10^5$	
		Risk characterization		
Risk		Probability of illness/person/day	$=1 - (1 + D/\beta)^{-\alpha}$	Park et al. [37]

By contrast, this was calculated to be 1.44×10^{-8} in drinking yogurt and 5.09×10^{-9} in regular yogurt with EHEC (Table 4). The risk of foodborne illness from both pathogens was higher from drinking yogurt due to its higher survival ability than regular yogurt. Additionally, the highest risk was found for EHEC in drinking yogurt due to the highest survival ability of EHEC in drinking yogurt (Figure 2), in which the highest delta value was noticed. As a result, the risk of EHEC is higher than LM in yogurt. Yogurt has an inhibition effect on pathogenic microorganisms due to organic acids such as lactic acid and acetic acid, which were produced by LAB [50], low pH below 4.1 [46], and bacteriocin or bacteriocin-like substances produced by LAB [51]. Yang et al. [51] isolated and identified bacteriocinogenic LAB from various cheeses and yogurts. They found that 20% of isolates (28 isolates) out of 138 LAB isolates had antimicrobial effects on all microorganisms tested, except for *E. coli*. In the present study, we found that EHEC shows better survival ability than LM in both types of yogurts. A similar trend was reported by Gulmez and Guven [52], who compared the inhibitory effects of LM, *E. coli* O157:H7, and *Yersinia enterocolitica* in yogurt and kefir samples during 24 h fermentation time and 10 days of storage. They found that *E. coli* O157:H7 showed the highest resistance during the yogurt's fermentation and storage time. The most recent study showed [53] that most of the bacteriocins produced by LAB isolates are active against Gram-positive bacteria, such as LM and *Staphylococcus aureus*, whereas Gram-negative bacteria, *E. coli*, and *Salmonella* Typhimurium, displayed considerable resistance.

Table 4. Probability of illness per day per person by *Listeria monocytogenes* (LM) and EHEC with consumption of yogurt with @Risk scenario.

			Probability of Illness/Person/Day				
Pathogens	**Sample**		**Min**	**25%**	**Mean**	**95%**	**Max**
LM	Drinking	Susceptible population	0	0	1.91×10^{-15}	8.44×10^{-15}	3.65×10^{-14}
		General population	0	0	0	0	0
	Regular	Susceptible population	0	0	2.87×10^{-16}	2.11×10^{-15}	3.63×10^{-14}
		General population	0	0	0	0	0
EHEC	Drinking		0	4.01×10^{-9}	1.44×10^{-8}	4.33×10^{-8}	1.75×10^{-7}
	Regular		0	4.39×10^{-10}	5.09×10^{-9}	2.12×10^{-8}	9.45×10^{-8}

3.7. Sensitivity Analysis

Sensitivity analysis was conducted to identify input variables with a major influence on results. If the result has a negative value, it has a negative correlation. As the input value increases, the output value decreases. If it is 0, there is no correlation. A positive value indicates a positive correlation, meaning that the output value increases as the input value increases [54]. Results of analysis of regression coefficients for the probability risk of foodborne illness caused by LM and EHEC due to yogurt consumption are shown in Figure 5. Both pathogens had a negative correlation with storage time at the market. The risk of foodborne illness decreased with increased storage time at the market. Both pathogens had the greatest positive correlation with the initial contamination level and consumption. As a result, it is considered that initial hygiene management before manufacture can reduce the risk of LM and EHEC. LM can survive longer in yogurt when LM is contaminated with higher concentrations during yogurt manufacture [55]. Kasımoğlu and Akgün [56] found that yogurt contaminated at 10^2 CFU/g level of *E. coli* O157:H7 has a lower elimination time than that contaminated at 10^6 CFU/g level. They suggested that the decline time of *E. coli* O157:H7 contaminated in the pre-fermentation stage could be affected by the initial contamination level. Therefore, initial hygiene management is important to inhibit the contamination and reduce the risk of pathogens in yogurt.

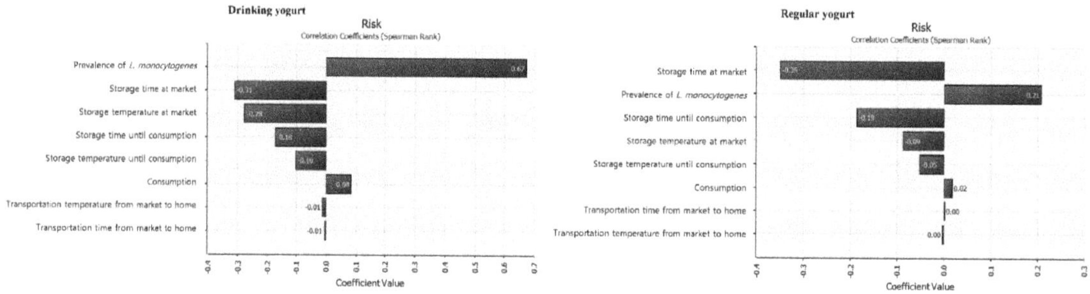

(**A**) *Listeria monocytogenens*

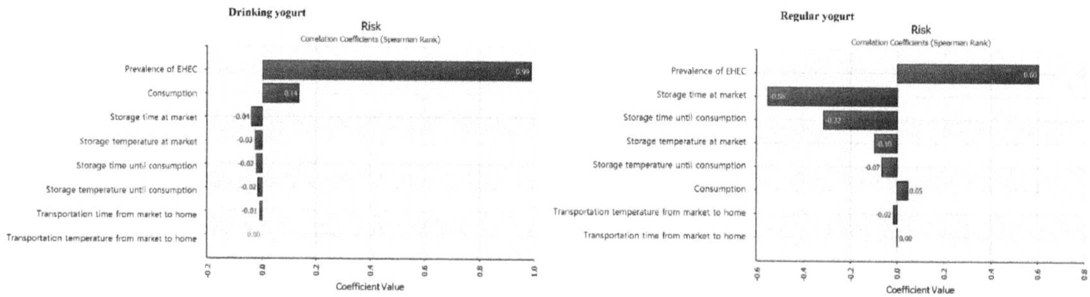

(**B**) EHEC

Figure 5. The correlation coefficient for sensitivity analysis affecting illness by *Listeria monocytogenes* (**A**) and EHEC (**B**) with consumption of yogurt with @Risk.

4. Conclusions

Results showed that the risk of serious illness from LM and EHEC due to drinking and regular yogurt consumption is very low. Yogurt does not permit the growth of LM and EHEC during storage at 4, 10, 17, 25, and 36 °C. The contamination level of both LM and EHEC decreased in yogurt from the market to home, and LM and EHEC died faster in regular yogurt than in drinking yogurt. However, controlling the initial contamination level of EHEC during yogurt manufacture should be emphasized because its survival ability in yogurt is higher in both drinking and regular yogurt than LM.

Author Contributions: Investigation, S.Y.Y.; methodology, S.Y.Y. and K.S.Y.; project administration, K.S.Y.; writing—original draft, S.Y.Y.; writing—review and editing, K.S.Y. All authors have read and agreed to the published version of the manuscript.

Funding: This research was funded by the Ministry of Food and Drug Safety in Korea, grant number 20162MFDS027.

Institutional Review Board Statement: Not applicable.

Informed Consent Statement: Not applicable.

Data Availability Statement: We did not report any additional data for this study.

Conflicts of Interest: The authors declare no conflict of interest.

References

1. Hamann, W.T.; Marth, E.H. Survival of Streptococcus thermophilus and *Lactobacillus bulgaricus* in commercial and experimental yogurts. *J. Food Prot.* **1984**, *47*, 781–786. [CrossRef] [PubMed]
2. Korea Health Supplements Association (KHSA). Consumer and Market Research Report. 2020. Available online: https://www.khsa.or.kr/user/info/InfoBoardUserView.do?_menuNo=374&boardSeqno=10039&postsSeqno=114886. (accessed on 18 August 2021).
3. Korea Dairy Committee. Distribution Consumption Statistics (2010–2020). 2021. Available online: https://www.dairy.or.kr/kor/sub05/menu_01_5_1.php (accessed on 18 August 2021).
4. Nyanzi, R.; Jooste, P.J.; Buys, E.M. Invited review: Probiotic yogurt quality criteria, regulatory framework, clinical evidence, and analytical aspects. *J. Dairy Sci.* **2021**, *104*, 1–19. [CrossRef] [PubMed]
5. Kaper, J.B.; Nataro, J.P.; Mobley, H.L. Pathogenic *Escherichia coli*. *Nat. Rev. Microbiol* **2004**, *2*, 123–140. [CrossRef] [PubMed]
6. Korea Disease Control and Prevention Agency (KDCA). The Prevalence of Pathogenic *Escherichia coli* Isolated by the Enteric Pathogens Active Surveillance Network (Enter-Net), 2010–2019. 2020. Available online: https://kdca.go.kr/board/board.es?mid=a20602010000&bid=0034&act=view&list_no=368538 (accessed on 23 August 2021).
7. Paton, J.C.; Paton, A.W. Pathogenesis and diagnosis of Shiga toxin-producing *Escherichia coli* infections. *Clin. Microbiol. Rev.* **1988**, *11*, 450–479. [CrossRef]
8. Kim, G.H.; Breidt, F.; Fratamico, P.; Oh, D.H. Acid resistance and molecular characterization of *Escherichia coli* O157: H7 and different non-O157 Shiga toxin-producing *E. coli* serogroups. *J. Food Sci.* **2015**, *80*, M2257–M2264. [CrossRef]
9. Morgan, D.; Newman, C.P.; Hutchinson, D.N.; Walker, A.M.; Rowe, B.; Majid, F. Verotoxin producing *Escherichia coli* O 157 infections associated with the consumption of yoghurt. *Epidemiol. Infect.* **1993**, *111*, 181–188. [CrossRef]
10. Cutrim, C.S.; de Barros, R.F.; da Costa, M.P.; Franco, R.M.; Conte-Junior, C.A.; Cortez, M.A.S. Survival of *Escherichia coli* O157: H7 during manufacture and storage of traditional and low lactose yogurt. *LWT* **2016**, *70*, 178–184. [CrossRef]
11. Moineau-Jean, A.; Guévremont, E.; Champagne, C.P.; Roy, D.; Raymond, Y.; LaPointe, G. Fate of *Escherichia coli* and *Kluyveromyces marxianus* contaminants during storage of Greek-style yogurt produced by centrifugation or ultrafiltration. *Int. Dairy J.* **2017**, *72*, 36–43. [CrossRef]
12. Farber, J.M.; Peterkin, P. Listeria monocytogenes, a food-borne pathogen. *Microbiol. Rev.* **1991**, *55*, 476–511. [CrossRef]
13. Lado, B.; Yousef, A.E. Characteristics of *Listeria monocytogenes* Important to Food Processors. In *Listeria, Listeriosis and Food Safety*, 3rd ed.; Ryser, E.T., Marth, E.H., Eds.; CRC Press: Boca Raton, FL, USA, 2007.
14. Radoshevich, L.; Cossart, P. Listeria monocytogenes: Towards a complete picture of its physiology and pathogenesis. *Nat. Rev. Microbiol.* **2018**, *16*, 32–46. [CrossRef]
15. Vázquez-Boland, J.A.; Kuhn, M.; Berche, P.; Chakraborty, T.; Domınguez-Bernal, G.; Goebel, W.; Gonzaılez-Zorn, B.; Wehland, J.; Kreft, J. Listeria pathogenesis and molecular virulence determinants. *Clin. Microbiol. Rev.* **2001**, *14*, 584–640. [CrossRef] [PubMed]
16. Farver, J.M.; Ross, W.H.; Harwig, J. Health risk assessment of listeria monocytogenes in Canada. *Int. J. Food Microbiol.* **1996**, *30*, 145–156. [CrossRef]
17. Food and Drug Administration (FDA). Listeria (Listeriosis). 2019. Available online: https://www.fda.gov/food/foodborne-pathogens/listeria-listeriosis (accessed on 28 July 2021).
18. Ulusoy, B.H.; Chirkena, K. Two perspectives of *Listeria monocytogenes* hazards in dairy products: The prevalence and the antibiotic resistance. *Food Qual. Saf.* **2019**, *3*, 233–241. [CrossRef]
19. Lammerding, A.M. An overview of microbial food safety risk assessment. *J. Food Prot.* **1997**, *60*, 1420–1425. [CrossRef]
20. World Health Organization (WHO). Microbiological Risk Assessment: Guidance for Food. 2021. Available online: https://www.who.int/publications/i/item/9789240024892 (accessed on 15 August 2021).
21. Codex. Principles and Guidelines for the Conduct of Microbiological Risk Assessment. 1999. Available online: http://www.fao.org/fao-who-codexalimentarius/sh-proxy/en/?lnk=1&url=https%253A%252F%252Fworkspace.fao.org%252Fsites%252Fcodex%252FStandards%252FCXG%2B30-1999%252FCXG_030e_2014.pdf (accessed on 14 July 2021).
22. FAO/WHO. Food Safety Risk Analysis: A Guide for National Food Safety Authorities. 2006. Available online: http://apps.who.int/iris/bitstream/handle/10665/43718/9789251056042_eng.pdf;jsessionid=689C4D54F8646357776C37326910DD2A?sequence=1 (accessed on 14 August 2021).
23. Makita, K.; Desissa, F.; Teklu, A.; Zewde, G.; Grace, D. Risk assessment of staphylococcal poisoning due to consumption of informally-marketed milk and home-made yoghurt in Debre Zeit, Ethiopia. *Int. J. Food Microbiol.* **2012**, *153*, 135–141. [CrossRef]
24. Ministry of Food and Drug Safety (MFDS). *Study on the Evaluation and Reduction of Risk of Listeria monocytogenes, Pathogenic* Escherichia coli *and Campylobacter spp. in Dairy Products*; Ministry of Food and Drug Safety (MFDS): Cheongju-si, Korea, 2021.
25. Ministry of Food and Drug Safety (MFDS). Food Code. Dairy Products. 2020. Available online: https://www.foodsafetykorea.go.kr/foodcode/01_03.jsp?idx=39 (accessed on 28 June 2021).
26. Vose, D.J. The application of quantitative risk assessment to microbial food safety. *J. Food Prot.* **1998**, *61*, 640–648. [CrossRef]
27. Sanaa, M.; Coroller, L.; Cerf, O. Risk assessment of listeriosis linked to the consumption of two soft cheeses made from raw milk: Camembert of Normandy and Brie of Meaux. *Risk Anal.* **2004**, *24*, 389–399. [CrossRef]
28. Ministry of Food and Drug Safety (MFDS). Risk Assessment and reduction of Listeria monocytogenes in Agriculture, Livestock and Fishery Products. 2019. Available online: https://scienceon.kisti.re.kr/srch/selectPORSrchReport.do?cn=TRKO202000029896 (accessed on 4 August 2021).

29. Mafart, P.; Couvert, O.; Gaillard, S.; Leguérinel, I. On calculating sterility in thermal preservation methods: Application of the Weibull frequency distribution model. *Int. J. Food Microbiol.* **2002**, *72*, 107–113. [CrossRef]
30. Geeraerd, A.H.; Valdramidis, V.P.; Van Impe, J.F. GInaFiT, a freeware tool to assess non-log-linear microbial survivor curves. *Int. J. Food Microbiol.* **2005**, *102*, 95–105. [CrossRef]
31. Mataragas, M.; Drosinos, E.H.; Vaidanis, A.; Metaxopoulos, I. Development of a predictive model for spoilage of cooked cured meat products and its validation under constant and dynamic temperature storage conditions. *J. Food Sci.* **2006**, *71*, M157–M167. [CrossRef]
32. Jung, H. Consumer Survey and Hazard Analysis for the Improvement of Food Hygiene and Safety in Purchase. Master's Thesis, Korea University, Seoul, Korea, 2011.
33. Ministry of Food and Drug Safety (MFDS). Food Hygiene in the Home for Consumer Awareness Survey. 2009. Available online: https://scienceon.kisti.re.kr/commons/util/originalView.do?cn=TRKO201000014868&dbt=TRKO&rn= (accessed on 30 August 2021).
34. Bahk, G.J. Statistical probability analysis of storage temperatures of domestic refrigerator as a risk factor of foodborne illness outbreak. *Korean J. Food Sci. Technol.* **2010**, *42*, 373–376.
35. Park, J.H.; Cho, J.L.; Joo, I.S.; Heo, J.J.; Yoon, K.S. Estimation of Amount and Frequency of Consumption of 50 Domestic Livestock and Processed Livestock Products. *J. Korean Soc. Food Sci Nutr.* **2016**, *45*, 1177–1191. [CrossRef]
36. Food and Agriculture Organization of the United Nations and the World Health Organization (FAO/WHO). Risk Assessment of Listeria Monocytogenes in Ready-to-Eat Foods: Technical Report. Microbiological Risk Assessment Series 5. 2004. Available online: https://apps.who.int/iris/handle/10665/42875 (accessed on 11 August 2021).
37. Park, M.S.; Cho, J.L.; Lee, S.H.; Bahk, G.J. A study on dose-response models for foodborne disease pathogens. *J. Food Hyg. Saf.* **2014**, *29*, 299–304. [CrossRef]
38. Cheigh, H.S.; Park, K.Y. Biochemical, microbiological and nutritional aspects of kimchi (Korean fermented vegetable products). *Crit. Rev. Food Sci. Nutr.* **1994**, *34*, 175–203. [CrossRef] [PubMed]
39. Zhao, S.; Han, J.; Bie, X.; Lu, Z.; Zhang, C.; Lv, F. Purification and characterization of plantaricin JLA-9: A novel bacteriocin against *Bacillus* spp. produced by *Lactobacillus plantarum* JLA-9 from Suan-Tsai, a traditional Chinese fermented cabbage. *J. Agric. Food Chem.* **2016**, *64*, 2754–2764. [CrossRef] [PubMed]
40. Panebianco, F.; Giarratana, F.; Caridi, A.; Sidari, R.; De Bruno, A.; Giuffrida, A. Lactic acid bacteria isolated from traditional Italian dairy products: Activity against Listeria monocytogenes and modelling of microbial competition in soft cheese. *LWT* **2021**, *137*, 110446. [CrossRef]
41. Adamberg, K.; Kask, S.; Laht, T.M.; Paalme, T. The effect of temperature and pH on the growth of lactic acid bacteria: A pH-auxostat study. *Int. J. Food Microbiol.* **2003**, *85*, 171–183. [CrossRef]
42. Choi, S.J.; Yang, S.Y.; Yoon, K.S. *Lactic acid* bacteria starter in combination with sodium chloride controls pathogenic *Escherichia coli* (EPEC, ETEC, and EHEC) in kimchi. *Food Microbiol.* **2021**, *100*, 103868. [CrossRef]
43. Bachrouri, M.; Quinto, E.J.; Mora, M.T. Survival of *Escherichia coli* O157: H7 during storage of yogurt at different temperatures. *J. Food Sci.* **2002**, *67*, 1899–1903. [CrossRef]
44. Szczawinski, J.; Szczawinska, M.E.; Lobacz, A.; Jackowska-Tracz, A. Modeling the effect of temperature on survival rate of *Listeria monocytogenes* in yogurt. *Pol. J. Vet. Sci.* **2016**, 19. [CrossRef]
45. Millet, L.; Saubusse, M.; Didienne, R.; Tessier, L.; Montel, M.C. Control of *Listeria monocytogenes* in raw-milk cheeses. *Int. J. Food Microbiol.* **2006**, *108*, 105–114. [CrossRef] [PubMed]
46. Guraya, R.; Frank, J.F.; Hassan, A.N. Effectiveness of salt, pH, and diacetyl as inhibitors for *Escherichia coli* O157: H7 in dairy foods stored at refrigeration temperatures. *J. Food Prot.* **1998**, *61*, 1098–1102. [CrossRef] [PubMed]
47. Faradonbeh, R.S.; Armaghani, D.J.; Monjezi, M. Development of a new model for predicting flyrock distance in quarry blasting: A genetic programming technique. *Bull. Eng. Geol. Environ.* **2016**, *75*, 993–1006. [CrossRef]
48. Coşansu, S. Survival kinetics of heat-stressed *Escherichia coli* O157: H7 and *Listeria monocytogenes* cells as post-fermentation contaminants in kefir during refrigerated storage. *LWT* **2018**, *98*, 635–641. [CrossRef]
49. Hu, C.H.; Ren, L.Q.; Zhou, Y.; Ye, B.C. Characterization of antimicrobial activity of three *Lactobacillus plantarum* strains isolated from Chinese traditional dairy food. *Food Sci. Nutr.* **2019**, *7*, 1997–2005. [CrossRef]
50. Ogawa, M.; Shimizu, K.; Nomoto, K.; Tanaka, R.; Hamabata, T.; Yamasaki, S.; Takeda, T.; Takeda, Y. Inhibition of in vitro growth of Shiga toxin-producing *Escherichia coli* O157: H7 by probiotic *Lactobacillus* strains due to production of lactic acid. *Int. J. Food Microbiol.* **2001**, *68*, 135–140. [CrossRef]
51. Yang, E.; Fan, L.; Jiang, Y.; Doucette, C.; Fillmore, S. Antimicrobial activity of bacteriocin-producing lactic acid bacteria isolated from cheeses and yogurts. *AMP Express.* **2012**, *2*, 48. Available online: https://amb-express.springeropen.com/articles/10.1186/2191-0855-2-48 (accessed on 28 July 2021). [CrossRef]
52. Gulmez, M.; Guven, A. Survival of *Escherichia coli* O157: H7, *Listeria monocytogenes* 4b and Yersinia enterocolitica O_3 in different yogurt and kefir combinations as prefermentation contaminant. *J. Appl. Microbiol.* **2003**, *95*, 631–636. [CrossRef]
53. Afrin, S.; Hoque, M.A.; Sarker, A.K.; Satter, M.A.; Bhuiyan, M.N.I. Characterization and profiling of bacteriocin-like substances produced by lactic acid bacteria from cheese samples. *Access Microbiol.* **2021**, *3*, 000234. [CrossRef]
54. Vose, D.J. *Understanding and Using the Results of a Risk Analysis. Risk Analysis: A Quantitative Guide*; John Wiley Sons: Hoboken, NJ, USA, 2008.

55. Akkaya, L.; Telli, R.; Sagdic, O. Growth-death kinetics of *Listeria monocytogenes* in strained yogurt. *Int. J. Food Prop.* **2009**, *12*, 705–712. [CrossRef]
56. Kasımoğlu, A.; Akgün, S. Survival of *Escherichia coli* O157: H7 in the processing and post-processing stages of acidophilus yogurt. *Int. J. Food Sci. Technol.* **2004**, *39*, 563–568. [CrossRef]

foods MDPI

Article

Bio-Mapping Indicators and Pathogen Loads in a Commercial Broiler Processing Facility Operating with High and Low Antimicrobial Intervention Levels

Juan F. De Villena, David A. Vargas, Rossy Bueno López, Daniela R. Chávez-Velado, Diego E. Casas, Reagan L. Jiménez and Marcos X. Sanchez-Plata *

International Center for Food Industry Excellence, Department of Animal and Food Sciences, Texas Tech University, Lubbock, TX 79409, USA; juan.devillena@ttu.edu (J.F.D.V.); andres.vargas@ttu.edu (D.A.V.); rossy.bueno@ttu.edu (R.B.L.); daniela.r.chavez@ttu.edu (D.R.C.-V.); diego.casas@ttu.edu (D.E.C.); reagan.brashears@ttu.edu (R.L.J.)

* Correspondence: marcos.x.sanchez@ttu.edu; Tel.: +1-806-834-6503

Abstract: The poultry industry in the United States has traditionally implemented non-chemical and chemical interventions against *Salmonella* spp. and *Campylobacter* spp. on the basis of experience and word-of-mouth information shared among poultry processors. The effects of individual interventions have been assessed with microbiological testing methods for *Salmonella* spp. and *Campylobacter* spp. prevalence as well as quantification of indicator organisms, such as aerobic plate counts (APC), to demonstrate efficacy. The current study evaluated the loads of both indicators and pathogens in a commercial chicken processing facility, comparing the "normal chemical", with all chemical interventions turned-on, at typical chemical concentrations set by the processing plant versus low-chemical process ("reduced chemical"), where all interventions were turned off or reduced to the minimum concentrations considered in the facility's HACCP system. Enumeration and prevalence of *Salmonella* spp. and *Campylobacter* spp. as well as indicator organisms (APC and Enterobacteriaceae—EB) enumeration were evaluated to compare both treatments throughout a 25-month sampling period. Ten locations were selected in the current bio-mapping study, including live receiving, rehanger, post eviscerator, post cropper, post neck breaker, post IOBW #1, post IOBW #2, prechilling, post chilling, and parts (wings). Statistical process control parameters for each location and processing schemes were developed for each pathogen and indicator evaluated. Despite demonstrating significant statistical differences between the normal and naked processes in *Salmonella* spp. counts ("normal" significantly lower counts than the "reduced" at each location except for post-eviscerator and post-cropper locations), the prevalence of *Salmonella* spp. after chilling is comparable on both treatments (~10%), whereas for *Campylobacter* spp. counts, only at the parts' location was there significant statistical difference between the "normal chemical" and the "reduced chemical". Therefore, not all chemical intervention locations show an overall impact on *Salmonella* spp. or *Campylobacter* spp., and certain interventions can be turned off to achieve the same or better microbial performance if strategic intervention locations are enhanced.

Keywords: poultry bio-mapping; chemical interventions; *Salmonella* enumeration; *Campylobacter* enumeration

Citation: De Villena, J.F.; Vargas, D.A.; Bueno López, R.; Chávez-Velado, D.R.; Casas, D.E.; Jiménez, R.L.; Sanchez-Plata, M.X. Bio-Mapping Indicators and Pathogen Loads in a Commercial Broiler Processing Facility Operating with High and Low Antimicrobial Intervention Levels. *Foods* **2022**, *11*, 775. https://doi.org/10.3390/foods11060775

Academic Editors: Antonio Afonso Lourenco, Catherine Burgess and Timothy Ells

Received: 28 January 2022
Accepted: 6 March 2022
Published: 8 March 2022

1. Introduction

The United States poultry industry is the largest producer and the second largest exporter of poultry meat in the world [1]. In 2020, the value of production combining broilers, eggs, and turkeys was USD 35.5 billion, with 61% from broilers, 24% from eggs, 15% from turkeys, and less than 1% from chickens (e.g., spent fowl) [2]. Moreover, consumption of poultry meat has been trending up in the last ten years, displacing a significant amount of red meat consumption perhaps in part because of favorable prices and health recommendations. According to the National Chicken Council, the per-capita consumption of poultry in the United States in 2020 was 113.4 lb, from which 97.6 lb were chicken, and 15.8 lb were turkey [3]. Furthermore, with almost 18% of total poultry production exported, the U.S. poultry industry is heavily influenced by currency fluctuations, trade negotiations, and economic growth in importing markets [2].

The Center for Disease Control and Prevention (CDC), in 2013, estimated that in the United States (U.S.), there are around 48 million people who suffer from foodborne illnesses every year: 128,000 required hospitalization, and 3000 died. Furthermore, the contribution of poultry and eggs to foodborne illnesses caused by bacteria is 22.8%, which is the second highest percentage overall for illnesses compared to land animals (meat: 23.2%) [4]. *Salmonella* spp. is one of the leading causes of foodborne illnesses, after Norovirus, accounting for approximately 1.1 million cases per year, with 19,336 hospitalizations and 378 deaths [5]. The CDC also notes that campylobacteriosis, caused by *Campylobacter* spp., is the most common bacterial cause of diarrheal illness in the U.S., with approximately 20 cases diagnosed annually for every 100,000 people [5]. The CDC estimates that *Campylobacter* spp. is responsible for infecting at least 1.5 million U.S. residents every year [6]. Therefore, the impact of these two pathogens on public health is a significant concern in the United States and globally [7,8].

The United States Department of Agriculture (USDA)—Food Safety and Inspection Service (FSIS) enforces microbial performance standards based on prevalence (positive or negative) in poultry-processing establishments. Whole birds and parts are collected after the chilling step, sent out to an official laboratory, and tested for *Salmonella* spp. as part of this verification system. FSIS established the *Salmonella* spp. performance standard of 5 positive results out of 51 samples collected (for whole birds) and 8 positive results out of 52 samples collected (for parts, e.g., wings). There is a *Campylobacter* spp. standard; however, it is not currently enforced. Whole-bird and/or part samples are collected one per week, and each result is entered into a 52-week moving window database that calculates individual plant performance and categorizes establishments in three categories. Category 1 is defined as establishments that have achieved 50% or less of the maximum allowable percent positive during the most recently completed 52-week moving window. Category 2 is for establishments that meet the maximum allowable percent positive but have results greater than 50% of the maximum allowable percent positive during the most recently completed 52-week moving window, and Category 3 is for establishments that have exceeded the maximum allowable percent positive during the most recently completed 52-week moving window [9]. Therefore, the focus remains in reducing the prevalence of *Salmonella* spp. through the implementation of sanitary dressing procedures, applying antimicrobial interventions, both chemical and non-chemical, to reduce cross contamination during processing and handling [10].

Most chicken processors in the U.S. proactively work to minimize pathogen contamination and comply with regulatory performance standards using process control and pathogen reduction initiatives based on Hazard Analysis and Critical Control Points (HACCP) systems to reduce consumer exposure to foodborne pathogens, such as *Salmonella* spp. and *Campylobacter* spp. [9]. The poultry industry has traditionally implemented non-chemical (e.g., physical removal of solids prior to the scalding step) and chemical interventions (e.g., chlorine and peroxyacetic acid rinses) against *Salmonella*spp. and *Campylobacter* spp., based on plant-to-plant experiences and word-of-mouth information shared among the industry. The validation of each intervention has been evaluated using traditional prevalence microbiological methods for *Salmonella* spp. and *Campylobacter* spp., which typically compares such prevalence before and after a particular intervention or a series of interventions is applied.

Typical chemical interventions that poultry processors utilize during first processing (e.g., evisceration) and second processing (e.g., deboning) include the use of sodium hypochlorite (chlorine) [11] and peroxyacetic acid (PAA) in equipment rinses, belt washers, inside-outside bird washers (IOBWs), on-line reprocessing (OLR) cabinets, pre-chillers, main chillers, shower heads, and dips/sprays. These chemicals and any chemical used as antimicrobial intervention in a federally inspected establishment must be listed under the USDA-FSIS safe and suitable ingredients used in production of meat, poultry, and egg products [12]. For instance, for PAA the maximum approved concentration is 2000 parts per million (ppm). Many chemical interventions have been studied for raw poultry products, and these must be approved for industry applications. Typically, laboratory validations are conducted to prove efficacy prior to field tests and/or application and chemicals should show at least, as a general rule, a 1 log CFU/mL reduction after the intervention application to be considered useful [13].

The use of PAA has increased in popularity among poultry processors, and research studies show its efficacy is greater than chlorine as well as other antimicrobials available for the poultry industry [14]. However, PAA has been associated with occupational concerns because of its corrosive and irritating effect on eyes, nasal passages, and skin [15]. OSHA has yet to establish occupational exposure limits for PAA; however, the American Conference of Governmental Industrial Hygienists (ACGIH) established an occupational exposure limit of 0.4 ppm as a short-term exposure limit for inhalable fraction and vapor [16]. Processors have been increasing PAA concentration levels at more locations to ensure compliance to regulatory standards; therefore, there is a need to re-assess the strategic use of PAA as an intervention in poultry processing to address occupational concerns and enhance microbial performance. The FSIS reported that between July 2020 and June 2021, the prevalence in raw chicken carcasses for *Salmonella* spp. was 3.42% (down slightly from the previous year) and for *Campylobacter* spp. was 16.45% (down significantly from the previous year). Similarly, the prevalence in raw chicken parts for *Salmonella* spp. was 6.53% (down from the previous year) and for *Campylobacter* spp. was 15.12% (down from the previous year) [17].

Despite poultry processors using a multi-hurdle approach to achieve the USDA-FSIS performance standards, there is minimal information regarding enumeration of *Salmonella* spp. and *Campylobacter* spp. levels in comparing individual chemical interventions or the contribution of these interventions in the multi-hurdle approach. This is the first biomapping study that incorporates ten sampling locations throughout carcass cleaning, evisceration, chilling, and deboning of chicken parts in comparing the microbial performance when all chemical interventions are turned on (normal chemical) versus the performance when the chemical interventions are turned off or reduced to the minimum allowed concentration (reduced chemical). The evaluation included indicator organisms, such as aerobic plate counts (AC) and Enterobacteriaceae (EB), as well as *Salmonella* spp. counts and *Campylobacter* spp. counts. Statistical process control parameters for each processing scheme and location were developed to assist the facility in continuous improvement of their food-safety system.

2. Materials and Methods

2.1. Sample Collection

The study was conducted on a commercial processing facility that processes on average 336,000 birds and runs in two lines at 175 birds per minute in the southern region of the United States. Samples were collected by trained plant personnel throughout a 25-month period of operations to account for flock-to-flock variability and day-to-day process variability. Whole chicken carcass and part rinses from a small birds (target 4.5 lb. live bird weight) were collected using 400 mL of buffered peptone water (BPW), (Millipore Sigma, Danvers, MA, USA). Rinses were immediately chilled and shipped overnight to the International Center for Food Industry Excellence (ICFIE) Food Microbiology laboratory at Texas Tech University for microbiological analysis.

2.2. Intervention Parameters

The normal processing conditions included chicken carcasses undergoing the standardized processing conditions of the operation with high levels of chemical interventions (CX—chemical treatments), including PAA, PAA + sodium hydroxide, and sodium hypochlorite, at various steps in the evisceration, chilling, and deboning processes, respectively. The reduced chemical treatment was planned to include no chemical interventions (just water) or reduced targeted chemical levels (RC—low chemical). The normal process interventions (CX) typically range from 100–400 ppm of PAA (in some cases in combination with sodium hydroxide to elevate the pH of the medium) and up to 50 ppm of total chlorine (sodium hypochlorite). For the low-chemical intervention process (RC), the chemical application was eliminated in several locations except for where needed as per the validated HACCP being verified by FSIS in the Public Health Information System (PHIS). Figure 1 shows a general flow chart of the process, identifying the CX and the RC processes and chemical concentrations along with the sampling locations. Ten locations throughout the processing line were sampled, including live receiving (LR)—where a warm and intact recently identified dead-on-arrival (DOA) was collected as the closest location to the actual live receiving step; rehanger (R); post eviscerator (M); post cropper (C); post neck breaker (NB); post inside-outside bird washer 1 (IOBW #1); post inside-outside bird washer 2 (IOBW #2); pre chilling (PRE); post chilling (POST); and parts (wings). At each location, at least ten rinses were taken per repetition for CX and RC treatments, five per shift. A total of 1309 samples were analyzed during the current study.

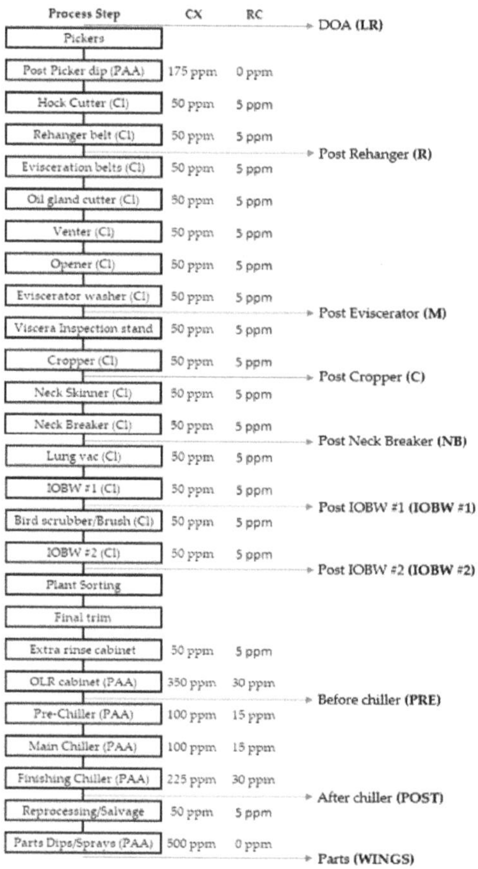

Figure 1. General flow chart of the process, identifying the CX and the RC processing schemes along with the sampling locations.

2.3. Microbial Indicators and *Campylobacter* spp. Enumeration

Rinses were homogenized by hand, and then, the TEMPO system (BioMérieux, Paris, France) was used for the enumeration of indicator microorganisms as well as *Campylobacter* spp. For aerobic plate counts (AC), the Association of Official Agricultural Chemists (AOAC) 121.204 was used, where TEMPO cards were incubated for 22–28 h at 35 ± 1 °C. For *Enterobacteriaceae* enumeration, the AOAC 050801 was used, where TEMPO cards were incubated for 22–28 h at 35 ± 1 °C. For *Campylobacter* spp. enumeration, the ISO 16140/AFNOR method was followed, where TEMPO cards were incubated for 44–48 h at 42 ± 1 °C under microaerophilic conditions using a gas pack generating system.

2.4. *Salmonella* spp. Enumeration and Prevalence

Rinses were homogenized by hand, and then, 30 mL of the rinses were combined with 30 mL of SalQuant solution (Hygiena, Camarillo, CA, USA). Samples were immediately incubated at 42 °C for 6 h for recovery. After incubation, the AOAC 081201 protocol for enumeration of *Salmonella* spp. using the BAX® System SalQuant™ (Hygiena, Camarillo, CA, USA) was followed. Subsequent to enumeration, samples were placed again in an incubator at 42 °C for 18 h for enrichment. After incubation, if samples were not positive for BAX® System SalQuant™, the BAX® System RT-*Salmonella* Assay for detection was followed.

2.5. Statistical Analysis

All data were analyzed using R (Version 4.04) statistical analysis software to evaluate the difference in reduction of microbial loads after following the normal process interventions when compared to low-chemical process interventions on each of the 10 locations analyzed. All counts were transformed to log CFU/mL of rinse with exception of *Salmonella* spp. counts, which were reported as log CFU/sample (Log CFU/400 mL), and a *t*-test was performed to compare the counts at each location with normal process interventions and low chemical process interventions. If parametric assumptions were not met, the Wilcoxon Sum Rank Test or Mann–Whitney test was used as a non-parametric alternative for the *t*-test. A *p*-value of 0.05 or less was used to determine significant differences.

3. Results

The log CFU/mL (or log CFU/Sample for *Salmonella* spp. counts) reductions from live receiving to rehanger locations were significant for all testing conducted on indicator and pathogen bacteria. For indicator organisms, the average reduction for AC was 2.92 log CFU/mL (p-value < 0.001) and 2.41 log CFU/mL (p-value < 0.001) for the CX and RC treatments, respectively, while for EB the average reduction was 2.43 log CFU/mL (p-value < 0.001) and 2.29 log CFU/mL (p-value < 0.001) for the CX and RC treatments, respectively.

For pathogen enumeration, the average reduction from live receiving to rehanger locations for *Campylobacter* spp. was 3.18 log CFU/mL (p-value < 0.001) and 3.23 log CFU/mL (p-value < 0.001) for the CX and RC treatments, respectively, while for *Salmonella* spp., the average reduction was 2.27 log CFU/mL (p-value < 0.001) and 1.94 log CFU/mL (p-value < 0.001) for the CX and RC treatments, respectively.

In the nine locations following the live receiving (LR) location, for indicators and pathogens enumeration, the variation of the data points for the low-chemical treatment (RC) treatments was greater than those for the normal chemical treatment (CX) treatments.

For each of the sampling locations and all indicators as well as pathogens counts, the standard error (SE) was calculated to show dispersion of sample means around the population mean. The mean plus three standard error of the mean (mean + 3SE) was also calculated in each treatment to show the upper control limit per the USDA FSIS recommendation on statistical process control [18].

3.1. *Aerobic Plate Counts (AC)*

The average incoming AC count measured at the live receiving area was 7.56 log CFU/mL (Table 1). These counts were prior to any (chemical) antimicrobial treatment at the processing plant. Subsequently, only feather removal and scalding, after hanging, stunning, and killing steps, were applied. There was a significant reduction from live receiving (7.56 CFU/mL) to rehanger location for both treatments: 4.64 log CFU/mL (CX with a p-value < 0.001) and 5.16 log CFU/mL (RC with a p-value < 0.001). The AC counts were not statistically different ($p > 0.05$) between CX and RC treatments at post-evisceration, post-cropper, post-IOBW #2, and post-chilling locations. Counts at the post rehanger, post neck breaker, post IOBW #1, pre chilling, and parts (wings) showed a statistically significant difference between treatments ($p < 0.05$), with the highest mean difference between treatments at the post-chilling location (0.45 log CFU/mL) and the lowest at the post-IOBW #2 location (0.15 log CFU/mL). For all locations, the low-chemical process (RC) shows greater counts than the normal process (CX) (see Figure 2). There was an increase in counts for both treatments from post-chilling to the parts (wings) location, where the CX treatment showed an average increase of 1.62 log CFU/mL; the RC treatment average increase was 2.01 log CFU/mL.

3.2. *Enterobacteriaceae (EB)*

The average incoming EB count measured at the live hanging area was 6.03 log CFU/mL (Table 2 and Figure 3). These counts were prior to any (chemical) antimicrobial treatment at the processing plant. The counts at the post-neck-breaker, post-IOBW #1, post-IOBW #2, pre-chilling. and parts (wings) locations had significant statistical differences ($p < 0.05$) between the CX and RC treatments, with the highest mean difference at the post-IOBW #2 location (1.01 log CFU/mL) and the lowest at the pre-chilling location (0.45 log CFU/mL). All locations showed higher counts with the RC treatments except for the post-evisceration and the post-cropper locations, where the RC treatment was lower than the CX treatment, with a mean difference of 0.04 log CFU/mL and 0.08 log CFU/mL, respectively. For the post-rehanger, post-evisceration, post-cropper, and post-chilling locations, there were no significant statistical differences ($p > 0.05$) between the CX and RC treatments.

Table 1. Aerobic plate counts (log CFU/mL) on each of the ten locations during the evisceration process under normal process interventions (CX) and low-chemical process interventions (RC) on chicken rinses.

Location	Aerobic Plate Counts (Log CFU/mL)					
	Chemical (CX)			Reduced Chemical (RC)		
	Mean ± SE [1]	Mean + 3SE	n	Mean ± SE	Mean + 3SE	n
Live Receiving [2]	7.56 ± 0.04 [a]	7.68	70	7.56 ± 0.04 [a]	7.68	70
Rehanger	4.64 ± 0.14 [b]	5.04	40	5.16 ± 0.13 [bc]	5.53	90
Post Eviscerator	4.71 ± 0.16 [b]	5.19	30	4.95 ± 0.11 [bc]	5.27	90
Post Cropper	4.75 ± 0.12 [b]	5.10	50	4.92 ± 0.12 [c]	5.29	90
Post NB	4.22 ± 0.11 [c]	4.56	50	5.25 ± 0.12 [b]	5.61	90
Post IOBW#1	4.03 ± 0.14 [c]	4.43	50	4.43 ± 0.12 [d]	4.77	84
Post IOBW#2	3.54 ± 0.08 [d]	3.77	50	3.68 ± 0.08 [e]	3.92	89
Pre Chilling	3.42 ± 0.06 [d]	3.61	50	3.84 ± 0.10 [e]	4.14	98
Post Chilling	1.39 ± 0.19 [f]	1.95	40	1.84 ± 0.08 [f]	2.09	106
Parts (Wings)	3.01 ± 0.10 [e]	3.31	50	3.84 ± 0.11 [e]	4.18	92

[1] Standard error of the mean; [2] For Live Receiving location, there was no treatment applied (CX nor RC); therefore, the same values are reported for each treatment on the table; [a–f] For each Location, with each treatment (CX and RC), Different Letters are Significantly Different according to ANOVA p-value < 0.01.

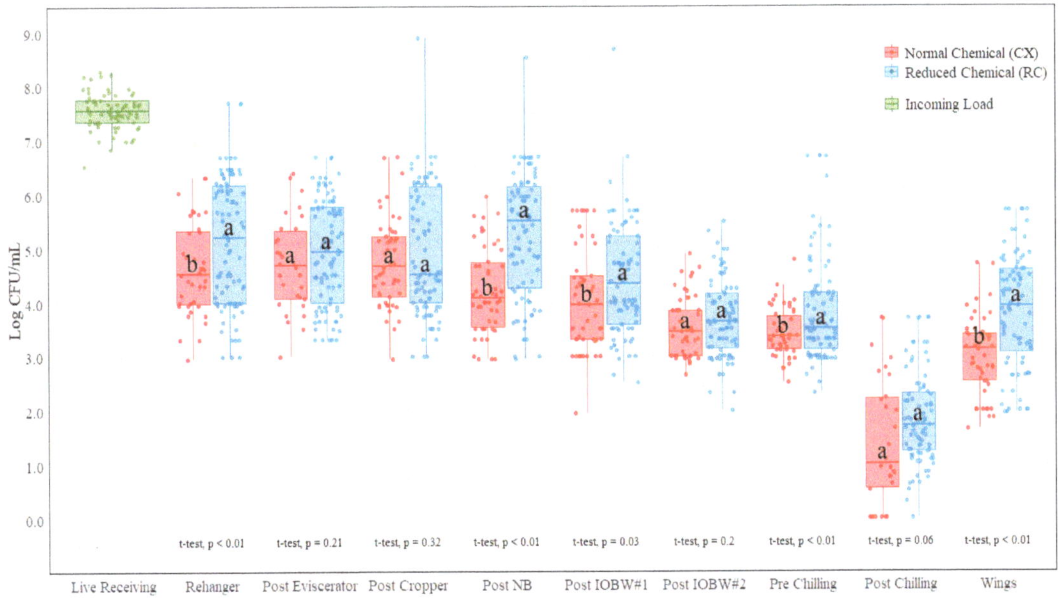

Figure 2. Aerobic plate counts (log CFU/mL) on each of the ten locations during the evisceration process under normal process interventions (CX) and lo- chemical process interventions (RC) on chicken rinses. In each boxplot, the horizontal line crossing the box represents the median, the top and bottom lines of the box represent the lower (0.25) and upper (0.75) quartiles, the vertical top lines represent 1.5 times the interquartile range, and the vertical bottom line represents 1.5 times the lower interquartile range. The dots represent the actual data points. [a,b] For each location, boxes with different letters are significantly different between treatments according to *t*-test analysis at *p*-value < 0.05.

Table 2. Enterobacteriaceae counts (log CFU/mL) on each of the ten locations during the evisceration process under normal process interventions (CX) and low-chemical process interventions (RC) on chicken rinses.

Location	Enterobacteriaceae Counts (Log CFU/mL)					
	Chemical (CX)			Reduced Chemical (RC)		
	Mean ± SE [1]	Mean + 3SE	*n*	Mean ± SE	Mean + 3SE	*n*
Live Receiving [2]	6.03 ± 0.07 [a]	6.25	70	6.03 ± 0.07 [a]	6.25	70
Rehanger	3.60 ± 0.17 [c]	4.10	40	3.74 ± 0.11 [cd]	4.07	90
Post Eviscerator	4.04 ± 0.17 [b]	4.56	30	4.00 ± 0.10 [c]	4.30	90
Post Cropper	3.67 ± 0.15 [bc]	4.10	50	3.59 ± 0.10 [d]	3.89	90
Post NB	3.53 ± 0.12 [c]	3.89	50	4.37 ± 0.11 [b]	4.69	90
Post IOBW#1	2.91 ± 0.10 [d]	3.22	50	3.48 ± 0.10 [de]	3.78	84
Post IOBW#2	2.24 ± 0.13 [e]	2.63	50	3.25 ± 0.11 [e]	3.59	89
Pre Chilling	2.24 ± 0.08 [e]	2.50	50	2.69 ± 0.08 [f]	2.92	98
Post Chilling	0.90 ± 0.08 [g]	1.15	40	0.92 ± 0.10 [g]	1.23	106
Parts (Wings)	1.64 ± 0.10 [f]	1.94	50	2.60 ± 0.11 [f]	2.91	92

[1] Standard error of the mean; [2] For Live Receiving location, there was no treatment applied (CX nor RC); therefore, the same values are reported for each treatment on the table; [a–g] For each Location, with each treatment (CX and RC), Different Letters are Significantly Different according to ANOVA *p*-value < 0.01.

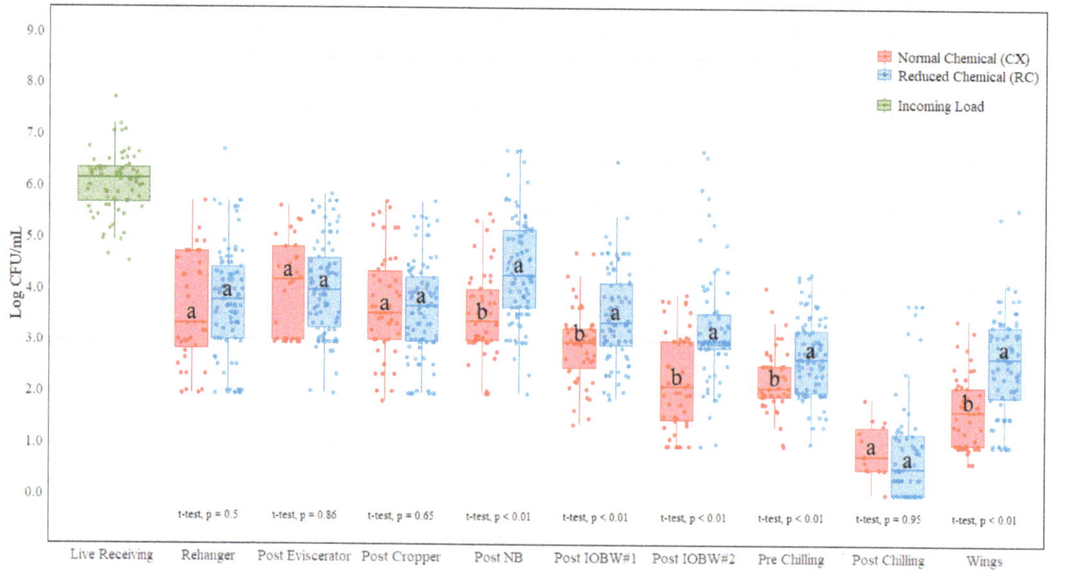

Figure 3. Enterobacteriaceae counts (log CFU/mL) on each of the ten locations during the evisceration process under normal process interventions (CX) and low-chemical process interventions (RC) on chicken rinses. In each boxplot, the horizontal line crossing the box represents the median, the top and bottom lines of the box represent the lower (0.25) and upper (0.75) quartiles, the vertical top lines represent 1.5 times the interquartile range, and the vertical bottom line represents 1.5 times the lower interquartile range. The dots represent the actual data points. [a,b] For each location, boxes with different letters are significantly different between treatments according to *t*-test analysis at *p*-value < 0.05.

3.3. *Salmonella* Detection and Enumeration

Salmonella spp. counts were substantially low when analyzed on a per-mL basis; thus, when transformed to log CFU/mL, some counts resulted in negative values (2.91% of the data with the CX treatment and 8.28% of the data with the RC treatment), making analysis and visualization more difficult for interpretation. Therefore, all data were transformed from to log CFU/sample equivalent to log CFU/400 mL to facilitate data visualization. The limit of quantification for SalQuant (LOQ) is 1 CFU/mL, but counts can be extrapolated below LOQ, as counts are obtained from a regression equation provided by the methodology, the reason why a new LOQ was established as 1% of the real LOQ (0.01 CFU/mL or 0.6 Log CFU/sample). Samples showing as <0.6 log CFU/sample were reported as 50% of the new LOQ (0.3 log CFU/sample). The same value was applied for samples that were not quantifiable but found positive for prevalence analysis. Samples that were not quantifiable nor detected were reported as 0 log CFU/sample. A summary of the parameters used for the data analysis can be found in Table 3.

The average incoming *Salmonella* spp. count measured at the live hanging area was 2.63 log CFU/sample (Table 4). These counts were prior to any (chemical) antimicrobial treatment at the processing plant. Counts were statistically different ($p < 0.05$) between treatments in all sampling locations except for the post-evisceration and post-cropper locations. The RC treatment had greater counts at each sampling location except for the post-cropper location, where the lowest average count was at the RC treatment (0.67 log CFU/Sample). The highest average difference between CX and RC treatments was at the post-neck-breaker location (0.61 log CFU/sample) and the lowest at the post-chilling location (0.01 log CFU/sample). In addition to enumeration (counts), prevalence (Table 5) was performed on non-quantifiable samples using BAX® system Real-Time *Salmonella* assays, and values are shown in Figure 4. The prevalence under the CX treatment is lower for all sampling locations except at the post-evisceration location.

Table 3. Observed and reported parameters established for *Salmonella* spp. quantification and prevalence analysis.

Observed SalQuant Result (Log CFU/Sample)	Observed Prevalence Result	Reported SalQuant Result (Log CFU/Sample)	Reported Prevalence Result
No Result	Negative	0	Negative
No Result	Positive	0.3	Positive
Less than 0.6	NA [1]	0.3	Positive
More or equal than 0.6	NA	Observed SalQuant result	Positive

[1] Not applicable, as prevalence test is not necessary in samples quantified and detected by SalQuant.

Table 4. *Salmonella* spp. counts (log CFU/sample) on each of the ten locations during the evisceration process under normal process interventions (CX) and low-chemical process interventions (RC) on chicken rinses.

Location	*Salmonella* spp. Counts (Log CFU/Sample)					
	Chemical (CX)			Reduced Chemical (RC)		
	Mean ± SE [1]	Mean + 3SE	*n*	Mean ± SE	Mean + 3SE	*n*
Live Receiving [2]	2.63 ± 0.21 [a]	3.26	70	2.63 ± 0.21 [a]	3.26	70
Rehanger	0.36 ± 0.13 [bc]	0.74	40	0.69 ± 0.13 [bc]	1.09	90
Post Eviscerator	0.63 ± 0.19 [b]	1.21	30	0.79 ± 0.14 [b]	1.21	90
Post Cropper	0.72 ± 0.24 [bc]	1.44	50	0.57 ± 0.12 [bc]	0.93	90
Post NB	0.09 ± 0.04 [cd]	0.21	50	0.66 ± 0.12 [bc]	1.03	90
Post IOBW#1	0.04 ± 0.01 [d]	0.08	50	0.43 ± 0.11 [bc]	0.75	84
Post IOBW#2	0.04 ± 0.02 [d]	0.10	50	0.13 ± 0.05 [bc]	0.27	89
Pre Chilling	0.02 ± 0.02 [d]	0.07	50	0.34 ± 0.08 [bc]	0.60	98
Post Chilling	0.00 ± 0.00 [d]	0.00	40	0.00 ± 0.00 [c]	0.00	106
Parts (Wings)	0.07 ± 0.05 [d]	0.22	50	0.15 ± 0.07 [bc]	0.35	92

[1] Standard error of the mean; [2] For Live Receiving location, there was no treatment applied (CX nor RC); therefore, the same values are reported for each treatment on the table; [a–d] For each Location, with each treatment (CX and RC), Different Letters are Significantly Different according to Krustal–Wallis test at p-value < 0.01.

Table 5. Prevalence of *Salmonella* spp. at each Sampling Location for each Treatment: Normal Chemical (CX) and Reduced Chemical (RC).

Location	Prevalence (%)	
	Normal Chemical (CX)	Reduced Chemical (RC)
Live Receiving *	94.29%	94.29%
Rehanger	42.50%	45.60%
Post Eviscerator	46.70%	40.00%
Post Cropper	28.00%	35.60%
Post Neck Breaker	16.00%	33.30%
Post IOBW #1	12.00%	30.00%
Post IOBW #2	10.00%	16.20%
Pre Chilling	4.00%	23.33%
Post Chilling	0.00%	1.11%
Parts (Wings)	10.00%	11.20%

* Percentages are the same under CX and RC because at Live Receiving location, no chemical treatment was applied.

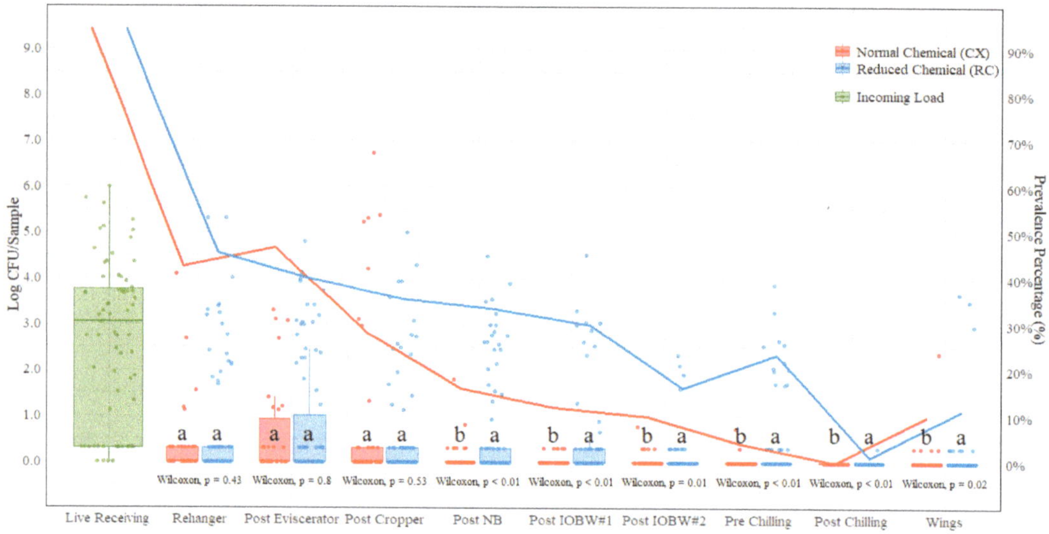

Figure 4. *Salmonella* spp. counts (log CFU/Sample) and prevalence (shown as solid lines) comparison on each of the ten locations during the evisceration process under normal process interventions (CX) and low-chemical process interventions (RC) on chicken rinses. In each boxplot, the horizontal line crossing the box represents the median, the top and bottom lines of the box represent the lower (0.25) and upper (0.75) quartiles, the vertical top lines represent 1.5 times the interquartile range, and the vertical bottom line represents 1.5 times the lower interquartile range. The dots represent the actual data points. [a,b] For each location, boxes with different letters are significantly different between treatments according to Wilcoxon test analysis at p-value < 0.05.

3.4. *Campylobacter* spp.

The average incoming *Campylobacter* spp. count measured at the live hanging area was 5.23 log CFU/mL (Table 6). These counts were prior to any (chemical) antimicrobial treatment at the processing plant. The only location with significant mean difference ($p < 0.05$) between CX and RC treatments was the parts (wings) location, where the difference between treatments was 0.30 log CFU/mL (CX treatment with lower counts than the RC treatment). However, higher counts were shown in the CX treatments for post-rehanger (2.05 log CFU/mL), post-cropper (2.34 log CFU/mL), post-neck-breaker (2.57 log CFU/mL), post-IOBW #1 (1.75 log CFU/mL), pre-chilling (1.23 log CFU/mL), and post-chilling (0.18 log CFU/mL) locations (Figure 5). The highest mean difference between treatments was shown at the post-cropper location (0.34 log CFU/mL higher on the CX treatment) and the lowest at the post-rehanger and post-chilling locations (0.05 log CFU/mL on both locations, higher on the CX treatment).

Prevalence was obtained from the TEMPO® quantification data, and values are shown in Table 7. The *Campylobacter* spp. incoming load measured at live receiving was 100.00% positive. After the slaughtering, bleeding, and defeathering (including scalding and picking) processing steps, the prevalence of *Salmonella* spp. was reduced to 90.00% (CX) and 86.70% (RC) positive, which represents a 10.00% (CX)/13.30% (RC) reduction without any chemical intervention applied other than under the RC treatment, where in some of the samples, the post-picker dip was kept at 175 ppm (PAA). After the rehanger, there was not a gradual reduction on counts; instead, the prevalence increased slightly from rehanger to the post-eviscerator location with both treatments: 93.33% positive with the CX treatment and 86.70% positive with the RC treatment. Furthermore, from the post-eviscerator to the post-cropper location, there was also an increase in prevalence with both treatments: 100.00% positive with the CX treatment and 90.00% positive with the RC treatment. At the post-neck-breaker location, with the CX treatment, the *Campylobacter* spp. prevalence stayed the same at 100% and with the RC treatment increased to 94.40%. There was a decrease in prevalence from the post-NB to the post-IOBW#1 location, and from the post-IOBW#1 to the post-IOBW#2 locations, *Campylobacter* spp. prevalence decreased from 98.00% to 94.00% positive with the CX treatment and from 86.90% to 75.30% positive with the RC treatment. There was also a decrease from the post-IOBW#2 (94.00% positive with CX and 75.30% with RC) and the pre-chilling location (92.00% positive with CX and 66.30% with RC).

Table 6. *Campylobacter* spp. counts (log CFU/mL) on each of the ten locations during the evisceration process under normal process interventions (CX) and low-chemical process interventions (RC) on chicken rinses.

Location	*Campylobacter* spp. Counts (Log CFU/mL)					
	Chemical (CX)			Reduced Chemical (RC)		
	Mean ± SE [1]	Mean + 3SE	*n*	Mean ± SE	Mean + 3SE	*n*
Live Receiving [2]	5.23 ± 0.16 [a]	5.72	70	5.23 ± 0.16 [a]	5.72	70
Rehanger	2.05 ± 0.18 [cd]	2.58	40	2.00 ± 0.12 [bc]	2.37	90
Post Eviscerator	2.18 ± 0.18 [c]	2.71	30	2.23 ± 0.12 [b]	2.59	90
Post Cropper	2.34 ± 0.12 [bc]	2.70	50	2.00 ± 0.11 [bc]	2.33	90
Post NB	2.57 ± 0.12 [b]	2.92	50	2.25 ± 0.11 [b]	2.57	90
Post IOBW#1	1.75 ± 0.12 [d]	2.10	50	1.54 ± 0.10 [cd]	1.85	90
Post IOBW#2	1.36 ± 0.10 [e]	1.67	50	1.38 ± 0.09 [cd]	1.65	89
Pre Chilling	1.23 ± 0.11 [e]	1.56	50	1.18 ± 0.10 [d]	1.47	98
Post Chilling	0.18 ± 0.07 [f]	0.39	40	0.13 ± 0.05 [f]	0.27	106
Parts (Wings)	0.27 ± 0.07 [f]	0.48	50	0.57 ± 0.06 [e]	0.76	92

[1] Standard error of the mean; [2] For Live Receiving location, there was no treatment applied (CX nor RC); therefore, the same values are reported for each treatment on the table; [a–f] For each Location, with each treatment (CX and RC), Different Letters are Significantly Different according to Kruskal–Wallis test at p-value < 0.01.

Table 7. Prevalence of *Campylobacter* spp. at each Sampling Location for each Treatment: Normal Chemical (CX) and Reduced Chemical (RC).

Location	Prevalence (%)	
	Normal Chemical (CX)	Reduced Chemical (RC)
Live Receiving *	100.00%	100.00%
Rehanger	90.00%	86.70%
Post Eviscerator	93.33%	87.80%
Post Cropper	100.00%	90.00%
Post NB	100.00%	94.44%
Post IOBW#1	98.00%	86.90%
Post IOBW#2	94.00%	75.30%
Pre Chilling	92.00%	66.30%
Post Chilling	17.50%	9.43%
Parts (Wings)	34.00%	50.00%

* Percentages are the same under CX and RC because at Live Receiving location, no chemical treatment was applied.

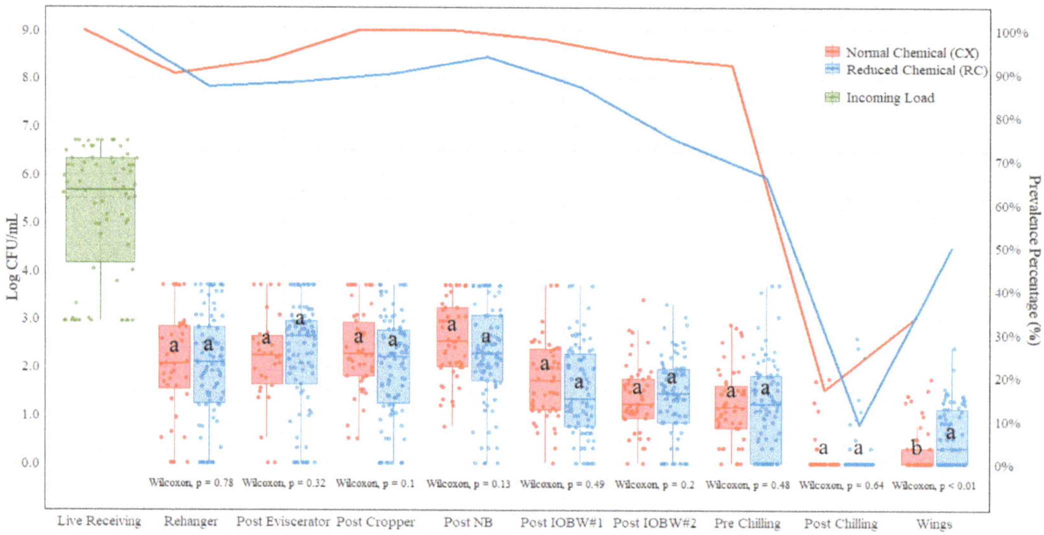

Figure 5. *Campylobacter* spp. counts (log CFU/mL) and prevalence (shown as solid lines) comparison on each of the ten locations during the evisceration process under normal process interventions (CX) and low-chemical process interventions (RC) on chicken rinses. In each boxplot, the horizontal line crossing the box represents the median, the top and bottom lines of the box represent the lower (0.25) and upper (0.75) quartiles, the vertical top lines represent 1.5 times the interquartile range, and the vertical bottom line represents 1.5 times the lower interquartile range. The dots represent the actual data points. [a,b] For each location, boxes with different letters are significantly different between treatments according to *t*-test analysis at *p*-value < 0.05.

4. Discussion

As observed in previous studies, the prevalence of *Salmonella* spp. was reduced from the pre-scalding to the post-chiller stages. These reductions were attributed to sequential washes and antimicrobial interventions applied during evisceration and in the pre- and post-chiller tanks [14,19–22]. Most of the research studies conducted on *Salmonella* spp. and *Campylobacter* spp. in poultry focus the microbiological methods on prevalence (%), whereas in the current study, we evaluated the quantification of indicator bacteria as well as pathogens (*Salmonella* spp. and *Campylobacter* spp.) in a processing operation running with chemical interventions and low levels of interventions, which makes the current research study unique. The sampling collection also occurred over a period of twenty-five months, capturing variability of flocks sampled and seasonality.

The significant log reductions from live receiving to the rehanger location for both indicator and pathogen loads provide validation data indicating that the scalding (washing effect and high temperature) and picking processes are key steps in bacterial reduction during poultry processing and a major pathogenic reduction stage for pathogen control if properly managed. The sample collected at the live receiving location included feathers, head, and feet, as well as any filth from the field, compared to the picked (plucked) bird at the rehanger location, where the feathers, head, and feet have been removed. As mentioned in previous studies [23], in general for industry professionals, a pathogen reduction of at least one logarithmic cycle from location to location is necessary to consider an intervention effective. In the current study, the average reduction from live receiving to rehanger across both treatments was 2.66 log CFU/mL (APC), 2.36 log CFU/mL (EB), 3.20 log CFU/mL (*Campylobacter* spp.), and 2.15 log CFU/sample (*Salmonella* spp.). At this particular processing plant, there is no chemical treatment applied in the scalding or the defeathering process. As indicated earlier, there is a post-picker dip with up to 175 ppm of

PAA immediately after the last picker, which showed to be statistically significant when comparing CX and RC treatments for AC and *Salmonella* spp. counts. Therefore, even without any pH adjustment treatment in the scalder tanks (one of the common antimicrobial interventions used in the poultry industry), the softening and removal of the feathers while keeping the bird warm during this process are definitely an important aid in bacterial reduction for the process.

The need to optimize the rather widespread use of PAA as interventions throughout the process is critical due to concerns on dose and time of contact variability [10] and the occupational concerns mentioned earlier [15]. Therefore, the current research study provides a standardized methodology to generate the evidence needed for the identification of focused intervention locations in the process, more specifically the use of PAA, in selected locations within first and second processing to maximize the efficacy and improved the microbial performance of the process.

In another study, it was determined that reductions in the AC and EB counts were not consistent between the post-scalding and post-defeathering locations [24] and did not provide a clear indication of what microorganisms could be affecting those results. We learned that the reduction from the live receiving to the rehanger location under the CX treatment on both AC (2.92 Log CFU/mL) and EB (2.43 Log CFU/mL) was consistent, and the counts remained somewhat constant between the rehanger and the post-neck-breaker location, suggesting that up to the post-neck-breaker location, there is no major reduction on AC and EB counts even with high levels of chemical interventions applied. In fact, the post-evisceration and post-cropper locations showed no significant statistical difference between the CX and RC treatments ($p > 0.05$).

Poultry processors have implemented various antimicrobial interventions to reduce cross contamination and minimize the presence of foodborne pathogens, such as *Salmonella* spp. and *Campylobacter* spp., during poultry processing. However, limited information on comprehensive biomapping conducted at a commercial poultry processing facility—which included enumeration of pathogens as well as prevalence—is available in the literature. Limited research studies are available, such as those using chicken parts, conducted in laboratory settings and in controlled environments. In the current study, whole birds and parts (wings) samples were collected over the course of twenty-five months and included quantification of indicators and pathogens in a plant setting, therefore making the current bio-mapping more representative of the process variability and allowing this processor to establish a facility-specific microbial baseline for decision making on the intervention's effectiveness.

The processing facility where the current research study was conducted is operating under the New Poultry Inspection System (NPIS) and has a line-speed waiver to process in evisceration, at line speeds of up to 175 birds per minute (BPM). The multi-hurdle approach for antimicrobial interventions at this processing facility, whether under the CX or the RC treatments, achieved post-chill pathogen counts of less than 0.27 and 0.57 log CFU/mL (*Campylobacter* spp.) or 0.07 and 0.15 log CFU/Sample (*Salmonella* spp.), respectively. These levels, according to the risk assessments of *Salmonella* spp. in broiler chickens [25], have a very low probability for causing illness, without even considering the effect of thermal processing on risk reduction from the raw poultry carcass or part evaluated. Furthermore, when comparing these results at the parts location (wings), the levels are below 1 log CFU/mL (*Campylobacter* spp.) or 1 log CFU/Sample (*Salmonella* spp.), which also represents a very low probability of illness. Therefore, the current data suggest that the increased evisceration line speed under NPIS does not affect or increase the risk of illness caused by foodborne pathogens, such as *Salmonella* spp. and *Campylobacter* spp. [26].

4.1. Aerobic Counts (AC)

There were significant statistical differences between CX and RC treatments observed at the rehanger location (0.51 CFU/mL with lower counts shown with the CX treatment). These results suggest that the use of the post-picker dip, located immediately after the last picker and containing up to 250 ppm PAA, may have an improved effect in the overall process for pathogen control. There was no statistical difference between treatments at the post-evisceration and post-cropper locations, indicating that neither the chlorinated washer located immediately after the removal of viscera from the birds nor the washer and brushes removing crops from the probes may have a reduction effect in the aerobic counts.

At the NB location, there was a statistically significant reduction in counts with the CX treatment, while the counts with the RC treatment appeared to increase. This suggests that chemical interventions are needed at this location to ensure proper sanitizing of the neck-breaker blades to reduce cross contamination. Because the birds are hung upside down, all the fluids draining from the cavity of the birds pass through the neck area. At this step of the process, the release of these fluids when breaking the necks may require a chemical treatment to reduce the AC load. Furthermore, the very next processing step, the first inside-outside bird washer, seems to have a beneficial effect when chemical interventions are used in reducing aerobic counts. The average reduction at the post-IOBW #1 location from the previous steps (excluding live receiving) was 0.55 log CFU/mL (CX treatment) and 0.64 log CFU/mL (RC treatment).

The brushes (after the IOBW #1) and the subsequent IOBW #2 seem to not have a major effect in AC levels with the addition of chemicals at those two steps, as there are no statistical differences between CX and RC treatments at the post-IOBW #2 location. This could be due to the reduction already achieved by the IOBW #1. However, at the pre-chilling location, which is after the on-line reprocessing (OLR) cabinet, there is a significant statistical difference between CX and RC treatment; however, the reduction is only 0.11 log CFU/mL. There was also an increase for the RC treatment from the post-IOBW#2 to the pre-chilling location (0.15 log CFU/mL). This suggests that the chemical effect at the OLR applied in this facility may not be an important antimicrobial intervention in the AC reduction and will need to be optimized. The typical chemical used at this location is PAA, at concentrations ranging from 300 ppm to 400 ppm under normal processing.

At the post-chill location, the difference between CX and RC treatments is not statistically significant, with a 2.04 log CFU/mL reduction from pre-chill to post-chill locations under the CX treatment and 2.00 log CFU/mL under the RC treatment. The lowest AC counts with both treatments occurred at the post-chilling location (lower with the CX treatment), indicating that the temperature reductions and chemical treatments in the pre-chiller, main-chiller, and post-chiller when combined are effective for reducing AC counts.

There is also a significant statistical difference between treatments at the parts location (wings), with CX treatment at 0.84 log CFU/mL, lower in average than the RC treatment. The overall reduction at this location has been previously reported at 1.27 log CFU/mL on a laboratory spray application setting on breast fillets [27]. Parts dips have become popular in commercial processing facilities, and they are currently widely used in the poultry industry, with concentrations of PAA up to 400 ppm to help in complying with parts performance standards. This antimicrobial intervention has proven to be very effective in reducing the loads of AC, as shown in the current research study.

4.2. *Enterobacteriaceae* Counts (EB)

Similar to what was found with the AC counts, the EB counts at the post-evisceration and post-cropper locations were not significantly different between the CX and RC treatments. In addition to these locations, the post-rehanger EB counts were also not significantly different between the treatments. However, there is a significant difference between the treatments at the post-neck-breaker location, with the RC treatment being higher than the CX treatment, on average 0.84 log CFU/mL. This difference could be due to the antimicrobial effect of the neck-breaker equipment washers. The use of chemicals during the process seem to have a positive impact when measuring EB at the post-neck-breaker location, which along with the removal of the viscera and crops, creates an opening around the neck area, helping drainage of contamination during processing. However, there was not much change among the counts from the rehanger to the neck-breaker locations, with an average

EB count of 3.71 log CFU/mL (CX) and 3.92 log CFU/mL (RC) in these four locations. As mentioned in previous studies, certain steps, such as those within the evisceration process, may contribute to higher levels of contamination [28], and in the current research study, we found that the EB counts do not seem to change much across these locations.

The chemical usage in the IOBWs as well as in the brushes between the washers seem to also have a positive impact in reducing EB counts, which is displayed in the significant differences between the treatments at these locations. Whereas no significant statistical differences were found at post-chill location, indicating that not much effect on EB was accomplished by the use of chemical in the chilling system, the lower temperature in the system may have a positive impact in the reduction of EB counts between pre-chill and post-chill locations. There is a significant statistical difference between treatments at the pre-chilling and parts (wings) locations, which reinforces the findings that parts dips with PAA have a positive impact in bacterial reduction in skin-on part samples.

4.3. *Salmonella* spp.

There was a statistical difference in *Salmonella* spp. counts between treatments at each sampling location (except for the post-evisceration and post-cropper locations), with CX being the lowest at each sampling location with the exception of post-cropper, where the CX treatment was lower (0.67 log CFU/mL at RC vs. 0.75 log CFU/mL at CX). The pattern for prevalence was very similar, with the highest prevalence of *Salmonella* spp. under the RC treatments except for samples collected at the post-evisceration location. At this location, the CX treatment had a slightly higher prevalence than the RC treatment.

The largest average difference between treatments was at the post-neck-breaker location, validating that cross-contamination control and adequate sanitary dressing in neck breaking are key steps in the reduction of *Salmonella* spp. Furthermore, chilling (pre main and post chiller) continued to be a crucial step in microbial control during poultry processing, which is validated by the 0% (CX) and 0.94% (RC) prevalence at the post-chill location, significantly lower than the performance standard limits.

The reduction in prevalence from the live receiving (>90%) to the rehanger (~40%) follows the same trend as with the quantification reduction at these two locations. Even though the prevalence reduction is close to 50%, in quantification, the average reduction from live receiving to rehanger locations (2.27 log CFU/Sample for CX and 1.94 log CFU/Sample for RC) was higher than 90% with the CX treatment and 75% with the RC treatment, and it can only be seen with quantification data. These discrepancies are a confirmation than prevalence alone is not a good indicator of food safety [29].

4.4. *Campylobacter* spp.

After the live receiving location, all locations except for parts (wings) show no significant difference between treatments CX and RC. Only the parts (wings) location, with an average difference of 0.30 log CFU/mL, showed minimal effect under the CX treatment, which is consistent with the AC and EB indicators as well as *Salmonella* spp. loads. This provides some evidence that parts interventions are effective in reducing pathogen loads. As previously reported, the use of antimicrobial interventions, such as post-chilling immersion tanks or spraying systems using high concentration of chemicals (with short contact times), have proven to be an added hurdle after primary chilling that further facilitates the reduction of pathogens on poultry carcasses [10].

As seen on the results, after the live receiving location, there seemed to be not much change in counts from the rehanger to the post-neck-breaker locations, which is a pattern observed with AC and EB counts. However, there is a reduction at the first IOBW #1, showing an average 0.82 log CFU/mL (CX) and 0.71 log CFU/mL (RC) from the previous location. Furthermore, between the post IOBW #1 and the pre-chilling locations, there is not much change in *Campylobacter* spp. counts until the post-chilling location. This provides strong evidence that the chilling of the birds is the primary step in pathogen reduction.

Prevalence of *Campylobacter* spp. under the CX treatment remains constant between 90% and 100% through the pre-chilling location; however, as discussed before, there is a 3.18 log CFU/mL reduction from live receiving to the rehanger location. This reduction is negligible when only looking at prevalence. Similarly, under the RC treatment, the prevalence of *Campylobacter* spp. remains between 85% and 100% through the post-IOBW #1 location, disregarding the reduction in counts from live receiving to rehanger of 3.23 log CFU/mL, which is a strong evidence that prevalence alone cannot be used as a sole representation of the microbial loads within a poultry-processing facility [29].

5. Conclusions

Pathogen quantification can result in improved risk assessment where chemical interventions can be targeted to stages with higher indicator and pathogen bacteria counts. The current research study provides evidence for the application of chemical treatments in strategic locations during poultry processing rather than the use of an array of interventions at different locations, thus assisting the processor to customize their antimicrobial intervention regimes and focus these efforts in higher-risk areas.

The development of biomapping baselines that include quantification of pathogens leads to the development of statistical process control parameters to support food safety management decision making. Nonparametric statistical process control can be approached to more representatively use pathogen prevalence and quantification data together, resulting in more educated decisions than using exclusively prevalence data.

In the current research study, it was evident that the scalding and picking processing steps leading up to the evisceration process are key steps in the reduction of indicator and pathogen bacteria. Furthermore, the reduction achieved between live receiving and rehanger is almost constant for both treatments (CX and RC) for any of the indicator and pathogen bacteria tested up to the neck-breaker location. After such step, the incorporation of chemicals (e.g., sodium hypochlorite) at the first inside-outside bird washer (IOBW #1), along with good sanitary dressing practices, seem to have the best performance. Therefore, the first step in the evisceration process that needs to have chemicals based on the results of the current study is the IOBW #1.

The on-line reprocessing (OLR) cabinet does not seem to have a major impact in bacterial reduction in this operation with either CX or the RC treatments; however, the chilling system, including the pre chiller, main chiller, and post-finishing chiller, were shown to be a major contributor to pathogen reductions (combining low water temperature and chemical usage, such as PAA) for bacterial reduction, thus indicating that the chiller process should be optimized as the second main location for chemical application in the process. The final antimicrobial intervention step, shown in the current study to have an impactful bacterial-reduction performance, is the parts dips, where PAA is mostly used.

The data generated from the current study demonstrate that the use of *Salmonella* spp. or *Campylobacter* spp. prevalence as a sole measurement of food-safety performance is not adequate or representative of the whole picture of contamination in a dynamic system. Pathogen prevalence is part of the equation, and several other variables, such as quantification, are necessary to make decisions that will improve the food-safety system's performance. There have been models published identifying risk factors for *Salmonella* control in poultry-processing operations [29], which support the conclusions of the current study. Published risk assessments support this approach, and the results of the current study can be used to conduct probabilistic quantitative microbial risk assessments similar to those conducted in prior publications (QMRA) [30]. Finally, this integrated approach to measure the performance of the pathogen-control system provides a risk-based approach to food-safety management and therefore is needed to establish a new performance standard for *Salmonella* spp. and *Campylobacter* spp. that is based on loads. A better performance-standard system can contribute in a better way to help achieve the Healthy People 2030 goals [31,32].

There is a significant amount of data generated by research conducted by poultry processors, who collect far more microbiological data than the official sampling programs of the USDA-FSIS sampling plans (e.g., *Salmonella* spp. 52-rolling window—one sample per week). Federally inspected establishments collect on a routine basis samples before and after chilling for every 22,000 birds processed. For example, if a single evisceration line processes 660,000 birds in one week, there would be a total of thirty samples (30) collected in one week for one of the indicator organisms compared to one (1) sample collected by USDA FSIS. These samples are in addition to other microbial samples collected by each establishment to evaluate the performance of some of their intervention schemes. Furthermore, poultry processors, through biomapping sampling, select more significant sampling locations that better represent the microbiological performance of the process. With more repetitions and extra sampling locations, the poultry industry can generate sufficient quantitative data on pathogen loads that, when statistically analyzed, would serve as a better measurement for the establishment's microbial performance and to generate actual risk-based performance standards. Therefore, it is important to consider outside data, such as that generated from the current research study, to evaluate large datasets from a variety of operations to establish a plant's microbial performance [33].

Author Contributions: Conceptualization, J.F.D.V. and M.X.S.-P. methodology, J.F.D.V., M.X.S.-P., D.A.V., D.E.C., D.R.C.-V., R.L.J. and R.B.L.; validation, J.F.D.V., D.A.V., M.X.S.-P. and D.R.C.-V.; formal analysis, J.F.D.V. and D.A.V.; investigation, J.F.D.V., D.A.V. and M.X.S.-P.; resources, J.F.D.V., R.B.L. and D.R.C.-V.; data curation, J.F.D.V. and D.A.V.; writing—review and editing, J.F.D.V., D.A.V. and M.X.S.-P.; supervision, M.X.S.-P.; project administration, J.F.D.V. and D.R.C.-V.; funding acquisition, J.F.D.V. and M.X.S.-P.; writing—original draft, J.F.D.V. All authors have read and agreed to the published version of the manuscript.

Funding: The current study was funded by the International Center for Food Industry Excellence (ICFIE) at Texas Tech University and was supported by in-kind contributions from Hygiena, BioMerieux, and Wayne Farms LLC for completion.

Institutional Review Board Statement: Not applicable.

Informed Consent Statement: Not applicable.

Data Availability Statement: Data available on request from the corresponding author. The data are not publicly available due to privacy from the poultry-processing partner that allowed the project to be conducted within their poultry-processing facility.

Acknowledgments: The authors would like to thank the ICFIE Food Microbiology Laboratory personnel for their contributions to completing the current project: Tanya Jackson, Gabriela Betancourt, Rigo Soler, Manoella Ajcet, Donald Molina, Estefania Pizzato, Loron Brown, Samuel Peabody, Paola Moncada, Sabrina Blandon, and Mariana Fernandez; and the following personnel for sample collection: Justin Pettis, Katelyn Youngblood, Shieriney Murphy, Charles Register, and Darryl Lane.

Conflicts of Interest: The authors declare no conflict of interest.

References

1. United States Department of Agriculture. Poultry Eggs. 2021. Available online: https://www.ers.usda.gov/topics/animal-products/poultry-eggs (accessed on 21 September 2021).
2. United States Department of Agriculture. Poultry-Production and Value Summary 2020. 2021. Available online: https://www.nass.usda.gov/Publications/Todays_Reports/reports/plva0421.pdf (accessed on 20 November 2021).
3. National Chicken Council. Per Capita Consumption of Poultry and Livestock, 1965 to Forecast 2022. 2021. (In Pounds). Available online: https://www.nationalchickencouncil.org/about-the-industry/statistics/per-capita-consumption-of-poultry-and-livestock-1965-to-estimated-2012-in-pounds/ (accessed on 20 November 2021).
4. Painter, J.A.; Hoekstra, R.M.; Ayers, T.; Tauxe, R.V.; Braden, C.R.; Angulo, F.J.; Griffin, P.M. Attribution of foodborne illnesses, hospitalizations, and deaths to food commodities by using outbreak data, United States, 1998–2008. *Emerg. Infect. Dis.* **2013**, *19*, 407–415. [CrossRef] [PubMed]
5. Center for Disease Control and Prevention Burden of Foodborne Illness: Findings. Available online: https://www.cdc.gov/foodborneburden/2011-foodborne-estimates.html (accessed on 11 November 2021).
6. Center for Disease Control and Prevention. Campylobacter. Available online: https://www.cdc.gov/campylobacter/index.html (accessed on 29 September 2021).

7. Zhao, C.; Ge, B.; De Villena, J.; Sudler, R.; Yeh, E.; Zhao, S.; White, D.G.; Wagner, D.; Meng, J. Prevalence of *Campylobacter* spp., Escherichia coli, and *Salmonella* Serovars in Retail Chicken, Turkey, Pork, and Beef from the Greater Washington, D.C., Area. *Appl. Environ. Microbiol.* **2001**, *67*, 5431–5436. [CrossRef] [PubMed]
8. Ricke, S.C. Strategies to Improve Poultry Food Safety, a Landscape Review. *Annu. Rev. Anim. Biosci.* **2021**, *9*, 379–400. [CrossRef] [PubMed]
9. Food Safety and Inspection Service. *Performance Standards: Salmonella Verification Program for Raw Poultry Products. Directive 10,250.2*; USDA FSIS: Washington, DC, USA, 2021.
10. Kataria, J.; Vaddu, S.; Rama, E.N.; Sidhu, G.; Thippareddi, H.; Singh, M. Evaluating the efficacy of peracetic acid on *Salmonella* and *Campylobacter* on chicken wings at various pH levels. *Poult. Sci.* **2020**, *99*, 5137–5142. [CrossRef] [PubMed]
11. Loretz, M.; Stephan, R.; Zweifel, C. Antimicrobial activity of decontamination treatments for poultry carcasses: A literature survey. *Food Control* **2010**, *21*, 791–804. [CrossRef]
12. USDA FSIS. FSIS Directive 7120.1. 2012; pp. 1–52. Available online: https://www.fsis.usda.gov/sites/default/files/media_file/2020-07/7120.1.pdf (accessed on 29 September 2021).
13. Cano, C.; Meneses, Y.; Chaves, B.D. Application of peroxyacetic acid for decontamination of raw poultry products and comparison to other commonly used chemical antimicrobial interventions—A Review. *J. Food Prot.* **2021**, *84*, 1772–1783. [CrossRef]
14. Wideman, N.; Bailey, M.; Bilgili, S.F.; Thippareddi, H.; Wang, L.; Bratcher, C.; Sanchez-Plata, M.; Singh, M. Evaluating best practices for *Campylobacter* and *Salmonella* reduction in poultry processing plants. *Poult. Sci.* **2016**, *95*, 306–315. [CrossRef]
15. *Committee on Acute Exposure Guideline Levels Acute Exposure Guideline Levels for Selected Airborne Chemicals*; The National Academies of Sciences, Engineering, and Medicine: Washington, DC, USA, 2010; Volume 8, pp. 6–13. ISBN 8436944542. Available online: http://www.nap.edu/catalog/12770.html (accessed on 22 November 2021).
16. Pechacek, N.; Osorio, M.; Caudill, J.; Peterson, B. Evaluation of the toxicity data for peracetic acid in deriving occupational exposure limits: A minireview. *Toxicol. Lett.* **2015**, *233*, 45–57. [CrossRef]
17. Food Safety and Inspection Service Data Collection and Reports: Sampling results for FSIS Regulated Products. Available online: https://www.fsis.usda.gov/science-data/sampling-program/sampling-results-fsis-regulated-products (accessed on 29 September 2021).
18. USDA/FSIS. Young Chicken Baseline Data Collection. 2006. Available online: https://www.fsis.usda.gov/sites/default/files/media_file/2020-07/Young_Chicken_Baseline_Data_Collection.pdf (accessed on 15 December 2021).
19. Ramirez-Hernandez, A.; Bugarel, M.; Kumar, S.; Thippareddi, H.; Brashears, M.M.; Sanchez-Plata, M.X. Phenotypic and genotypic characterization of antimicrobial resistance in *Salmonella* strains isolated from chicken carcasses and parts collected at different stages during processing. *J. Food Prot.* **2019**, *82*, 1793–1801. [CrossRef]
20. Sánchez, M.X.; Fluckey, W.M.; Brashears, M.M.; McKee, S.R. Microbial profile and antibiotic susceptibility of *Campylobacter* spp. and *Salmonella* spp. in broilers processed in air-chilled and immersion-chilled environments. *J. Food Prot.* **2002**, *65*, 948–956. [CrossRef]
21. Fluckey, W.M.; Sanchez, M.X.; McKee, S.R.; Smith, D.; Pendleton, E.; Brashears, M.M. Establishment of a microbiological profile for an air-chilling poultry operation in the united states. *J. Food Prot.* **2003**, *66*, 272–279. [CrossRef]
22. Ramirez-Hernandez, A.; Carrascal-Camacho, A.K.; Varón-García, A.; Brashears, M.M.; Sanchez-Plata, M.X. Genotypic characterization of antimicrobial resistant *Salmonella* spp. Strains from three poultry processing plants in colombia. *Foods* **2021**, *10*, 491. [CrossRef] [PubMed]
23. Brashears, M.M.; Chaves, B.D. The diversity of beef safety: A global reason to strengthen our current systems. *Meat Sci.* **2017**, *132*, 59–71. [CrossRef] [PubMed]
24. Wages, J.A.; Feye, K.M.; Park, S.H.; Kim, S.A.; Ricke, S.C. Comparison of 16S rDNA next sequencing of microbiome communities from post-scalder and post-picker stages in three different commercial poultry plants processing three classes of broilers. *Front. Microbiol.* **2019**, *10*, 972. [CrossRef]
25. Skovgaard, N. Risk assessments of *Salmonella* in eggs and broiler chickens. Interpretative summary. *Int. J. Food Microbiol.* **2004**, *91*, 223. [CrossRef]
26. Cox, L.A. Higher line speed in young chicken slaughter establishments does not predict increased *Salmonella* contamination risks. *Poult. Sci.* **2021**, *100*, 635–642. [CrossRef]
27. Kumar, S.; Singh, M.; Cosby, D.E.; Cox, N.A.; Thippareddi, H. Efficacy of peroxy acetic acid in reducing *Salmonella* and *Campylobacter* spp. populations on chicken breast fillets. *Poult. Sci.* **2020**, *99*, 2655–2661. [CrossRef] [PubMed]
28. Buncic, S.; Sofos, J. Interventions to control *Salmonella* contamination during poultry, cattle and pig slaughter. *Food Res. Int.* **2012**, *45*, 641–655. [CrossRef]
29. Oscar, T. *Salmonella* Prevalence Alone Is Not a Good Indicator of Poultry Food Safety. *Risk Anal.* **2021**, *41*, 110–130. [CrossRef] [PubMed]
30. Lambertini, E.; Ruzante, J.M.; Kowalcyk, B.B. The Public Health Impact of Implementing a Concentration-Based Microbiological Criterion for Controlling *Salmonella* in Ground Turkey. *Risk Anal.* **2021**, *41*, 1376–1395. [CrossRef]
31. U.S. Department of Health and Human Services. Campylobacter Goals (Healthy People 2030). Available online: https://health.gov/healthypeople/objectives-and-data/browse-objectives/foodborne-illness/prevent-increase-proportion-macrolide-antibiotic-resistant-campylobacter-infections-fs-06 (accessed on 15 December 2021).

32. U.S. Department of Health and Human Services. Salmonella Goals (Healthy People 2030). Available online: https://health.gov/healthypeople/objectives-and-data/browse-objectives/foodborne-illness/prevent-increase-proportion-nontyphoidal-salmonella-infections-are-resistant-multiple-drug-classes-fs-05 (accessed on 15 December 2021).
33. Lambertini, E.; Ruzante, J.M.; Chew, R.; Apodaca, V.L.; Kowalcyk, B.B. The public health impact of different microbiological criteria approaches for *Salmonella* in chicken parts. *Microb. Risk Anal.* **2019**, *12*, 44–59. [CrossRef]

foods

Article

Antimicrobial Resistance and Virulence of Non-Typhoidal *Salmonella* from Retail Foods Marketed in Bangkok, Thailand

Thida Kong-Ngoen [1], Sirijan Santajit [2,3], Witawat Tunyong [1], Pornpan Pumirat [1], Nitat Sookrung [4], Wanpen Chaicumpa [5] and Nitaya Indrawattana [1,*]

1 Department of Microbiology and Immunology, Faculty of Tropical Medicine, Mahidol University, Bangkok 10400, Thailand; thida.kon@mahidol.ac.th (T.K.-N.); witawat.tun@mahidol.ac.th (W.T.); pornpan.pum@mahidol.ac.th (P.P.)
2 Department of Medical Technology, School of Allied Health Sciences, Walailak University, Nakhon Si Thammarat 80160, Thailand; sirijan.sa@wu.ac.th
3 Research Center in Tropical Pathobiology, Walailak University, Nakhon Si Thammarat 80160, Thailand
4 Biomedical Research Incubator Unit, Department of Research, Faculty of Medicine Siriraj Hospital, Mahidol University, Bangkok 10700, Thailand; nitat.soo@mahidol.ac.th
5 Center of Research Excellence on Therapeutic Proteins and Antibody Engineering, Department of Parasitology, Faculty of Medicine Siriraj Hospital, Mahidol University, Bangkok 10700, Thailand; wanpen.cha@mahidol.ac.th
* Correspondence: nitaya.ind@mahidol.ac.th; Tel.: +66-2-354-9100 (ext. 1598)

Abstract: Nontyphoidal-*Salmonella* bacteria cause foodborne gastroenteritis that may lead to fatal bacteremia, osteomyelitis, and meningitis if not treated properly. The emergence of multidrug-resistant *Salmonella* strains is a global public health threat. Regular monitoring of genotypes and phenotypes of *Salmonella* isolated from humans, animals, foods, and environments is mandatory for effective reduction and control of this food-borne pathogen. In this study, antimicrobial-resistant and virulent genotypes and phenotypes of *Salmonella* isolated from retail food samples in Bangkok, Thailand, were investigated. From 252 raw food samples, 58 *Salmonella* strains that belonged only to serotype Enteritidis were isolated. Disc diffusion method showed that all isolates were still sensitive to amikacin and carbapenems. More than 30% of the isolates were resistant to ampicillin, tetracycline, and ciprofloxacin. Twenty isolates resist at least three antibiotic classes. Minimum inhibitory concentration tests showed that 12.07% of the isolates produced extended-spectrum β-Lactamase. Polymerase chain reaction indicated that 32.76, 81.03, 39.66, and 5.17% of the isolates carried $bla_{\text{TEM-1}}$, *tet*A, *sul*2, and *dfr*A7, respectively. All isolates were positive for invasion-associated genes. Effective prevention and control of *Salmonella* (as well as other food-borne pathogens) is possible by increasing public awareness and applying food hygienic practices. Active and well harmonised "One Health" co-operation is required to effectively control food-borne zoonosis.

Keywords: food-borne salmonellosis; *Salmonella* Enteritidis; multi-drug resistance; invasion genes bacterial virulence

Citation: Kong-Ngoen, T.; Santajit, S.; Tunyong, W.; Pumirat, P.; Sookrung, N.; Chaicumpa, W.; Indrawattana, N. Antimicrobial Resistance and Virulence of Non-Typhoidal *Salmonella* from Retail Foods Marketed in Bangkok, Thailand. *Foods* **2022**, *11*, 661. https://doi.org/10.3390/foods11050661

Academic Editors: Antonio Bevilacqua, Antonio Afonso Lourenco, Catherine Burgess and Timothy Ells

Received: 21 January 2022
Accepted: 22 February 2022
Published: 24 February 2022

1. Introduction

Salmonella causes food-borne gastroenteritis (salmonellosis) with high and increasing prevalence worldwide [1–3]. The bacteria are ubiquitously present in the environment and throughout the food chain, i.e., farm-to-folk. Humans become infected through the consumption of contaminated water or foods mainly of animal origins, such as poultry meat, eggs, pork, beef, dairy products, and ready-to-eat produce [4,5]. *Salmonella* serovars with human host preference include *S.* Typhimurium and *S.* Enteritidis [6,7]. Clinical symptoms of salmonellosis usually begin 6–8 h to 7 days after infection and are characterised by abdominal cramp, fever, and diarrhoea [8]. The diseases can be self-limited in healthy individuals but may be severe, which requires prompt medical attention and may also be life-threatening if the bacteria invade beyond the gastrointestinal tract [9]. According to the World Health Organization (WHO), *Salmonella* is one of the key causative agents of diarrheal disease, which inflicts not only huge medical intervention expenses but also loss of productivity [10].

Pathogenesis of *Salmonella* is related to the abundance of the virulence genes in the chromosomally located *Salmonella* pathogenicity islands (SPIs) [11,12]. Among the virulence-associated genes are *inv*A, which encodes the type III secretion system, and the *hil*A, which encodes an OmpR/ToxR family transcriptional regulator that activates the expression of invasion genes required for *Salmonella* invasion into host intestinal epithelial cells [13–15]. Besides, *Salmonella* bacteria also harbour plasmids carrying a myriad of antimicrobial resistance genes, such as $bla_{TEM\text{-}1}$ (class A broad-spectrum β-lactamase, TEM-1), $bla_{CMY\text{-}2}$ (class C β-lactamase CMY-2), *tet*A (tetracycline efflux major facilitator superfamily (MFS) transporter, TetA), *tet*C (tetracycline resistance-associated transcriptional repressor, TetC), *sul*2 (sulfonamide-resistance gene), and *dfr*A7 (dihydrofolate reductase, a single gene cassette within the class 1 integrons). These genes contribute to drug-resistant phenotypes, which are currently the major global public health worrisome [16–22].

Antibiotic resistance among bacteria is a global phenomenon. Regular monitoring of serotypes and drug-resistant phenotypes and genotypes of *Salmonella* that contaminate foods may help track the cause of the food-borne diseases and may lead to appropriate food safety policy for intervention, prevention, and/or effective treatment measures of food-borne illnesses. Therefore, in this study, we assessed the prevalence of antimicrobial phenotypes and drug resistance-associated and virulence genes in *Salmonella* isolated from retail food samples in the Bangkok metropolitan area.

2. Materials and Methods

2.1. Sample Collection and Bacterial Isolation and Identification

Five different food categories (chicken, n = 44; pork and beef, n = 28; seafood, n = 60; fruits and vegetables, n = 60; and dairy products, n = 60) comprising 252 samples were collected from 19 wet markets and 2 supermarkets between October and December 2017. All markets are located in the central and peripheral districts of the Bangkok Metropolitan area. Food samples were maintained in sterile bags on ice and transferred to the laboratory within 2 h.

Food samples were processed according to the international standard, five-step method of the ISO protocol: 6579: 2002 Microbiology of Food and Animal Feeding Stuffs-Horizontal Method for the Detection of *Salmonella* spp. [23,24]. Firstly, individual samples were pre-enriched in a non-selective medium. Twenty-five grams of each sample was placed in a sterile 500 mL flask containing 225 mL of Trypticase Soy Broth and incubated at 37 °C for 18–24 h. Then, 0.1 mL of each overnight culture was inoculated into 10 mL of selective enrichment medium, Rappaport-Vassiliadis Soya broth (Merck, Darmstadt, Germany), and incubated at 42 °C for 24 h. The cultures (0.1 mL aliquots) were spread onto selective agar plates, i.e., xylose lysine deoxycholate agar (XLD) and *Salmonella–Shigella* agar (SS) selective plates, and the plates were incubated at 37 °C for 18–24 h. Suspected *Salmonella* colonies (small red colonies with/without central black dots on XLD agar and translucent colourless colonies with/without central black dots on SS agar) were subjected to conventional biochemical assays for *Salmonella* verification, including triple sugar iron (TSI) agar utilisation, deamination of lysine, ornithine decarboxylation, citrate and urease productions, and indole formation, as well as motility testing [25].

2.2. Serotyping of the Salmonella Isolates

All *Salmonella* isolates were serotyped using polyvalent O and H antisera by slide agglutination technique (Kauffmann–White–Le Minor scheme) [26]. The assay was performed according to the manufacturer's instructions (Serosystem, Clinag, Bangkok, Thailand). Briefly, individual *Salmonella* colonies were suspended in normal saline solution on glass slides. They were mixed separately with 9 polyvalent *Salmonella* antisera reagents in a 1:1

ratio, and the slides were rocked in a circular motion for 30 s. Bacterial agglutination was visually observed. Strains giving negative or positive agglutinations were recorded.

2.3. Determination of Intestinal Cell Invasion by Salmonella Isolates

The ability of the isolated *Salmonella* strains to invade human colon carcinoma cells (Caco-2 cell line) was investigated. Confluent Caco-2 cell monolayer was established in 24-well tissue culture plates (approximately 2×10^5 cells/well) containing Dulbecco's modified Eagle's medium (DMEM) (Gibco, NY, USA) supplemented with 10% fetal bovine serum and 50 μg/mL gentamicin at 37 °C in 5% CO_2 atmosphere. The monolayers were rinsed twice in phosphate-buffered saline, pH 7.4 (PBS). Cells were infected with individual *Salmonella* strains at a multiplicity of infection (MOI) 1:50 [27]. Plates were incubated at 37 °C in 5% CO_2 incubator for 4 h. The cells were rinsed to remove extracellular bacteria and replenished with DMEM containing gentamicin (50 μg/mL) for 1.5 h. Cells were then rinsed with PBS and stained with Giemsa reagent. *Salmonella* invasion into the Caco-2 cells was observed under inverted microscopy (200 and 400× magnifications) (Zeiss, Jena, Germany). Alternatively, the infected cells were lysed by adding 1% Triton X-100 (Sigma); the lysate was spread on an LB plate and incubated at 37 °C for 24 h. The presence of bacterial colonies on the cultured plate indicates the invasive ability of the bacterial isolate.

2.4. Antimicrobial Resistance Profiles

Antimicrobial susceptibility was evaluated based on Clinical and Laboratory Standards Institute 2017 (CLSI 2017) guidelines using the disc diffusion method. Briefly, *Salmonella* isolates were aerobically cultured in 10 mL of Mueller–Hinton (MH) broth (Oxoid, Hampshire, UK) at 37 °C for 24 h. Overnight cultures were adjusted to an optical density of 0.5 MacFarland units. The bacterial suspensions were aseptically spread onto MH agar plates, and the plates were allowed to dry for 2–4 min. Individual antimicrobial discs were placed on the surface using a disc dispenser, and the plates were incubated at 37 °C for 24 h. The tested antibiotics were ampicillin (10 μg), ampicillin/sulbactam (10 μg/10 μg), piperacillin/tazobactam (100 μg/10 μg), cefepime (30 μg), cefotaxime (30 μg), ceftazidime (30 μg), ceftriaxone (30 μg), gentamicin (10 μg), amikacin (30 μg), ertapenem (10 μg), meropenem (10 μg), imipenem (10 μg), tetracycline (30 μg), ciprofloxacin (5 μg), and trimethoprim/sulfamethoxazole (1.75/23.25 μg) (Oxoid). Extended-spectrum β-lactamase (ESBL) production was also determined using the combination disc test comprising ceftazidime with and without clavulanate and cefotaxime with and without clavulanate (Oxoid). A positive test was defined as a $\geq$5 mm difference in zone diameter between the respective two discs. The CLSI 2017 criteria were followed for the interpretation of the antimicrobial susceptibility results.

2.5. Polymerase Chain Reaction for Determination of Drug Resistance and Virulence Genes of the Salmonella Isolates

All *Salmonella* isolates were screened for the presence of virulence genes (*inv*A and *hil*A) and antimicrobial resistance genes (*tet*A, *tet*C, $bla_{\text{TEM-1}}$, $bla_{\text{CMY-2}}$, *sul*2, and *dfr*A7) by using PCR. Genomic DNA was extracted from each *Salmonella* culture using the conventional boiling method [27]. Two millilitres of each bacterial culture were centrifuged at 14,000× *g* for 5 min. Sterile distilled water (600 μL) was added to the pellet and re-centrifuged. The supernatant was discarded, and 200 μL of sterile distilled water was added to the pellet. The sample was then placed in a 100 °C heat-block for 10 min, immediately cooled on ice for 5 min, and centrifuged at 14,000× *g* for 5 min. The supernatant was used as a PCR template.

PCR was conducted using primers listed in Table 1. The PCR reaction mixture (25 μL) contained 3 μL of DNA template, 2.5 μL of 10× *Taq* buffer, 2 mM $MgCl_2$, 0.2 mM dNTP, 1 μM each primer, and 1 U of *Taq* DNA polymerase (Thermo Fisher Scientific, Waltham, MA, USA). The thermal cycles were initial denaturation at 94 °C for 5 min, 35 cycles of denaturation at 94 °C for 45 s, annealing at 52–60 °C for 40 s, extension at 72 °C for 40 s and a final extension at 72 °C for 7 min. *Salmonella* Enteritidis ATCC 13076 and constructed plasmids containing the antibiotic-resistant genes served as positive controls, while buffer alone (without DNA template) served as a negative control. The PCR products were electrophoresed on 1.5% (*w*/*v*) agarose gels in 100 mL of 1× TAE buffer and stained with ethidium bromide. DNA bands were visualised using the ChemiDoc MP imaging system (Bio-Rad, Hercules, CA, USA).

2.6. Statistical Analysis

The statistical analysis and data comparison were performed using one-way ANOVA in GraphPad Prism version 9 (La Jolla, CA, USA). The *p*-value < 0.05 was considered statistically significant.

Table 1. PCR primers used for amplification of different drug resistance-associated and virulence genes.

Gene Name	Oligonucleotide Sequence (5′-3′)	Product Size (bp)	Annealing Temperature (°C)	Reference
*inv*A	Forward: ACAGTGCTCGTTTACGACCTGAAT Reverse: AGACGACTGGTACTGATCGATAAT	244	60	[28]
*hil*A	Forward: CGTGAAGGGATTATCGCAGT Reverse: GTCCGGGAATACATCTGAGC	296	56	[29]
$bla_{\text{TEM-1}}$	Forward: TTGGGTGCACGAGTGGGT Reverse: TAATTGTTGCCGGGAAGC	504	56	[30]
$bla_{\text{CMY-2}}$	Forward: ATAACCACCCAGTCACGC Reverse: CAGTAGCGAGACTGCGCA	631	52	[31]
*sul*2	Forward: CGGCATCGTCAACATAACC Reverse: GTGTGCGGATGAAGTCAG	405	60	[31]
*tet*A	Forward: GCTACATCCTGCTTGCCTTC Reverse: CATAGATCGCCGTGAAGAGG	210	52	[32]
*tet*C	Forward: CTTGAGAGCCTTCAACCCAG Reverse: ATGGTCGTCATCTACCTGCC	418	52	[32]
*dfr*A7	Forward: GGTAATGGCCCTGATATCCC Reverse: TGTAGATTTGACCGCCACC	265	50	[33]

3. Results

3.1. Prevalence and Serotypes of Salmonella in Retail Food Samples

Fifty-eight *Salmonella* isolates (23%) were recovered from a total of 252 retail food samples. All of them belonged to serovar Enteritidis. The isolated bacteria were from chicken (36 isolates, 62.07%), pork (16 isolates, 27.59%), and beef (6 isolates, 10.34%). The comparative prevalence of S. Enteritidis isolated from chicken and pork, chicken and beef, chicken and fruits, chicken and vegetables, pork and fruits, and pork and vegetables were different ($p < 0.001$). The *Salmonella* prevalence in pork and beef samples was also different ($p < 0.05$). Nevertheless, no difference was found between samples of beef and fruits, beef and vegetables, and fruits and vegetables ($p > 0.05$). The isolates were further classified into six different groups, i.e., B ($n = 17$; 29.31%), C ($n = 22$; 37.93%), E ($n = 15$; 25.86%), G ($n = 1$; 1.72%), and I ($n = 2$; 3.45%), and non-A–I ($n = 1$; 1.72%). Group C was predominant in this study (Table 2).

Table 2. Serotypes, antibiotic resistance profiles, virulence genes, and drug resistance-associated genes of *Salmonella* Enteritidis isolates of this study.

Salmonella Isolates	Source	Antibiotic-Resistant Profile	*Salmonella* Serotype	Virulence Gene		Drug Resistance Associated Gene					
				*inv*A	*hil*A	*tet*A	*tet*C	bla_{TEM-1}	bla_{CMY-2}	*sul2*	*dfr*A7
Sal1	pork	AMP, TE, and SXT	*B*	+	+	+	–	+	–	+	–
Sal2	pork	AMP, TE, and SXT	*B*	+	+	+	–	+	–	+	–
Sal3	pork	AMP and SXT	*E*	+	+	+	–	+	–	+	+
Sal4	pork	AMP, CTX, CRO, FEP, GN, and TE	*E*	+	+	+	–	–	–	+	–
Sal5	pork	AMP, CTX, CRO, FEP, GN, and TE	*E*	+	+	+	–	–	–	+	–
Sal6	pork	AMP, TE, CIP, and SXT	*E*	+	+	+	–	+	–	+	+
Sal7	pork	AMP, CTX, CRO, FEP, GN, and TE	*E*	+	+	–	–	–	–	+	–
Sal8	pork	AMP and TE	*C*	+	+	+	–	+	–	+	–
Sal9	pork	–	*E*	+	+	+	–	–	–	–	–
Sal10	pork	AMP, CTX, CRO, FEP, GN, and TE	*E*	+	+	+	–	–	–	–	–
Sal11	pork	–	*E*	+	+	+	–	–	–	–	–
Sal12	pork	AMP and TE	*B*	+	+	+	–	+	–	+	–
Sal13	pork	AMP	*C*	+	+	+	–	–	–	–	–
Sal14	pork	AMP, TE, CIP, and SXT	*B*	+	+	–	–	+	–	–	–
Sal15	pork	AMP, CTX, CRO, FEP, GN, and TE	*E*	+	+	+	–	–	–	–	–
Sal16	pork	AMP, SAM, CAZ, CTX, CRO, FEP, GN, and TE	*B*	+	+	+	–	+	–	+	–
Sal17	chicken	AMP, SAM, TE, and SXT	*B*	+	+	+	–	+	–	–	–
Sal18	chicken	–	*I*	+	+	+	–	–	–	–	–
Sal20	chicken	–	*I*	+	+	+	–	–	–	–	–
Sal21	chicken	–	*C*	+	+	+	–	–	–	–	–
Sal22	chicken	–	*C*	+	+	–	–	–	–	–	–
Sal23	chicken	CIP	*C*	+	+	+	–	–	–	–	–
Sal24	chicken	CIP	*C*	+	+	+	–	–	–	–	–
Sal25	chicken	–	*E*	+	+	+	–	–	–	–	–
Sal26	chicken	TE and CIP	*B*	+	+	+	–	–	–	–	–
Sal27	chicken	CIP	*C*	+	+	+	–	–	–	–	–
Sal28	chicken	–	*C*	+	+	+	–	–	–	–	–
Sal29	chicken	–	*Non A-I*	+	+	+	–	–	–	–	–
Sal30	chicken	AMP, TE, CIP, and SXT	*B*	+	+	+	–	+	–	–	–
Sal31	chicken	AMP, TE, CIP, and SXT	*B*	+	+	+	–	+	–	–	–
Sal32	chicken	TE	*C*	+	+	+	–	–	–	+	–
Sal33	chicken	CIP	*C*	+	+	+	–	–	–	–	–
Sal34	chicken	TE and CIP	*C*	+	+	+	–	–	–	+	–
Sal35	chicken	TE and CIP	*C*	+	+	–	–	–	–	+	–
Sal36	chicken	AMP, TE, and SXT	*B*	+	+	+	–	+	–	–	–
Sal37	chicken	TE	*C*	+	+	+	–	–	–	+	–
Sal38	chicken	–	*C*	+	+	+	–	–	–		–
Sal39	chicken	AMP, TE, and SXT	*B*	+	+	+	–	+	–	+	–
Sal40	chicken	AMP, SAM, TE, and CIP	*C*	+	+	+	–	+	–	+	–
Sal42	chicken	–	*C*	+	+	+	–	–	–	–	–
Sal43	chicken	TE	*B*	+	+	+	–	–	–	+	–

Table 2. *Cont.*

Salmonella Isolates	Source	Antibiotic-Resistant Profile	*Salmonella* Serotype	Virulence Gene		Drug Resistance Associated Gene					
				*inv*A	*hil*A	*tet*A	*tet*C	$bla_{TEM\text{-}1}$	$bla_{CMY\text{-}2}$	*sul2*	*dfr*A7
Sal44	chicken	GN, TE, CIP, and SXT	*B*	+	+	+	–	+	–	+	–
Sal45	chicken	CIP and SXT	*E*	+	+	+	–	–	–	–	+
Sal46	chicken	AMP, TE, and SXT	*B*	+	+	+	–	+	–	–	–
Sal47	chicken	AMP and CIP	*C*	+	+	+	–	–	–	–	–
Sal48	chicken	–	*G*	+	+	+	–	–	–	–	–
Sal50	chicken	AMP, TE, and CIP	*E*	+	+	–	–	+	–	+	–
Sal52	chicken	TE	*C*	+	+	+	–	–	–	+	–
Sal53	chicken	TE and CIP	*C*	+	+	+	–	–	–	+	–
Sal54	chicken	CIP	*C*	+	+	+	–	–	–	+	–
Sal55	chicken	AMP and TE	*C*	+	+	+	–	+	–	+	–
Sal56	chicken	AMP, CTX, CRO, FEP, GN, TE, and CIP	*B*	+	+	+	–	+	–	–	–
Sal57	beef	–	*B*	+	+	–	–	–	–	–	–
Sal58	beef	–	*B*	+	+	–	–	–	–	–	–
Sal59	beef	–	*E*	+	+	–	–	–	–	–	–
Sal60	beef	–	*E*	+	+	–	–	–	–	–	–
Sal62	beef	–	*E*	+	+	–	–	–	–	–	–
Sal63	beef	–	*C*	+	+	–	–	–	–	–	–
		Number of isolates (%)		58 (100)	58 (100)	0 (0)	19 (32.76)	0 (0)	23 (39.66)		3 (5.17)

+ represent as "present "; – represent as "not present".

3.2. Antimicrobial and Virulence Genotypes of the Salmonella Isolates

PCR was used to determine drug resistance and virulence genes of the *Salmonella* isolates. The drug resistance and virulence genes that were detected included *invA*, *hilA*, *tetA*, bla_{TEM-1}, *sul2*, and *dfrA7*, of which their PCR amplicon sizes were 244, 296, 210, 504, 405, and 265 base pairs (bp), respectively (Figure 1). The invasion operon genes, *invA* and *hilA*, were detected in all isolates. The bla_{TEM-1} (n = 19; 32.76%), *tetA* (n = 47; 81.03%), *sul2* (n = 23; 39.66%) and *dfrA7* (n = 3; 5.17%) genes were carried by the resistance strains, a clear difference was noticed in the occurrence of these genes among the isolates. None of the isolates was positive for bla_{CMY-2} and *tetC* genes. The pork and chicken isolates were positive for at least one antimicrobial resistance-associated gene. The *tetA* was the most prevalent gene among the *Salmonella* isolated from pork and chicken, followed by *sul2*. None of the beef isolates carried the antimicrobial resistance-associated gene, and all of them were not resistant to any of the antibiotics tested (Table 2).

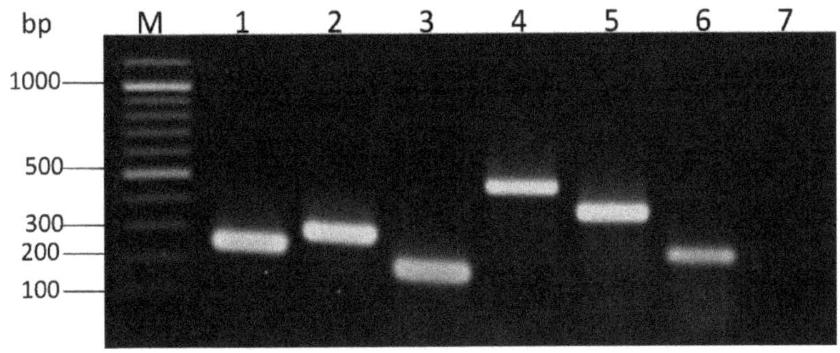

Figure 1. Molecular detection of virulence and drug-resistance associated genes of *Salmonella* isolates using PCR methods. Lane M: 100 bp plus DNA ladder; Lane 1: the representative *invA* amplicon; Lane 2: the representative *hilA* amplicon; Lane 3: the representative *tetA* amplicon; Lane 4: the representative bla_{TEM-1} amplicon; Lane 5: the representative *sul2* amplicon; Lane 6: the representative *dfrA7* amplicon, and Lane 7: negative control.

3.3. Antimicrobial Phenotypes of the Salmonella Isolates

Antibiotic sensitivity testing was performed for the 58 *Salmonella* isolates, and the results are shown in Table 3. All isolates were sensitive to ertapenem and amikacin. Twenty-six isolates (44.83%) were resistant to ampicillin (penicillin group); 3 isolates (5.17%) were resistant to ampicillin/sulbactam (β-lactam combination agents); 7 isolates (12.07%) each were resistant to cefepime, cefotaxime, and ceftriaxone, and 1 isolate resisted ceftazidime (cephalosporin group); 7 isolates (12.07%) resisted gentamicin (aminoglycoside group); 32 isolates (55.17%) resisted tetracycline (tetracycline group); 20 isolates (34.48%) resisted ciprofloxacin (fluoroquinolone group); and 12 isolates (20.69%) resisted trimethoprim/sulfamethoxazole (folate pathway antagonist group). Seven isolates (12.07%) were ESBL producing *S.* Enteritidis. Among 58 isolates, 20 (34.48%) were multi-drug resistant (MDR); *Salmonella* group B were resistant to at least three antibiotic classes (Table 3). A heatmap of the distribution of antimicrobial resistance genes and their phenotypes is illustrated in Figure 2. The isolates with phenotypic resistance to at least one antibiotic are displayed.

Table 3. The antibiotic resistance phenotypes of the *Salmonella* isolates.

Antimicrobial Agent	Number of Isolates Tested	Anti-Biogram Phenotypes of *Salmonella* Isolates Number of Isolates (%)		
		Sensitive	Intermediate	Resistant
Group Penicillin				
ampicillin (AMP)	58	32 (55.17)	0 (0)	26 (44.83)
Group Combined β-lactam agents				
ampicillin/sulbactam (SAM)	58	49 (84.49)	6 (10.34)	3 (5.17)
piperacillin/tazobactam (TZP)	58	56 (96.55)	2 (3.45)	0 (0)
Group Cephalosporin				
cefepime (FEP)	58	51 (87.93)	0 (0)	7 (12.07)
cefotaxime (CTX)	58	47 (81.03)	4 (6.90)	7 (12.07)
ceftazidime (CAZ)	58	52 (89.66)	5 (8.62)	1 (1.72)
ceftriaxone (CRO)	58	51 (87.93)	0 (0)	7 (12.07)
Group Aminoglycoside				
gentamicin (GN)	58	51 (87.93)	0 (0)	7 (12.07)
amikacin (AK)	58	58 (100)	0 (0)	0 (0)
Group Carbapenem				
ertapenem (ERT)	58	58 (100)	0 (0)	0 (0)
meropenem (MEM)	58	46 (79.11)	12 (20.89)	0 (0)
imipenem (IPM)	58	54 (93.10)	4 (6.90)	0 (0)
Group Tetracycline				
tetracycline (TE)	58	26 (44.83)	0 (0)	32 (55.17)
Group Fluoroquinolone				
ciprofloxacin (CIP)	58	4 (6.90)	34 (58.62)	20 (34.48)
Group Folate pathway antagonist				
trimethoprime/sulfamethoxazole (SXT)	58	46 (79.31)	0 (0)	12 (20.69)
ESBL	**Number of isolates tested**	**Number of positive isolates (%)**	**Number of negative isolates (%)**	
ceftazidime	58	7 (12.07)	51 (87.93)	
cefotaxime	58	7 (12.07)	51 (87.93)	

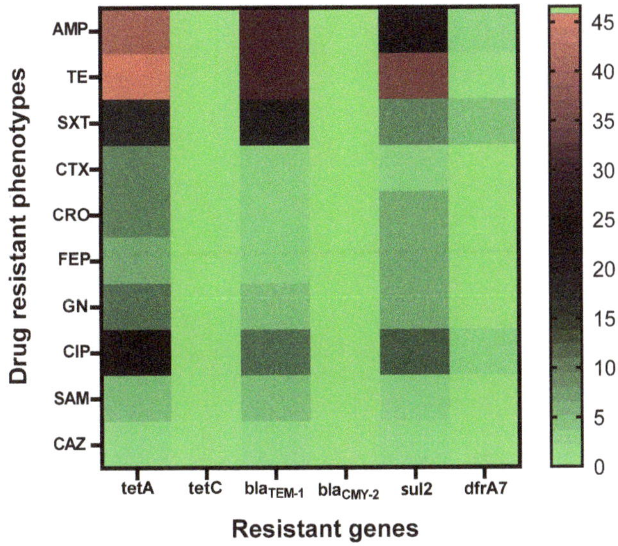

Figure 2. Heatmap of percent distribution for drug-resistant phenotypes and genotypes of *S.* Enteritidis isolates that were present in at least one isolate with antibiotic-resistant phenotype. The colored strip depicts the percentage of genes associated with a particular antibiotic-resistant phenotype. Created using GraphPad Prism version 9 (La Jolla, CA, USA).

3.4. Caco-2 Invasion Assay on Isolates

The ability of *S.* Enteritidis isolates to invade human intestinal epithelial (Caco-2) cells was determined. All 58 isolates, which carried *inv*A and *hil*A genes, could invade the Caco-2 cells. The cell invasion of the representative isolate is shown in Figure 3.

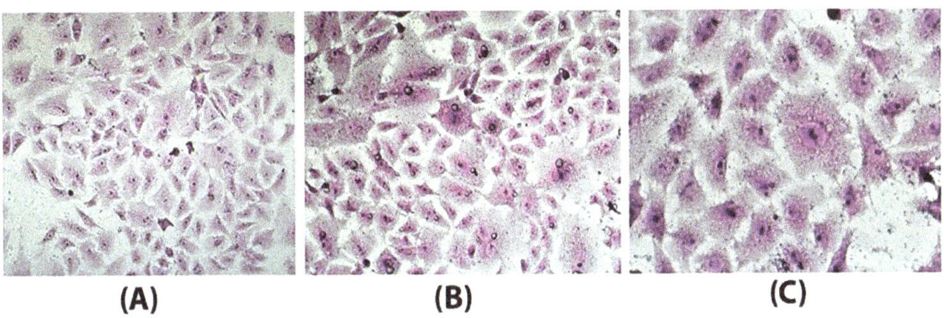

Figure 3. Microscopic appearance of Giemsa's stained CaCo-2 cells: (**A**) before (**B**,**C**) and after infecting with the representative *Salmonella* Enteritidis isolate no. 44 (Sal44). Bacteria are predominantly seen in the CaCo-2 cells' cytoplasm (original magnification 200× and 400×, respectively).

4. Discussion

Regular monitoring of serotypes, antimicrobial-resistant characteristics, and virulence of food-borne pathogenic bacteria, particularly *Salmonella enterica*, can provide useful epidemiological information on food-borne bacterial infections in a locality [34]. In recent decades, *S.* Enteritidis has been identified as the predominant causative agent of salmonellosis in Thailand [35,36]. In this study, 23% of the raw food samples collected from open markets in the Bangkok metropolitan region were found to be contaminated with *Salmonella*. The contaminated food samples were solely meat (chicken > pork > beef), while seafood, fruits, vegetables, and dairy products were not contaminated. All contaminated *Salmonella* isolates belonged to serovar Enteritidis, of which group C was predominant. When compared with the prevalence of *S.* Enteritidis from raw foods in other countries, e.g., abattoirs in Iran and butcher shops and supermarkets in Pakistan where the prevalence rates were 43 and 37.5%, respectively, the bacterial prevalence in our study was less [37,38].

Drug susceptibility testing data revealed that even though the *S.* Enteritidis isolated in this study were resistant to many groups of antibiotics, including penicillin, combined β-lactam agents, cephalosporins, aminoglycosides, tetracyclines, fluoroquinolones, and folate pathway antagonists, most of these MDR *Salmonella* strains were still sensitive to amikacin and carbapenems. Even though the isolates of this study showed high resistance to ampicillin, tetracycline, and ciprofloxacin, the prevalence of resistant isolates was still less compared to those isolated in Brazil, Iran, and China [39–41].

Invasion into cultured epithelial cells has been routinely used for determining *Salmonella* virulence [42–46]. Genotypic and phenotypic analysis of the *S.* Enteritidis isolates of this study revealed that the bacteria carried invasion genes (*inv*A and *hil*A). Nevertheless, they showed different degrees of invasiveness when tested by the invasion assay using intestinal epithelial (Caco-2) cells. The results conformed to those reported previously by others [47–51]. Most MDR *Salmonella* isolates were found to carry the antimicrobial-associated genes, namely, $bla_{\text{TEM-1}}$, *tet*A, *sul*2, and *dfr*A7 [28,52]. The prevalence of drug resistance genes was highest for *tetA*, followed by *sul*2, $bla_{\text{TEM-1}}$, and *dfr*A7. No isolate carried *tet*C and $bla_{\text{CMY-2}}$. Detail analysis of the entire genomes of the isolates by using next-generation sequencing should be performed further to provide the insight information for guiding appropriate treatment decisions and allow rapid tracking of transmission of the drug-resistant clones.

Epidemics of human salmonellosis are generally associated with a particular prevalent serovar and serotype of *S. enterica*. Epidemic tracking of the bacterial pathogens, e.g., through identification of the causative strain origin as well as the antimicrobial susceptibility pattern and their virulence characteristics in an outbreak, can be readily performed either phenotypically or genotypically, or both [29]. It is also noteworthy that retail food products undergo extensive processing and handling during production, which potentially enhance the risk of contamination [30]. Appropriate food hygienic education for end-consumers must be regularly implemented. Since the majority of food-borne diseases, including salmonellosis, are zoonotic, thus, improving food hygiene through health education and "One Health" approach should be practiced at all levels, i.e., from a locale to a nation-wide and global responsible practices.

5. Conclusions

In conclusion, the findings of this study supported the notion of the divergence of *Salmonella* serotypes isolated from a variety of raw food samples from the opened market and hypermarket in Bangkok and its periphery, Thailand. The findings also provided insight into the molecular characterisation of virulence- and drug-resistance traits, as well as the antimicrobial susceptibility pattern of the bacterial pathogen. The spread of MDR strains of *Salmonella* isolates with the cell invasion potential was become growing continuously. This requires good planning and effective control programs to prevent and manage infections for their spreading to community and public health.

Author Contributions: Conceptualization, N.I. and W.C.; methodology, T.K.-N., S.S. and W.T.; software, T.K.-N. and S.S.; validation, P.P. and N.S.; formal analysis, T.K.-N.; investigation, N.I.; resources, N.I. and N.S.; writing—original draft preparation, T.K.-N., N.I. and S.S.; writing—review and editing, T.K.-N. and N.I.; visualization, T.K.-N.; supervision, N.I.; funding acquisition, N.I. All authors have read and agreed to the published version of the manuscript.

Funding: This research was funded by the Agricultural Research Development Agency (Public Organization), grant number CRP5605021810.

Data Availability Statement: Not applicable.

Conflicts of Interest: The authors declare no conflict of interest.

References

1. Majowicz, S.E.; Musto, J.; Scallan, E.; Angulo, F.J.; Kirk, M.; O'brien, S.J.; Jones, T.F.; Fazil, A.; Hoekstra, R.M. International Collaboration on Enteric Disease "Burden of Illness" Studies. The global burden of nontyphoidal *Salmonella* gastroenteritis. *Clin. Infect. Dis.* **2010**, *50*, 882–889.
2. Morgado, M.E.; Jiang, C.; Zambrana, J.; Upperman, C.R.; Mitchell, C.; Boyle, M.; Sapkota, A.R.; Sapkota, A. Climate change, extreme events, and increased risk of salmonellosis: Foodborne diseases active surveillance network (FoodNet), 2004–2014. *Environ. Health* **2021**, *20*, 1–10.
3. Popa, G.L.; Papa, M.I. *Salmonella* spp. infection A continuous threat worldwide. *Germs* **2021**, *11*, 88.
4. Pouokam, G.B.; Foudjo, B.U.; Samuel, C.; Yamgai, P.F.; Silapeux, A.K.; Sando, J.T.; Atonde, G.F.; Frazzoli, C. Contaminants in foods of animal origin in cameroon: A one health vision for risk management "from Farm to Fork". *Front. Public Health* **2017**, *5*, 197.
5. Golden, C.E.; Rothrock, M.J., Jr.; Mishra, A. Mapping foodborne pathogen contamination throughout the conventional and alternative poultry supply chains. *Poult. Sci.* **2021**, *100*, 101157.
6. Fàbrega, A.; Vila, J. *Salmonella enterica* serovar Typhimurium skills to succeed in the host: Virulence and regulation. *Clin. Microbiol. Rev.* **2013**, *26*, 308–341.
7. Foley, S.L.; Johnson, T.J.; Ŗicke, S.C.; Nayak, R.; Danzeisen, J. *Salmonella* pathogenicity and host adaptation in chicken-associated serovars. *Microbiol. Mol. Biol. Rev.* **2013**, *77*, 582–607.
8. CDC. *Antibiotic Resistance Threats in the United States*; U.S. Department of Health and Human Services, CDC: Atlanta, GA, USA, 2019.
9. Chen, H.M.; Wang, Y.; Su, L.H.; Chiu, C.H. Nontyphoid *Salmonella* infection: Microbiology, clinical features, and antimicrobial therapy. *Pediatr. Neonatol.* **2013**, *54*, 147–152.
10. World Health Organization (WHO). Salmonella (Non-Typhoidal). 2018. Available online: https://www.who.int/news-room/fact-sheets/detail/salmonella-(non-typhoidal) (accessed on 19 September 2020).
11. Collazo, C.M.; Galán, J.E. The invasion-associated type-III protein secretion system in *Salmonella*–A review. *Gene* **1997**, *192*, 51–59.
12. Hensel, M. *Salmonella* pathogenicity island 2. *Mol. Microbiol.* **2000**, *36*, 1015–1023.
13. Murray, R.A.; Lee, C.A. Invasion genes are not required for *Salmonella enterica* serovar Typhimurium to breach the intestinal epithelium: Evidence that *Salmonella* pathogenicity island 1 has alternative functions during infection. *Infect. Immun.* **2000**, *68*, 5050–5055.
14. Boddicker, J.D.; Knosp, B.M.; Jones, B.D. Transcription of the *Salmonella* invasion gene activator, *hil*A, requires HilD activation in the absence of negative regulators. *J. Bacteriol.* **2003**, *185*, 525–533.
15. Golubeva, Y.A.; Sadik, A.Y.; Ellermeier, J.R.; Slauch, J.M. Integrating global regulatory input into the *Salmonella* pathogenicity island 1 type III secretion system. *Genetics* **2012**, *190*, 79–90.
16. Swamy, S.C.; Barnhart, H.M.; Lee, M.D.; Dreesen, D.W. Virulence determinants *inv*A and *spv*C in salmonellae isolated from poultry products, wastewater, and human sources. *Appl. Environ. Microbiol.* **1996**, *62*, 3768–3771.
17. Guerra, B.; Soto, S.M.; Argüelles, J.M.; Mendoza, M.C. Multidrug resistance is mediated by large plasmids carrying a class 1 integron in the emergent *Salmonella enterica* serotype [4, 5, 12: I:−]. *Antimicrob. Agent. Chemother.* **2001**, *45*, 1305–1308.
18. Cardona-Castro, N.; Restrepo-Pineda, E.; Correa-Ochoa, M. Detection of *hil*A gene sequences in serovars of *Salmonella enterica* subspecies enterica. *Memórias Inst. Oswaldo Cruz.* **2002**, *97*, 1153–1156.
19. Martin, L.C.; Weir, E.K.; Poppe, C.; Reid-Smith, R.J.; Boerlin, P. Characterization of bla_{CMY-2} plasmids in *Salmonella* and *Escherichia coli* isolates from food animals in Canada. *Appl. Environ. Microbiol.* **2012**, *78*, 1285–1287.
20. Glenn, L.M.; Lindsey, R.L.; Folster, J.P.; Pecic, G.; Boerlin, P.; Gilmour, M.W.; Harbottle, H.; Zhao, S.; McDermott, P.F.; Fedorka-Cray, P.J.; et al. Antimicrobial resistance genes in multidrug-resistant *Salmonella enterica* isolated from animals, retail meats, and humans in the United States and Canada. *Microb. Drug Resist.* **2013**, *19*, 175–184.
21. Sabry, M.A.; Abdel-Moein, K.A.; Abdel-Kader, F.; Hamza, E. Extended-spectrum β-lactamase-producing *Salmonella* serovars among healthy and diseased chickens and their public health implication. *J. Glob. Antimicrob. Resist.* **2020**, *22*, 742–748.
22. Pavelquesi, S.L.S.; de Oliveira Ferreira, A.C.A.; Rodrigues, A.R.M.; de Souza Silva, C.M.; Orsi, D.C.; da Silva, I.C.R. Presence of Tetracycline and Sulfonamide Resistance Genes in *Salmonella* spp.: Literature Review. *Antibiotics* **2021**, *10*, 1314.
23. *ISO 6579:2002*; Microbiology of Food and Animal Feeding Stuffs–Horizontal Method for the Detection of *Salmonella* spp. International Organization for Standardization (ISO): Geneva, Switzerland, 2002.
24. Assaf, A.; Cordella, C.B.; Thouand, G. Raman spectroscopy applied to the horizontal methods ISO 6579: 2002 to identify *Salmonella* spp. in the food industry. *Anal. Bioanal. Chem.* **2014**, *406*, 4899–4910.
25. Aslanzadeh, J. Biochemical profile-based microbial identification systems. In *Advanced Techniques in Diagnostic Microbiology*; Springer: Boston, MA, USA, 2006; pp. 84–116.
26. Grimont, P.A.; Weill, F.X. Antigenic formulae of the *Salmonella* serovars. *WHO Collab. Cent. Ref. Res. Salmonella* **2007**, *9*, 1–166.
27. Gal-Mor, O.; Suez, J.; Elhadad, D.; Porwollik, S.; Leshem, E.; Valinsky, L.; McClelland, M.; Schwartz, E.; Rahav, G. Molecular and cellular characterization of a *Salmonella enterica* serovar Paratyphi A outbreak strain and the human immune response to infection. *Clin. Vaccine Immunol.* **2012**, *19*, 146–156.

28. Alam, S.B.; Mahmud, M.; Akter, R.; Hasan, M.; Sobur, A.; Nazir, K.H.; Noreddin, A.; Rahman, T.; El Zowalaty, M.E.; Rahman, M. Molecular detection of multidrug resistant *Salmonella* species isolated from broiler farm in Bangladesh. *Pathogens* **2020**, *9*, 201.
29. Zou, M.; Keelara, S.; Thakur, S. Molecular characterization of *Salmonella enterica* serotype Enteritidis isolates from humans by antimicrobial resistance, virulence genes, and pulsed-field gel electrophoresis. *Foodborne Pathog. Dis.* **2012**, *9*, 232–238.
30. Thung, T.Y.; Mahyudin, N.A.; Basri, D.F.; Radzi, C.W.M.; Nakaguchi, Y.; Nishibuchi, M.; Radu, S. Prevalence and antibiotic resistance of *Salmonella* Enteritidis and *Salmonella* Typhimurium in raw chicken meat at retail markets in Malaysia. *Poult. Sci. J.* **2016**, *95*, 1888–1893.
31. Zambrana-Torrelio, C.; Murray, K.A.; Daszak, P. One health and hotspots of food-borne EIDs. In *Improving Food Safety through a One Health Approach: Workshop Summary*; National Academies Press: Washington, DC, USA, 2012.
32. Poppe, C.; Martin, L.C.; Gyles, C.L.; Reid-Smith, R.; Boerlin, P.; McEwen, S.A.; Prescott, J.F.; Forward, K.R. Acquisition of resistance to extended-spectrum cephalosporins by *Salmonella enterica* subsp. enterica serovar Newport and *Escherichia coli* in the turkey poult intestinal tract. *Appl. Environ. Microbiol.* **2005**, *71*, 1184–1192.
33. Fonseca, E.L.; Mykytczuk, O.L.; Asensi, M.D.; Reis, E.M.; Ferraz, L.R.; Paula, F.L.; Ng, L.K.; Rodrigues, D.P. Clonality and antimicrobial resistance gene profiles of multidrug-resistant *Salmonella enterica* serovar Infantis isolates from four public hospitals in Rio de Janeiro, Brazil. *J. Clin. Microbiol.* **2006**, *44*, 2767–2772.
34. Chai, L.C.; Robin, T.; Ragavan, U.M.; Gunsalam, J.W.; Bakar, F.A.; Ghazali, F.M.; Radu, S.; Kumar, M.P. Thermophilic *Campylobacter* spp. in salad vegetables in Malaysia. *Int. J. Food Microbiol.* **2007**, *117*, 106–111.
35. Dominguez, M.; Jourdan-Da Silva, N.; Vaillant, V.; Pihier, N.; Kermin, C.; Weill, F.X.; Delmas, G.; Kerouanton, A.; Brisabois, A.; de Valk, H. Outbreak of *Salmonella enterica* serotype Montevideo infections in France linked to consumption of cheese made from raw milk. *Foodborne Pathog. Dis.* **2009**, *6*, 121–128.
36. Chotinan, S.; Tadee, P. Epidemiological Survey of S. Enteritidis Pulsotypes from Salmonellosis Outbreak in Chiang Mai and Samut Songkhram Provinces, Thailand. *Vet. Integr. Sci.* **2015**, *13*, 73–80.
37. Afshari, A.; Baratpour, A.; Khanzade, S.; Jamshidi, A. *Salmonella* Enteritidis and *Salmonella* Typhimorium identification in poultry carcasses. *Iran. J. Microbiol.* **2018**, *10*, 45.
38. Altaf Hussain, M.; Wang, W.; Sun, C.; Gu, L.; Liu, Z.; Yu, T.; Ahmad, Y.; Jiang, Z.; Hou, J. Molecular Characterization of Pathogenic *Salmonella* Spp. From Raw Beef in Karachi, Pakistan. *Antibiotics* **2020**, *9*, 73.
39. Cardoso, M.O.; Ribeiro, A.R.; Santos, L.R.; Pilotto, F.; de Moraes, H.L.; Salle, C.T.; Rocha, S.L.; Nascimento, V.P. Antibiotic resistance in *Salmonella* Enteritidis isolated from broiler carcasses. *Braz. J. Microbiol.* **2006**, *37*, 368–371.
40. Ghazaey, S.; Mirmomeni, M.H. Microbial-resistant *Salmonella* Enteritidis isolated from poultry samples. *Rep. Biochem. Mol. Biol.* **2012**, *1*, 9.
41. Lin, D.; Chen, K.; Chan, E.W.C.; Chen, S. Increasing prevalence of ciprofloxacin-resistant food-borne *Salmonella* strains harboring multiple PMQR elements but not target gene mutations. *Sci. Rep.* **2015**, *5*, 14754.
42. Darwin, K.H.; Miller, V.L. Molecular basis of the interaction of *Salmonella* with the intestinal mucosa. *Clin. Microbiol. Rev.* **1999**, *12*, 405–428.
43. Dibb-Fuller, M.P.; Allen-Vercoe, E.; Thorns, C.J.; Woodward, M.J. Fimbriae-and flagella-mediated association with and invasion of cultured epithelial cells by *Salmonella* Enteritidis. *Microbiology* **1999**, *145*, 1023–1031.
44. Van Asten, F.J.; Hendriks, H.G.; Koninkx, J.F.; Van der Zeijst, B.A.; Gaastra, W. Inactivation of the flagellin gene of *Salmonella enterica* serotype Enteritidis strongly reduces into differentiated Caco-2 cells. *FEMS Microbiol. Lett.* **2000**, *185*, 175–179.
45. van Asten, F.J.; Hendriks, H.G.; Koninkx, J.F.; van Dijk, J.E. Flagella-mediated bacterial motility accelerates but is not required for *Salmonella* serotype Enteritidis invasion of differentiated Caco-2 cells. *Int. J. Med. Microbiol.* **2004**, *294*, 395–399.
46. Sharma, I.; Das, K. Detection of *inv*A gene in isolated *Salmonella* from marketed poultry meat by PCR assay. *J. Food Process. Technol.* **2016**, *7*, 2.
47. Solano, C.; García, B.; Valle, J.; Berasain, C.; Ghigo, J.M.; Gamazo, C.; Lasa, I. Genetic analysis of *Salmonella* Enteritidis biofilm formation: Critical role of cellulose. *Mol. Microbiol.* **2002**, *43*, 793–808.
48. Pang, J.C.; Lin, J.S.; Tsai, C.C.; Tsen, H.Y. The presence of major world-wide clones for phage type 4 and 8 *Salmonella enterica* serovar Enteritidis and the evaluation of their virulence levels by invasiveness assays in vitro and in vivo. *FEMS Microbiol. Lett.* **2006**, *263*, 148–154.
49. Pan, Z.; Carter, B.; Núñez-García, J.; AbuOun, M.; Fookes, M.; Ivens, A.; Woodward, M.J.; Anjum, M.F. Identification of genetic and phenotypic differences associated with prevalent and non-prevalent *Salmonella* Enteritidis phage types: Analysis of variation in amino acid transport. *Microbiology* **2009**, *155*, 3200–3213.
50. Borges, K.A.; Furian, T.Q.; Borsoi, A.; Moraes, H.L.; Salle, C.T.; Nascimento, V.P. Detection of virulence-associated genes in *Salmonella* Enteritidis isolates from chicken in South of Brazil. *Pesqui. Vet. Bras.* **2013**, *33*, 1416–1422.
51. Thung, T.Y.; Radu, S.; Mahyudin, N.A.; Rukayadi, Y.; Zakaria, Z.; Mazlan, N.; Tan, B.H.; Lee, E.; Yeoh, S.L.; Chin, Y.Z.; et al. Prevalence, virulence genes and antimicrobial resistance profiles of *Salmonella* serovars from retail beef in Selangor, Malaysia. *Front. Microbiol.* **2018**, *8*, 2697.
52. Mthembu, T.P.; Zishiri, O.T.; El Zowalaty, M.E. Molecular detection of multidrug-resistant *Salmonella* isolated from livestock production systems in South Africa. *Infect. Drug Resist.* **2019**, *12*, 3537.

foods

MDPI

Article

Performance Testing of *Bacillus cereus* Chromogenic Agar Media for Improved Detection in Milk and Other Food Samples

Eva Fuchs [1], Christina Raab [1], Katharina Brugger [2], Monika Ehling-Schulz [3], Martin Wagner [1,4] and Beatrix Stessl [1,*]

1 Unit of Food Microbiology, Institute of Food Safety, Food Technology and Veterinary Public Health, University of Veterinary Medicine Vienna, 1210 Vienna, Austria; fuchs_e@hotmail.com (E.F.); christina.raab@hotmail.com (C.R.); Martin.Wagner@vetmeduni.ac.at (M.W.)

2 Unit of Veterinary Public Health and Epidemiology, Institute of Food Safety, Food Technology and Veterinary Public Health, University of Veterinary Medicine Vienna, 1210 Vienna, Austria; katharina.brugger@vetmeduni.ac.at

3 Functional Microbiology Group, Institute of Microbiology, University of Veterinary Medicine Vienna, 1210 Vienna, Austria; Monika.Ehling-Schulz@vetmeduni.ac.at

4 Austrian Competence Center for Feed and Food Quality, Safety and Innovation (FFOQSI GmbH), 3430 Tulln an der Donau, Austria

* Correspondence: Beatrix.Stessl@vetmeduni.ac.at

Abstract: In this study, the performance of four alternative selective chromogenic *B. cereus* agar was compared to the reference mannitol-yolk polymyxin (MYP) agar (ISO 7932) using inclusion and exclusion test strains (n = 110) and by analyzing naturally contaminated milk and other food samples (n = 64). Subsequently, the *panC* group affiliation and toxin gene profile of *Bacillus cereus senso lato* (*s.l.*) isolates were determined. Our results corroborate that the overall best performing media CHROMagar™ *B. cereus* (93.6% inclusivity; 82.7% exclusivity) and BACARA® (98.2% inclusivity, 62.7% exclusivity) are more sensitive and specific compared to Brilliance™ *B. cereus*, MYP and ChromoSelect *Bacillus* Agar. Both media allow unequivocal detection of *B. cereus* with low risks of misidentification. Media containing ß-D-glucosidase for the detection of presumptive *B. cereus* may form atypical colony morphologies resulting in a false negative evaluation of the sample. Naturally contaminated samples presented high numbers of background flora, while numbers of presumptive *B. cereus* were below the detection limit (<10 CFU g^{-1} or mL^{-1}). Recovery after freezing resulted in the highest detection of *B. cereus s.l.* on BACARA® (57.8%), CHROMagar™ *B. cereus* (56.3%) and MYP agar (54.7%). The *panC*/toxin profile combination IV/A was the most abundant (33.0%), followed by III/F (21.7%) and VI/C (10.4%). More *panC* and toxin combinations were present in 15.6% of samples when reanalyzed after freezing. In order to improve detection and confirmation of *B. cereus s.l.* in food samples, we recommend the parallel use of two complementary selective media followed by molecular characterization (e.g., *panC* typing combined with toxin gene profiling). When determining psychrotolerant or thermophilic members of the *B. cereus* group, the selective agar media should additionally be incubated at appropriate temperatures (5 °C, ≥45 °C). If high-risk toxin genes (e.g., *ces* or *cytK-1*) are detected, the strain-specific ability to produce toxin should be examined to decisively assess risk.

Keywords: *Bacillus cereus* group; food safety; chromogenic media; performance testing; toxin gene profiling; *panC* sequencing

Citation: Fuchs, E.; Raab, C.; Brugger, K.; Ehling-Schulz, M.; Wagner, M.; Stessl, B. Performance Testing of *Bacillus cereus* Chromogenic Agar Media for Improved Detection in Milk and Other Food Samples. *Foods* **2022**, *11*, 288. https://doi.org/10.3390/foods11030288

Academic Editors: Timothy Ells, Catherine Burgess and Antonio Afonso Lourenco

Received: 17 December 2021
Accepted: 10 January 2022
Published: 21 January 2022

1. Introduction

Bacillus senso latu (*s.l.*) consists of Gram-positive, rod-shaped, aerobic or facultative anaerobic spore-forming bacteria that are widespread in the environment and commonly found in soil, plant material and in the gut of insects [1–3]. As toxin producers and food spoiling bacteria, they pose a health risk and cause economic damage when entering and persisting in the food chain [4–6].

The *B. cereus s.l.* group is represented by *B. cereus s.s.*, *B. anthracis*, *B. cytotoxicus*, *B. mycoides*, *B. pseudomycoides*, *B. thuringiensis*, *B. toyonensis* and *B. weihenstephanensis* [2,7–10].

In order to protect consumer health, a process criterion for presumptive *B. cereus* in infant formulas ($\leq$500 colony forming units [CFU] g^{-1}) was set within the Commission Regulation (EC) No 1441/2007 (EC, 2007; https://eur-lex.europa.eu/legal-content/EN/TXT/HTML/?uri=CELEX:32007R1441&from=EN; accessed on 17 December 2021). In addition, warning values are available, for example, for dried herbs and spices; tofu and bakery products (4 log CFU g^{-1}); or fruits and vegetables, cereals, pasta, mayonnaises, dressings, soups and ready-to-eat instant products (3 log CFU g^{-1}) (https://www.dghm-richt-warnwerte.de/de/dokumente; accessed on 17 December 2021).

In current practice, presumptive *B. cereus s.l.* is detected and enumerated on classical culture media as for example mannitol egg yolk polymyxin (MYP) agar. Chromogenic reactions rely on enzymatic cleavage (e.g., by β-D-glucosidase) of a particular substrate and the release of a chromogen, which is more specific than conventional microbiological growth media. Some chromogenic media additionally detect PLC activity in order to facilitate unambiguous identification. Selectivity is achieved in both media types by the addition of antibiotic substances (e.g., polymyxin B or trimethoprim), which inhibit the growth of undesirable Gram-positive and Gram-negative bacteria [11,12].

Apart from *B. cereus s.l.* counts, strain-specific properties such as toxin gene profiles or other virulence factors need to be investigated for risk characterization efforts. A broad range of phenotypical (e.g., biochemical profile, growth behavior and σ-endotoxin crystal staining) and genotypical methods (e.g., Multilocus Sequence Typing (MLST) and *panC*-typing) are required to accurately identify and group *B. cereus s.l.* on the species level [13–15], rendering species differentiation difficult under routine laboratory conditions.

Fourier transform infrared (FTIR) spectroscopy and Matrix-Assisted Laser Desorption/Ionization Time of Flight (MALDI-TOF) Mass Spectrometry have been developed to speed up the identification and characterization of *B. cereus s.l.* and cereulide. However, database richness is decisive for accurate species identification, and an enhanced cloud-based exchange of spectral data would be necessary for propagation [16–20].

As many innovative methods are established exclusively in expert's laboratories, there is still the need for rapid and unambiguous isolation and differentiation methods applicable in food and dairy plant laboratories. Contemporary chromogenic media may represent a useful tool to facilitate identification of *B. cereus s.l.* and accelerate the time to result by easier visual evaluation of morphology and color changes of media.

This study was initiated to assess and compare the performance of the ISO standard medium MYP agar with four alternative chromogenic selective plating media for detection and enumeration of food-intoxication and spoilage-associated *B. cereus* group members by using a bacterial test strain panel and analyzing naturally contaminated samples under everyday conditions. Furthermore, an in-depth molecular-biological characterization of inclusivity test strains and sample isolates was performed to explore strain-specific features.

2. Materials and Methods

2.1. *Performance Testing of Selective Media*

2.1.1. Test Media

Within the scope of this study, the performance of commercially available chromogenic selective media ChromoSelect *Bacillus* (HI; Merck KgaA, Darmstadt, Germany; formerly branded HiCrome™ *Bacillus*), CHROMagar™ *B. cereus* (CH; CHROMagar, Paris, France), Brilliance™ *B. cereus* (BRI; Thermo Fisher Scientific Inc., Oxoid, Waltham, MA, USA) and BACARA® agar (BA; *B. cereus* Rapid Agar; bioMérieux, Marcy l'Etoile, France) was evaluated in comparison to the ISO recommended standard medium MYP [21] (Thermo Fisher Scientific Inc., Oxoid, Waltham, MA, USA). Information on media composition—as indicated in the media manufacturer's specifications—is listed in Supplementary Table S1.

2.1.2. Inclusivity and Exclusivity Test Strains

In order to evaluate the performance of *B. cereus* selective media, a bacterial test strain panel (n = 220) consisting of *B. cereus* target organisms (for inclusivity testing, n = 110) and non-target *Bacillus* spp. (for exclusivity testing, panel included spoilage-associated microbes and Gram–positive and Gram–negative competitors; n = 110) was compiled. *B. cereus s.l.* strains originated from emetic and diarrheal outbreaks (Institute for Microbiology strain collection, University of Veterinary Medicine Vienna), environmental samples, fruits and vegetables, cereals, fish, tea, herbs, spices, milk and dairy products (isolate collection Unit of Food Microbiology, University of Veterinary Medicine Vienna; Supplementary Table S2). Exclusivity strains were selected according to their relevance and frequency as food contaminants and covered, among others, isolates deriving from fruits and vegetables, meat and meat products, dried spices and seeds, milk products and dairy processing environments (Supplementary Table S3). All strains are preserved as cryogenic cultures (Corning, VWR, Vienna, Austria) in a volume of 1.5 mL brain heart infusion broth (BHI; Merck KGaA, Darmstadt, Germany) with 15% glycerol (Merck KGaA) at −80 °C (GFL Gesellschaft für Labortechnik GmbH, Großwedel, Germany) in the strain collection of the Unit of Food Microbiology.

After activation of test strains from glycerol stocks and subculturing on trypto-casein-soy agar plus 0.6% yeast (TSA-Y; Biokar Diagnostics, Beauvais, France), selective media were inoculated. In order to obtain a few well-defined bacterial colonies, an isolated single colony from the working culture was transferred onto selective *B. cereus* media by fractioned three loop inoculation. After incubation (Ehret GmbH & Co. KG, Emmendingen, Germany) at the specified conditions (Supplementary Table S1), the presences of bacterial growth and colony morphology were recorded for all media. By qualitative classification into "typically growing," "atypically growing", or "non-growing" strains, media benefits and limitations were determined.

2.1.3. Naturally Contaminated Food Samples

In order to evaluate media reliability under routine laboratory conditions, food samples (n = 64) from 18 producers were collected from the production chain and retail level. Food samples (20.3%, n = 13/64; producer A–F) belonged to the source categories "fruits and vegetables", "nuts, nut products and seeds", "fish and fishery products", "herbs" and "cocoa and cocoa preparations, coffee, and tea" (Supplementary Table S4A). Milk samples (79.7%, n = 51/64; producer G-R) were heat treated, except for one raw milk sample (bactofugation) provided by a local dairy (Supplementary Table S4B).

Important information regarding processing was gathered, including the type of milk with reference to animal species (cow or small ruminant), agricultural system (organic or conventional farming), processing and predicted shelf-life (homogenized, pasteurized or high pasteurized) (Supplementary Table S4B). The majority of milk samples (80.4%, n = 41/51) were produced organically. Milk samples were examined after 24 h provocation at 30 °C for enrichment to ensure detection of *B. cereus s.l.* All food samples were analyzed before and after freezing at −20 °C to trigger outgrowth of spores.

In order to prepare sample homogenates, 25 mL or 25 g of food product was diluted 1:10 in sterile buffered peptone water (BPW; Fisher Scientific Inc., Oxoid); food samples were additionally mixed for 180 s in a paddle blender (Stomacher®; Seward Ltd., West Sussex, UK). Ten-fold serial dilutions in sterile Ringer's solution (B. Braun Melsungen AG, Melsungen, Germany) were plated in duplicate up to 10^{-5} on selective agar media by using the spatula method. Following incubation, growth was assessed, and colonies displaying characteristic morphology were enumerated to determine the extent of *B. cereus s.l.* contamination. Randomly picked colonies with typical and atypical morphology were isolated, subjected to confirmation and characterized with regard to *panC* group affiliation and toxin gene profile.

2.2. Molecular and Phenotypical Characterization

2.2.1. DNA-Extraction

Bacterial DNA was extracted from *B. cereus s.l.* cultures grown overnight on TSA-Y at 30 °C (Biokar Diagnostics and Merck KGaA) using the Chelex® 100 resin (Bio-Rad Laboratories, Inc., Hercules, CA, USA) method as described by Walsh et al. [22]. After extraction, 100 µL DNA of each isolate was kept at −20 °C until use in characterization experiments.

2.2.2. Confirmation of Group Affiliation

Bacillus cereus s.l. strains from culture collections and presumptive isolates from naturally contaminated food products and milk were confirmed as group members by PCR method targeting the gyrase B gene (*gyrB*) as described by Dzieciol et al. [23].

2.2.3. Toxin Gene Screening and Profiling

Confirmed *B. cereus s.l.* strains were screened for their toxin gene content by conventional PCR assays. Amplification was performed according to Ehling-Schulz et al. [24] with minor modifications, addressing the most widespread toxin genes. Two genes of the NHE-complex and two genes of the HBL-complex were taken into consideration: the enterotoxin genes *nheA*, *nheB*, *hblA* and *hblD*. Furthermore, PCR pre-screening assay were applied for *cytK-1/cytK-2*. Detection of the emetic toxin cereulide gene *ces* was performed after Dzieciol et al. [23] with slight adjustments. Strain-specific toxin gene profiles were assigned based on prevailing toxin gene combinations as in Ehling-Schulz et al. [24] (Figure 5, Supplementary Tables S2 and S6).

2.2.4. Partial *panC* Sequencing

Amplification, purification and sequencing (LGC, Berlin) of a fragment of the pantothenate synthetase (*panC*) gene were conducted as previously reported [13]. In order to assign *B. cereus s.l.* strains to one of the seven major phylogenetic groups (i.e., I-VII) defined by Guinebretière et al. [13,25], sequences were matched with deposited sequences in a web-based database (https://www.tools.symprevius.org/Bcereus/english.php, accessed on 17 December 2021) (Figure 5, Supplementary Tables S2 and S6).

2.2.5. Assessment of Hemolytic Activity

β-hemolytic activity of inclusivity test strains was determined on Columbia agar plates containing 5% sheep blood (COS; bioMérieux) after overnight incubation at 30 °C [21] (Supplementary Table S2).

2.3. Evaluation Criteria and Statistics

In order to differentiate the phenotypic appearance of test strains and evaluate their potential for misidentification, the growth of inclusivity and exclusivity strains was classified in typical and atypical according to their reaction(s) and colony morphology on selective media (Figure 1).

A mosaic plot was used for visualizing the results of growth and phospholipase C reactions of inclusivity and exclusivity test strains for each of the tested media MYP, HI, BRI, CH and BA (Figure 1). Detectability of *B. cereus s.l.* in naturally contaminated samples was illustrated in a bar plot (Figure 5). The relative frequency of *panC* group (II–VI) and toxin gene profile (A–F) combinations among *B. cereus s.l.* isolates (n = 106) associated with naturally contaminated samples was depicted as pie chart (Figure 6). Graphics were created with open-source statistical computer environment R version 4.1.0 [26].

3. Results

3.1. Inclusivity and Exclusivity Testing

The detailed strain properties of the inclusivity test strains are presented in Supplementary Table S2. The majority of inclusivity test strains (n = 110) were assigned to toxin profile C (*nhe+/hbl+*; 33.6%, n = 37) and A (*nhe+/hbl+/cytK+*; 27.3%, n = 30). The *ces* gene was present in six (5.5%) test strains derived from foodborne outbreaks. Among the target test strains, *panC* group III (30.9%, n = 34), IV (30.0%, n = 33) and VI (20.0%, n = 22) were the most abundant. The most frequent combination of *panC* group and toxigenic profile in the entire panel of inclusivity test strains was IV/A (21.8%, n = 24), obtained from milk and dried products (such as tea, spices and mushrooms). Other common combinations were VI/C (17.3%, n = 19) isolated from milk, soil and salad, as well as III/D (10.9%, n = 12) mainly detectable in strains isolated from protein-rich food (e.g., feta, dried fish and mushrooms).

Examination of target strains showed >99% inclusivity on all media (n = 109–110/110); one *B. pseudomycoides* strain did not grow on three selective media (Figure 1). The highest rates of atypical β-D-glucosidase negative colonies were observed on BRI (12.7%, n = 14), HI (6.4%, n = 7) and CH (5.5%, n = 6), resulting in an atypical white phenotype (Figure 2 and Supplementary Table S2). Such colony morphologies were largely related to the milk- or soil-derived *panC*-type/toxin profile VI/C. On chromogenic media (CH and BA), the PLC reaction was more distinct in comparison to MYP agar (Figure 2).

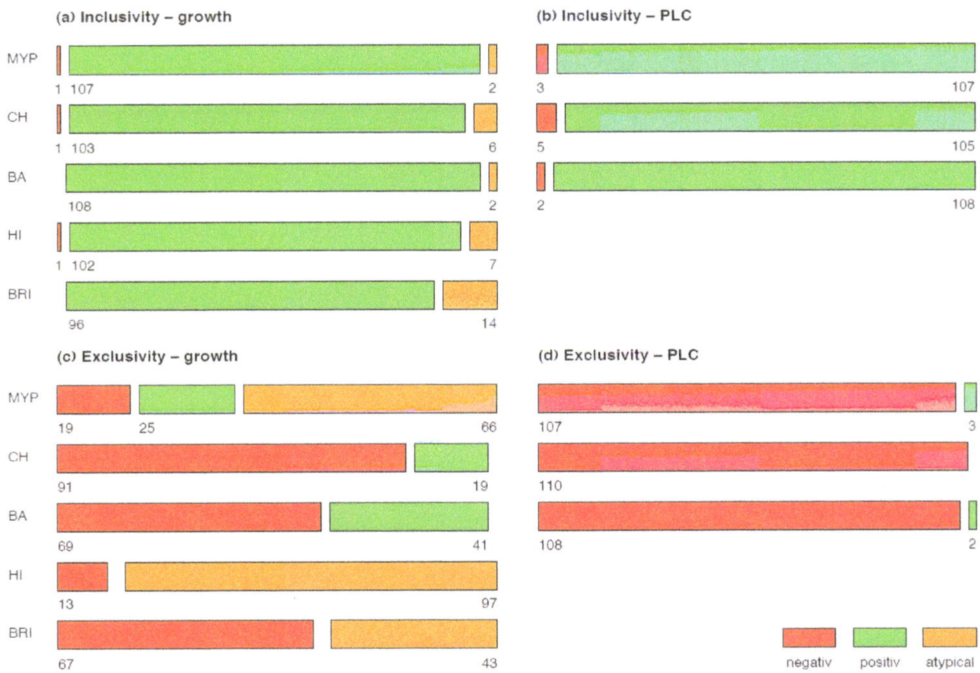

Figure 1. Growth (**a**,**c**) and phospholipase C reaction (**b**,**d**) of inclusivity (n = 110) and exclusivity (n = 110) test strains. Negative is no-growth, positive is typical growth and atypical is not presumptive *Bacillus cereus sensu lato* morphology on selective agar media. Abbreviations: PLC—phospholipase C; MYP—mannitol egg yolk polymyxin agar; CH—CHROMagar™ *B. cereus*; BA—BACARA® agar; HI—ChromoSelect *Bacillus* agar; BRI—Brilliance™ *B. cereus* agar.

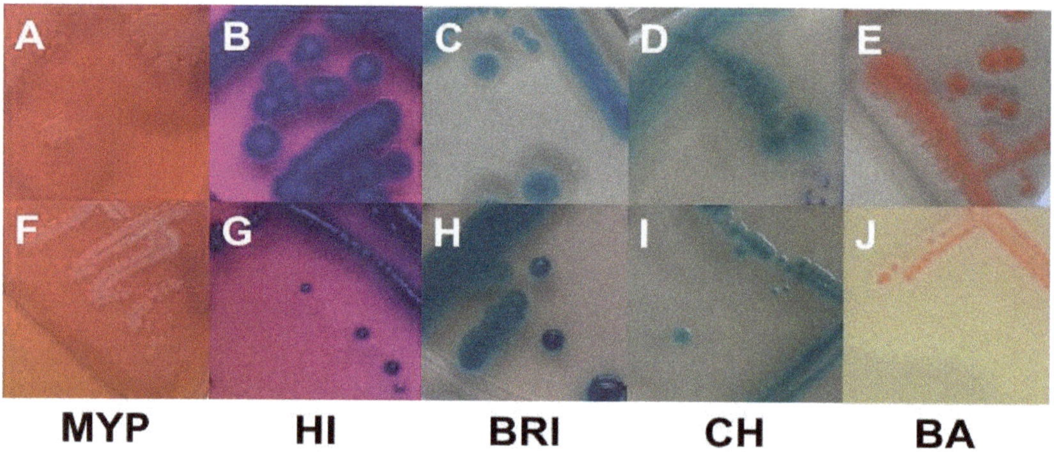

Figure 2. Typical (**A–E**) and atypical colonies (**F–J**) of *Bacillus cereus sensu lato* on MYP agar (A (BCG 6), F (BC 66)), ChromoSelect *Bacillus* agar (B (BC 30), G (BC 20)), Brilliance™ *B. cereus* agar (C (BC 2), H (BC 19)), CHROMagar™ *B. cereus* (D (BC 63), I (BC 2)) and BACARA® agar (E (BC 50), J (BC 34)). Bluish-green colonies are the result of β-D-glucosidase reaction. Precipitation zones surrounding typical colonies are caused by phospholipase C reaction, while lack of mannitol fermentation results in pink background. Abbreviations: MYP—mannitol egg yolk polymyxin agar; HI—ChromoSelect *Bacillus* agar; BRI—Brilliance™ *B. cereus* agar; CH—CHROMagar™ *Bacillus cereus*; BA—BACARA® agar.

Best performing media in terms of exclusivity (Supplementary Table S3) were CH (82.7%, n = 91/110) and BA (62.7%, n = 69/110). Several non-target organisms were not effectively suppressed by polymyxin B in MYP and (82.7%, n = 91/110) and HI (88.2%, n = 97/110) (Figure 3 and Supplementary Table S3). Comparatively low inhibition of exclusivity strains (n = 110) was also observed on BRI (60.9%, n = 67), although we only observed colony morphologies that could not be misidentified as presumptive *B. cereus* due to their atypical pin-point growth (Figure 3).

PLC reaction typical for the target organisms was observed in three and two exclusivity tests strains on MYP and BA, respectively (*Listeria monocytogenes*, *Paenibacillus polymyxa* and *Serratia marcescens*).

3.2. Naturally Contaminated Samples

Milk (n = 51) and food (n = 13) samples analyzed were contaminated with presumptive *B. cereus* at the limit of detection, resulting in quantitative data below 10 and 100 CFU g^{-1}, respectively. Further details on sample characteristics can be found in Supplementary Table S4A,B. Typical and atypical *B. cereus s.l.* colonies grown on selective media test panel are shown in Figure 4.

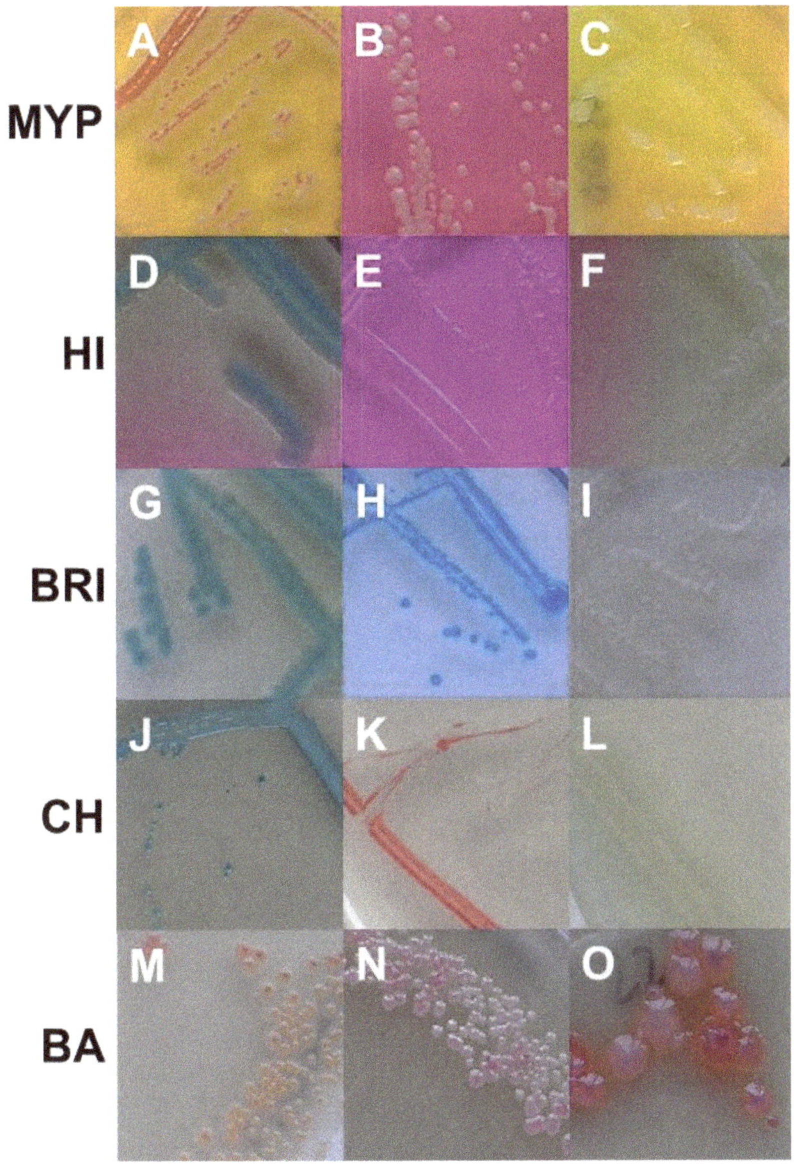

Figure 3. Growth of non-target organisms on *Bacillus cereus* selective media (from upper left to lower right). MYP agar: (**A**)—*Serratia marcescens* (EGN 54); (**B**)—*Brochothrix thermospacta* (EGP 2); (**C**)—*Bacillus stratosphericus* (BG 28). ChromoSelect *Bacillus* agar: (**D**)—*Aeromonas hydrophila* (EGN 5); (**E**)—*Acinetobacter baumannii* (EGN 2); (**F**)—*Citrobacter freundii* (EGN 9). Brilliance™ *B. cereus* agar: (**G**)—*Staphylococcus sciuri* (EGP 13); (**H**)—*Serratia marcescens* (EGN 54); (**I**)—*Pseudomonas fluorescens* (EGN 45). CHROMagar™ *B. cereus*: (**J**)—*Morganella morganii* (EGN 39); (**K**)—*Enterococcus faecalis* (EGP 5); (**L**)—*Providencia rettgeri* (EGN 42). BACARA® agar: (**M**)—*Staphylococcus haemolyticus* (EGP 11); (**N**)—*Staphylococcus chromogenes* (EGP 9); (**O**)–*Listeria monocytogenes* (EGP 22). Abbreviations: MYP—mannitol egg yolk polymyxin agar; HI—ChromoSelect *Bacillus* agar; BRI—Brilliance™ *B. cereus* agar; CH—CHROMagar™ *Bacillus cereus*; BA—BACARA® agar.

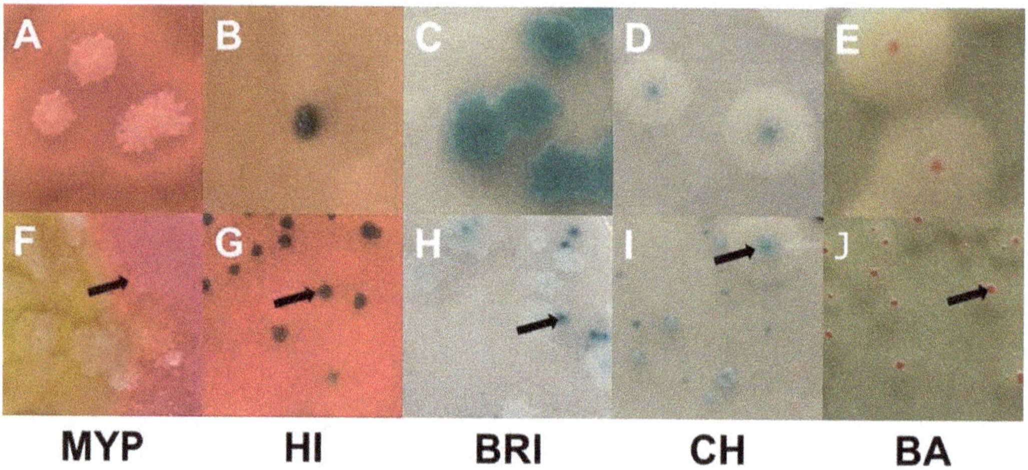

Figure 4. Examples of typical *Bacillus cereus sensu lato* colonies obtained by sampling of naturally contaminated food ans milk samples. Demarcated colonies with typical morphology (top row): (**A**)—MYP agar (ESL-milk); (**B**)—HI, ChromoSelect *Bacillus* agar (ESL-goat milk); (**C**)—Brilliance™ *B. cereus* agar (ESL-milk); (**D**)—CHROMagar™ *B. cereus* (ESL-milk); (**E**)—BACARA® agar (ESL-milk). Colonies masked by high growth of background flora and atypical morphologies (bottom row; arrows point on typical *B. cereus* colonies): (**F**)—coalescing *B. cereus* colonies surrounded by mannitol-positive background-flora (*B. licheniformis*) on MYP agar (dried fish snack); (**G**)—mixed culture on ChromoSelect *Bacillus* agar, growth of mannitol-positive background-flora (*Staphylococcus* spp.) intersparsed with typical colonies (raw milk); (**H**)—atypical light colonies with weak β-D-glucosidase activity and typical colonies on Brilliance™ *B. cereus* agar (ESL-milk); (**I**)—atypical PLC-negative and weakly β-D-glucosidase positive colonies lacking the distinctive halo together with typical colony on CHROMagar™ *B. cereus* (Chinese water spinach); (**J**)—atypical small colonies with weak PLC acitivity on BACARA® agar (dried fish snack). Abbreviations: MYP—mannitol egg yolk polymyxin agar; HI—ChromoSelect *Bacillus* agar; BRI—Brilliance™ *B. cereus* agar; CH—CHROMagar™ *B. cereus*; BA—BACARA® agar.

Figure 5 shows the *B. cereus* group containing samples with respect to the distribution of the *panC* group in combination with toxin profiles. The samples were negative in PCR confirmation of the emetic toxin gene (*ces*); in consequence, the toxin profiles B (*nhe, hbl, ces* gene combination positive) and E (*nhe* and *ces* gene combination positive) were not detected. The *panC*/toxin profile combination IV/A was the most abundant in the sample set (33.0%), followed by III/F (21.7%) and VI/C (10.4%). Representatives of *panC* group IV are described as highly cytotoxic and do generally grow at temperatures ≥10 °C. Toxin profile A represents *nhe, hbl* and *cytK* gene (*cytK-2*) positive isolates. The enterotoxin genes *nhe, hbl* and *cytK*-2 are located in the chromosome of different species of the *B. cereus* group, whereas the *cytK-1* gene is harbored exclusively by thermophilic species *B. cytotoxicus* (*panC* group VII). Representatives of *panC* group VII were not detected in any sample. *panC* group III is considered highly cytotoxic and is representative of *B. cereus* group grown at temperatures of ≥15 °C. The carriage of *nhe* gene (non-hemolytic enterotoxin) characterizes toxin profile F. Strains affiliated to *panC* group VI, which low cytotoxic and grown at ≥5 °C. Toxin profile C is characterized by the presence of *nhe* and *hbl* genes [24] (https://www.tools.symprevius.org/bcereus/english.php; accessed on 17 December 2021).

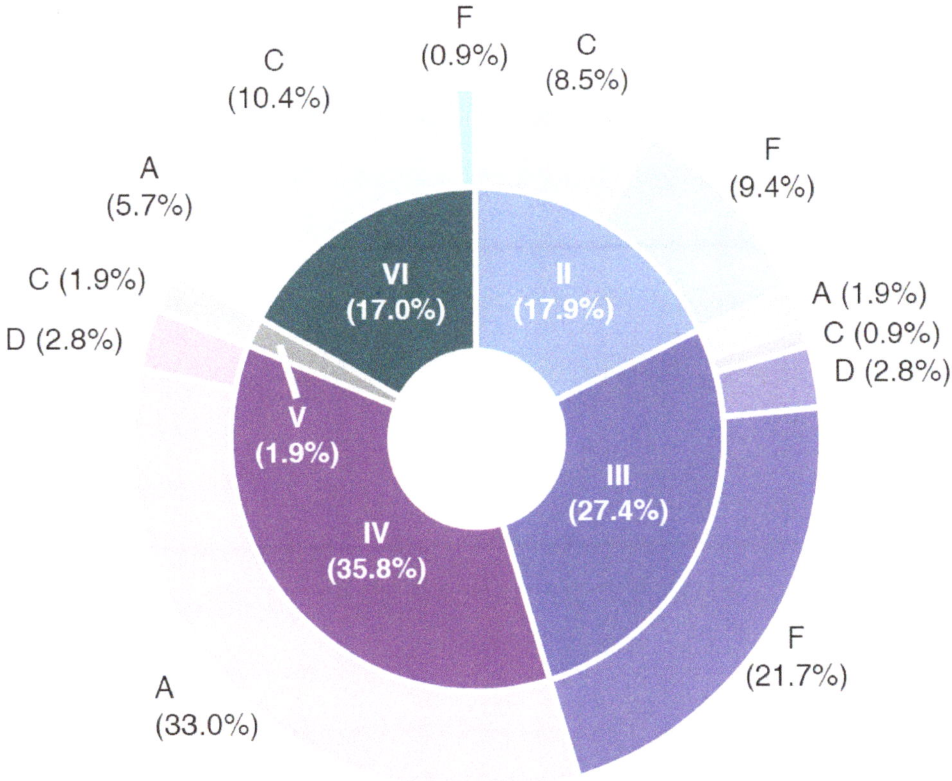

Figure 5. Combinations of *panC* groups (II–VI) and toxin gene profiles (A–F) of *Bacillus cereus sensu lato* isolates obtained from 64 naturally contaminated samples. Abbreviations: A—toxin profile A (*nhe*+, *hbl*+ and *cytK*+); C—toxin profile C (*nhe*+ and *hbl*+), D—toxin profile D (*nhe*+ and *cytK*+); F—toxin profile F (*nhe*+); II—*panC* group II (cytotoxic, growth ≥7 °C); III—*panC* group III (cytotoxic-highly cytotoxic, growth ≥15 °C); IV—*panC* group IV (highly cytotoxic, growth ≥10 °C); V—*panC* group V (low cytotoxic, growth ≥8 °C); VI—*panC* group VI (non or low cytotoxic; growth ≥5 °C).

Naturally contaminated samples were initially pre-screened for the presence of presumptive *B. cereus s.l.* on MYP agar prior to freezing (67.2%, *n* = 43 positive). Recovery after freezing was tested using the selective media test panel, and it resulted in the highest recovery on BA (57.8%, *n* = 37), CH (56.3%, *n* = 36) and MYP (54.7%, *n* = 35) (Figure 6). Supplementary Table S5 indicates that samples tested negative on MYP before freezing were positive for some of the selective media after freezing. The highest accordance (*n* = 6) for presumptive *B. cereus s.l.* recovery before and after freezing was observed for milk samples of different origin. *panC* group and toxin gene profile combinations of *B. cereus s.l.* detected before and after freezing are provided in Supplementary Table S6. In 13 of 64 samples (20.3%), *panC* group and toxin profile combinations were identical before and after freezing. In 12 (18.8%) and 16 samples (25.0%), respectively, *B. cereus s.l.* was detectable either only before or after freezing. In 13 samples (20.3%), different *panC* and toxin combinations were detectable after freezing in comparison to analysis before freezing.

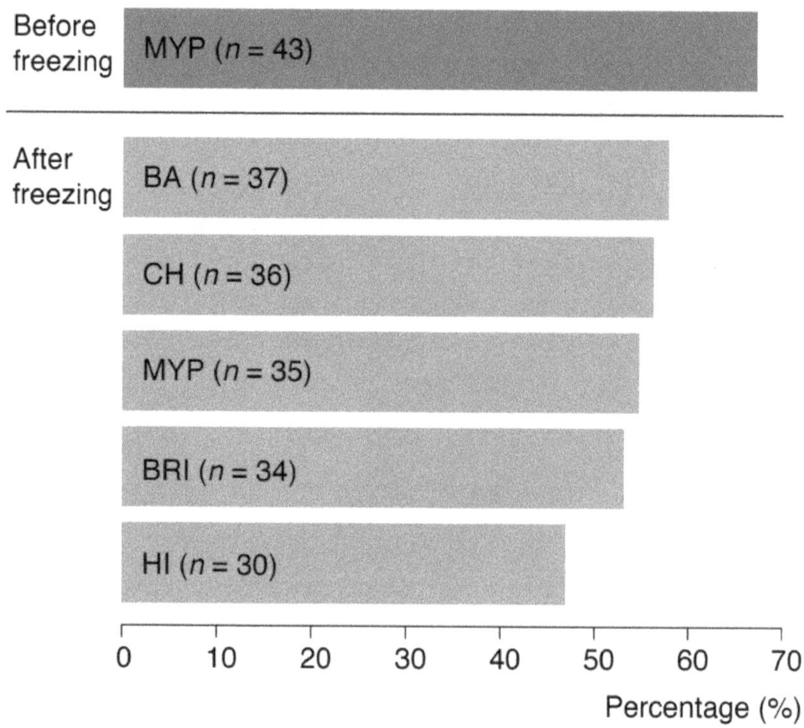

Figure 6. Detectability of *Bacillus cereus* sensu lato in 64 naturally contaminated food samples before freezing on MYP agar and after freezing on MYP agar and chromogenic media. Abbreviations: MYP—mannitol egg yolk polymyxin agar; BA—BACARA® agar; CH—CHROMagar™ *Bacillus cereus*; BRI—Brilliance™ *B. cereus* agar; HI—ChromoSelect *Bacillus* agar.

4. Discussion

B. cereus s.l. is documented among the most prevalent foodborne pathogens, causing one third of food poisoning events in Europe [27].

The presence of *B. cereus s.l.* in food depends mainly on the contamination of the raw material, as well as on recontamination during processing and extrinsic and intrinsic growth conditions during storage. This results in an increased likelihood of disease-relevant concentrations in minimally processed foods consumed either raw or unheated or in inadequately stored extended shelf-life (ESL) products (e.g., in case of cold storage interruption or accidental household refrigerator temperature abuse) [28–31]. In addition, the availability of nutrients and other extrinsic factors can influence toxin levels formed in the food matrix. In particular, high starch, carbohydrate, vitamin, trace element content, neutral pH and moderate to high water activity have been shown to be associated with increased risk of cereulide formation [32].

The detection of presumptive *B. cereus* requires microbiology-trained personnel and is labor-intensive if samples are comprehensively assessed. Most commonly, detection and confirmation are performed using selective culture media such as MYP agar according to ISO 7932 [21]. In industry, samples are often plated on MYP or PEMBA agar and a further discrimination is pursued. Sample analysis is challenged if a high level of accompanying flora jeopardizes outreads since other microbes will stain the agar yellow due to mannitol consumption. As a result, individual colonies of presumptive *B. cereus s.l.* are missed in the yellow-stained agar, and the sample is often considered false negative by the investigator.

This study focused on the comparison of alternative chromogenic selective nutrient media to identify the best performer for *B. cereus s.l.* detection and enumeration. For this purpose, inclusivity and exclusivity were elicited, and group diversity was determined by using naturally contaminated samples before and after freezing.

The study of *B. cereus s.l.* strains showed an inclusivity of >99% for all media, which is in general very promising. Nevertheless, atypical colony morphologies may occur. The highest rates of atypical β-D-glucosidase negative colonies were observed on BRI (12.7%, n = 14), HI (6.4%, n = 7) and CH (5.5%, n = 6) agar, resulting in a white phenotype. Atypical morphologies were largely related to the milk-derived or soil-derived *panC*-type/toxin profile combination VI/C. These atypical *B. cereus s.l.* phenotypes appear to be niche specific and may possibly be associated with specific *panC* types with variable exploitability of starch and various carbohydrates in the genetic clade. For instance, *panC* group IV comprises strains isolated from vegetables indicated limited substrate utilization pathways. Furthermore, a sub-branch within *panC* group III showed the least carbohydrate fermentation capacity due to a lack of aryl-6-phospho-β-glucosidase-encoding genes in the genome [33].

Previous studies focusing on agar evaluations also reported ß-D-glucosidase-negative *B. cereus s.l.* colonies on chromogenic *B. cereus* media manufactured by Oxoid or BMC-Biosynth, which is a concern for a proper evaluation [11,12,34]. In contrast, Chon et al. [35] showed increased specificity and selectivity of BRI agar in foods with high background microbial load and particularly recommended this agar for quantitative analysis. In a comparative analysis of BA and BRI agar, these two culture media were clearly superior to conventional culture media, with BRI agar being more efficient and selective for *B. cereus s.l.* isolation in this setting [36]. In a more recent comparison of the standard media MYP, PEMBA, BRI and a novel—yet not commercially listed—chromogenic agar medium (BCCA), atypical colony morphologies were also described on BRI agar (dark blue color) [37]. BCCA was based again on the detection of ß-D-glucosidase comparable to the BRILLIANCE agar and was fortified by polymyxin B (100,000 IU), trimethoprim (10 mg), ceftazidime (16 mg) and egg yolk emulsion (50 mL). This alternative medium seemed to be more selective in comparison to MYP and PEMBA and circumvented the false negative diagnosis of atypically grown presumptive *B. cereus* colonies by additional lecithinase reaction. All this research shows that *B. cereus s.l.* analysis is demanding and that current media are not sufficiently selective to analyze the diversity of the group.

According to literature, the presence of PC-specific or PI-specific PLC is widespread among *B. cereus s.l.* Almost all group isolates were PLC-positive in the literature: 96% [38] and 93% [39] or 100% PLC and 83% PI-PLC positive isolates [40]. The best performer in the detection of PLC reaction mediated by phosphatidyl-inositol (PI) or phosphatidyl-cholin (PC) was BA (98.2%), followed by MYP (97.3%) and CH (95.5%) (Figure 1). Interestingly, four of five inclusivity test strains lacking PLC reaction also showed atypical colony color due to a lack in ß-D-glucosidase on BRI, HI, or CH agar. *Bacillus pseudomycoides* (*panC* group I/toxin profile C) growth was inhibited on MYP, HI and CH (Figure 1). An explanation for this rare atypical observation was provided by Slamti et al. [41], who observed 2% PC-PLC-negative and non-hemolytic test strains due to the absence of PlcR-regulated proteins.

Cross-reactivity for PI-PLC, PC-PLC and β-D-glucosidase was previously observed for *Staphylococcus aureus* and pathogenic *Listeria* [37]. In our study, *Paenibacillus polymyxa* caused PLC cross-reactivity on MYP and BA agar and *L. monocytogenes* grew on BA. ß-D-glucosidase-positive reaction was observed for a broader range of exclusivity test strains (e.g., enterococci, *Listeria*, staphylococci, bacilli, *Microbacterium* and other Gram-negative bacteria) on the tested media (Figures 1 and 3). On BRI, the only agar investigated based on solely one differentiation system (ß-D-glucosidase), several Gram-positive (e.g., *Bacillus*, staphylococci and enterococci) and Gram-negative (e.g., *Enterobacter cloacae, Aeromonas hydrophila* and *Brevundimonas dimenuta*) non-target strains grew despite the addition of polymyxin B in combination with trimethoprim. Lower growth of cocci was observed on

BRI in contrast to MYP and HI. However, the proprietary antibiotic mixtures of CH and BA were even superior in selectivity compared to other media.

The detection and differentiation of presumptive *B. cereus s.l.* can be improved by the parallel use of two complementary selective agars, as it is already standard practice in the detection of *L. monocytogenes* [42] and *Salmonella* spp. [43]. The combination of agar media operating on different biochemical principles and characterized by different sensitivity and selectivity (e.g., the highly selective BA or CH with the less selective MYP, BRI, or HI) could allow for a more accurate detection of a broad spectrum of group members in food samples. Since other aerobic spore-formers are also relevant as hygiene indicators in food industry, BRI or HI could be supplemented with egg yolk to detect a broader spectrum of bacilli and improve initial differentiation. Parallel incubation of selective agar plates under mesophilic, psychrophilic, or thermophilic conditions would be recommendable depending on the food type (Figure 7). Incubation at 5–7 °C for the investigation of dairy products may support the assessment of a potential proliferation of bacilli even if the cold chain is maintained [44]. Starch-containing foods as well as herbs and spices have been contaminated with the thermotolerant *B. cytotoxicus*, as shown in previous reports [45,46]. Therefore, thermophilic (≥45 °C) and mesophilic (30 °C) incubation should be considered for these food categories.

In our study, all naturally contaminated samples contained levels of presumptive *B. cereus s.l.* at the limit of detection. In principle, this finding is reassuring, but when assessing the safety of a product throughout the food production chain, including storage to the end of shelf-life, particularly nutrient-rich products contaminated with low levels of *B. cereus s.l.* lacking competitive flora cannot be considered completely safe. On the one hand, one can assume low level contaminations in the case of fresh produce, which, however, can result in rapid multiplication and accumulation of emetic and enterotoxins when temperature deviations occur. Moreover, low contamination levels of highly processed foods do not preclude the presence of the heat-stable and acid-stable toxin cereulide at the time of consumption posing a health risk to the consumer [32]. Naturally contaminated food samples from different manufacturers and batches presented very heterogeneous *B. cereus s.l.* populations. In particular, the diversity of milk isolates between manufacturers was distinctive, which could be attributed to processing methods applied (such as microfiltration and high-temperature treatment) and/or the presence of persister cells in the production environment (Figure 5, Supplementary Tables S5 and S6).

Naturally contaminated samples were pre-screened for the presence of presumptive *B. cereus s.l.* prior to freezing. Recovery after freezing was tested using the selective media test panel and resulted in the highest recovery on BA (57.8%), CH (56.3%) and MYP (54.7%) (Figure 6). Identical *panC* group and toxin gene combinations before and after freezing were detected in 20.3% of samples. Sampling before and after freezing revealed shifts in *panC* groups and toxin gene profiles, but within samples from the same producer the distributions were consistent. In 15.6% of samples, divergent *panC* and toxin combinations were detected after freezing. This phenomenon can be explained by the non-uniform distribution of *B. cereus s.l.* contamination at the detection limit (Poisson distribution) and by the influence of matrix components during initial testing [47]. Group species and their toxins may be bound to lipid globules (e.g., in the case of dairy products) and only become detectable following rougher digestion after more stringent sample treatment process procedures, e.g., using such as beads beating [48,49]. In our investigation, freezing samples resulted in the detection of an extended spectrum of *panC* and toxin profile combinations.

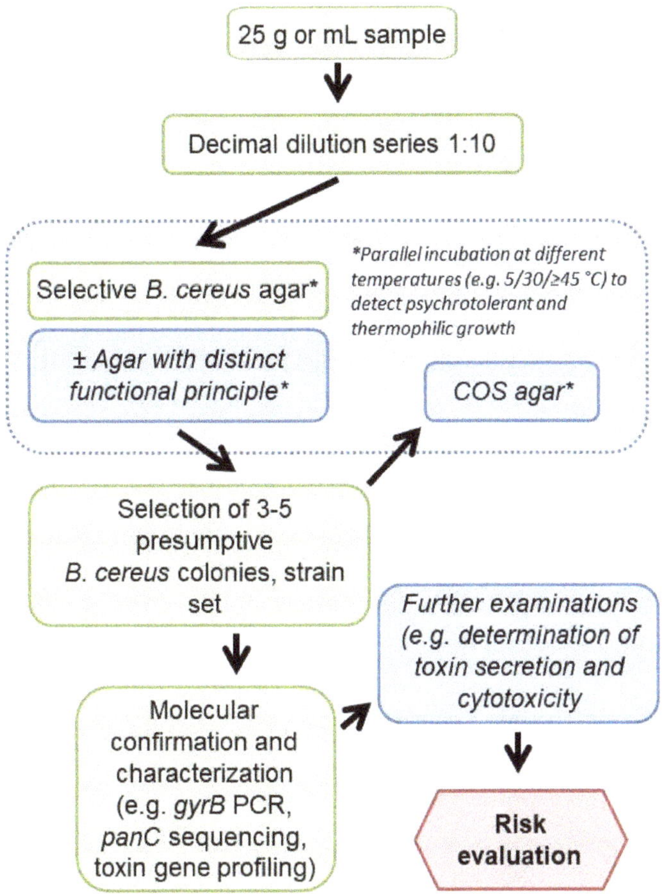

Figure 7. Proposed workflow for *Bacillus cereus sensu lato* analysis performable in routine food analysis (Italic—optional steps). COS, Columbia agar plus 5% sheep blood. *, parallel incubation at different temperatures to detect psychrotolerant and thermophilic growth.

The predominant *panC*/toxin profile combination among target strains and naturally contaminated sample isolates was IV/A (21.8% and 33.0%, respectively), followed by VI/C (17.3% and 10.4%), III/F and II/F (21.7% and 10.4% in naturally contaminated samples) and III/D (10.9% target strains) (Figure 5 and Supplementary Table S2).

Phylogenetic groups II, III and IV comprise moderately to highly cytotoxic strains, most likely posing a potential health risk. In addition, *panC* group III strains may carry the *ces* gene encoding for emetic toxin cereulide [25]. *B. cereus s.l.* strains assigned to *panC* group VI were often isolated from raw milk (target strain set) and were highly abundant among heat-treated milk samples (Supplementary Tables S2 and S6) [50].

Recently, the connection of biopesticidal *B. thuringiensis* strains to foodborne outbreaks in France was investigated. In 39% of outbreaks, *B. thuringiensis panC* group IV was suspected to be the causative organism [51]. In our study, *panC* group IV was highly abundant among isolates from salads, vegetables, herbs and spices that may also include biopesticidal *B. thuringiensis* strains. Furthermore, *cytK*-2 was highly abundant among *panC* group IV strains [18,52]. This is concordant with our results as we identified *cytK-2* highly abundant in *panC*/toxin gene profile combination IV/A (21.8% and 33.0% among target strains and sample isolates). Since other studies have found the use of *B. thuringiensis* biopesticides to be safe or of low risk to public health [53,54], future research should address the contribution of extensively used biopesticide strains to the contamination of raw materials, such as vegetables and fresh produce processed into ready-to-eat foods.

5. Conclusions

This study dealt with culture-based *B. cereus s.l.* diagnosis, which is especially practiced in routine analysis. We tested a selective media panel using test strains and naturally contaminated samples at the detection limit, which is relevant for practice. The results show that it is necessary to include more than one selective medium in the analysis, comparable to *Listeria monocytogens* and *Salmonella* diagnostics in food and animal feed. In order to be able to make a statement about contamination with presumptive *B. cereus s.l.* at all, it is recommended to perform, e.g., PCR, FTIR or MALDI-based confirmation and subtyping (e.g., *panC* and toxin gene profiling) and to assess growth behavior (e.g., psychrotolerance) (Figure 7).

Supplementary Materials: The following supporting information can be downloaded at: https://www.mdpi.com/article/10.3390/foods11030288/s1, Table S1: Media specifications according to the respective manufacturer's descriptions; Table S2: Inclusivity test strains (n = 110); Table S3: Exclusivity test strains (n = 110); Table S4A: Background information on naturally contaminated food samples; Table S4B: Background information on naturally contaminated milk samples; Table S5: Accordance in the detection of *Bacillus cereus* group in naturally contaminated samples (n = 64) before and after freezing on selective agar media; Table S6: *panC* group and toxin gene profile combinations of *Bacillus cereus* group detected before and after freezing (n = 64 samples).

Author Contributions: Conceptualization, E.F. and B.S.; methodology, E.F., C.R. and B.S.; validation, E.F. and B.S.; formal analysis, E.F., B.S. and K.B.; investigation, E.F., C.R. and B.S.; resources, M.W. and B.S.; data curation, E.F., C.R. and B.S.; writing—original draft preparation, B.S. and E.F.; writing—review and editing, B.S., E.F., K.B., M.E.-S. and M.W.; visualization, B.S., K.B. and E.F.; supervision, M.W., M.E.-S. and B.S.; project administration, M.W. and B.S.; funding acquisition, M.W. and B.S. All authors have read and agreed to the published version of the manuscript.

Funding: This research was funded by the Austrian Research Promotion Agency (FFG; Vienna, Austria) collaborative K-project "ADDA-Advancement of Dairying in Austria" under the COMET program (grant number 843543) and the Austrian Economic Chamber (WKO; Vienna, Austria) project "Microbial safety of ethnics foods".

Acknowledgments: The authors wish to express gratitude to Monika Ehling-Schulz (Institute of Microbiology, University of Veterinary Medicine, Vienna) for kindly providing outbreak strains, oligonucleotide primers and PCR protocols for toxin gene screening and intellectual input. Furthermore, we thank Sonja Muri-Klinger BSc. for excellent technical assistance in the laboratory. Open Access Funding by the University of Veterinary Medicine Vienna.

Conflicts of Interest: The authors declare no conflict of interest.

References

1. Kumari, S.; Sarkar, P.K. *Bacillus cereus* hazard and control in industrial dairy processing environment. *Food Control* **2016**, *69*, 20–29. [CrossRef]
2. Logan, N.A.; De Vos, P. Bacillus. In *Bergey's Manual of Systematics of Archaea and Bacteria*; Whitman, W.B., Rainey, F., Kämpfer, P., Trujillo, M., Chun, J., de Vos, P., Hedlund, B., Dedysh, S., Eds.; John Wiley & Sons: Hoboken, NJ, USA, 2015.
3. Messelhäußer, U.; Ehling-Schulz, M. *Bacillus cereus*—A Multifaceted Opportunistic Pathogen. *Curr. Clin. Microbiol. Rep.* **2018**, *5*, 120–125. [CrossRef]
4. Ehling-Schulz, M.; Fricker, M.; Scherer, S. *Bacillus cereus*, the causative agent of an emetic type of food-borne illness. *Mol. Nutr. Food Res.* **2004**, *48*, 479–487. [CrossRef]
5. Ehling-Schulz, M.; Lereclus, D.; Koehler, T.M. The *Bacillus cereus* Group: *Bacillus* Species with Pathogenic Potential. *Microbiol. Spect.* **2019**, *7*. [CrossRef]
6. Dietrich, R.; Jessberger, N.; Ehling-Schulz, M.; Märtlbauer, E.; Granum, P.E. The Food Poisoning Toxins of *Bacillus cereus*. *Toxins* **2021**, *13*, 98. [CrossRef]

7. Lechner, S.; Mayr, R.; Francis, K.P.; Pruess, B.M.; Kaplan, T.; Wießner-Gunkel, E.; Steward, G.S.; Scherer, S. *Bacillus weihenstephanensis* sp. nov. is a new psychrotolerant species of the *Bacillus cereus* group. *Int. J. Syst. Evol. Microbiol.* **1998**, *48*, 1373–1382. [CrossRef]
8. Nakamura, L.K. *Bacillus pseudomycoides* sp. nov. *Int. J. Syst. Evol. Microbiol.* **1998**, *48*, 1031–1035. [CrossRef] [PubMed]
9. Guinebretière, M.H.; Auger, S.; Galleron, N.; Contzen, M.; De Sarrau, B.; De Buyser, M.L.; Lamberet, G.; Fagerlund, A.; Granum, P.E.; Lereclus, D.; et al. *Bacillus cytotoxicus* sp. nov. is a novel thermotolerant species of the *Bacillus cereus* Group occasionally associated with food poisoning. *Int. J. Syst. Evol. Microbiol.* **2013**, *63*, 31–40. [CrossRef] [PubMed]
10. Jiménez, G.; Urdiain, M.; Cifuentes, A.; López-López, A.; Blanch, A.R.; Tamames, J.; Kämpfer, P.; Kolstø, A.B.; Ramón, D.; Martínez, J.F.; et al. Description of *Bacillus toyonensis* sp. nov., a novel species of the *Bacillus cereus* group, and pairwise genome comparisons of the species of the group by means of ANI calculations. *Syst. Appl. Microbiol.* **2013**, *36*, 383–391. [CrossRef] [PubMed]
11. Fricker, M.; Reissbrodt, R.; Ehling-Schulz, M. Evaluation of standard and new chromogenic selective plating media for isolation and identification of *Bacillus cereus*. *Int. J. Food Microbiol.* **2008**, *121*, 27–34. [CrossRef]
12. Tallent, S.M.; Kotewicz, K.M.; Strain, E.A.; Bennett, R.W. Efficient Isolation and Identification of *Bacillus cereus* Group. *J. AOAC Int.* **2012**, *95*, 446–451. [CrossRef]
13. Guinebretière, M.H.; Thompson, F.L.; Sorokin, A.; Normand, P.; Dawyndt, P.; Ehling-Schulz, M.; Svensson, B.; Sanchis, V.; Nguyen-The, C.; Heyndrickx, M.; et al. Ecological diversification in the *Bacillus cereus* Group. *Environ. Microbiol.* **2008**, *10*, 851–865. [CrossRef]
14. Ehling-Schulz, M.; Messelhäußer, U. Bacillus "next generation" diagnostics: Moving from detection toward subtyping and risk-related strain profiling. *Front. Microbiol.* **2013**, *4*, 32. [CrossRef]
15. Nair, K.; Al-Thani, R.; Al-Thani, D.; Al-Yafei, F.; Ahmed, T.; Jaoua, S. Diversity of *Bacillus thuringiensis* Strains From Qatar as Shown by Crystal Morphology, δ-Endotoxins and *cry* Gene Content. *Front. Microbiol.* **2018**, *9*, 708. [CrossRef] [PubMed]
16. Doellinger, J.; Schneider, A.; Stark, T.D.; Ehling-Schulz, M.; Lasch, P. Evaluation of MALDI-ToF mass spectrometry for rapid detection of cereulide from Bacillus cereus cultures. *Front. Microbiol.* **2020**, *11*, 2483. [CrossRef]
17. Ehling-Schulz, M.; Svensson, B.; Guinebretière, M.H.; Lindbäck, T.; Andersson, M.; Schulz, A.; Fricker, M.; Christiansson, A.; Granum, P.E.; Märtlbauer, E.; et al. Emetic toxin formation of *Bacillus cereus* is restricted to a single evolutionary lineage of closely related strains. *Microbiology* **2005**, *151*, 183–197. [CrossRef] [PubMed]
18. Johler, S.; Kalbhenn, E.M.; Heini, N.; Brodmann, P.; Gautsch, S.; Bağcioğlu, M.; Contzen, M.; Stephan, R.; Ehling-Schulz, M. Enterotoxin Production of *Bacillus thuringiensis* Isolates From Biopesticides, Foods, and Outbreaks. *Front. Microbiol.* **2018**, *9*, 1915. [CrossRef]
19. Bağcıoğlu, M.; Fricker, M.; Johler, S.; Ehling-Schulz, M. Detection and Identification of *Bacillus cereus, Bacillus cytotoxicus, Bacillus thuringiensis, Bacillus mycoides* and *Bacillus weihenstephanensis* via Machine Learning Based FTIR Spectroscopy. *Front. Microbiol.* **2019**, *10*, 902. [CrossRef]
20. Manzulli, V.; Rondinone, V.; Buchicchio, A.; Serrecchia, L.; Cipolletta, D.; Fasanella, A.; Parisi, A.; Difato, L.; Iatarola, M.; Aceti, A.; et al. Discrimination of *Bacillus cereus* Group Members by MALDI-TOF Mass Spectrometry. *Microorganisms* **2021**, *9*, 1202. [CrossRef]
21. *ISO 7932*; Microbiology of Food and Animal Feeding Stuffs-Horizontal Method for the Enumeration of Presumptive Bacillus cereus-Colony-Count Technique at 30 °C; International Organization for Standardization: Geneva, Switzerland, 2004.
22. Walsh, P.S.; Metzger, D.A.; Higuchi, R. Chelex 100 as a medium for simple extraction of DNA for PCR-based typing from forensic material. *Biotechniques* **2013**, *10*, 506–513. [CrossRef]
23. Dzieciol, M.; Fricker, M.; Wagner, M.; Hein, I.; Ehling-Schulz, M. A novel diagnostic real-time PCR assay for quantification and differentiation of emetic and non-emetic *Bacillus cereus*. *Food Control* **2013**, *32*, 176–185. [CrossRef]
24. Ehling-Schulz, M.; Guinebretière, M.H.; Monthán, A.; Berge, O.; Fricker, M.; Svensson, B. Toxin gene profiling of enterotoxic and emetic *Bacillus cereus*. *FEMS Microbiol. Lett.* **2006**, *260*, 232–240. [CrossRef] [PubMed]
25. Guinebretière, M.H.; Velge, P.; Couvert, O.; Carlin, F.; Debuyser, M.L. Ability of *Bacillus cereus* Group Strains To Cause Food Poisoning Varies According to Phylogenetic Affiliation (Groups I to VII) Rather than Species Affiliation. *J. Clin. Microbiol.* **2010**, *48*, 3388–3391. [CrossRef]
26. R Development Core Team. *R: A Language and Environment for Statistical Computing*; R Foundation for Statistical Computing: Vienna, Austria, 2021; ISBN 3-900051-07-0. Available online: http://www.R-project.org (accessed on 17 December 2021).
27. Glasset, B.; Herbin, S.; Granier, S.A.; Cavalié, L.; Lafeuille, E.; Guérin, C.; Ruimy, R.; Casagrande-Magne, F.; Levast, M.; Chautemps, N.; et al. *Bacillus cereus*, a serious cause of nosocomial infections: Epidemiologic and genetic survey. *PLoS ONE* **2018**, *13*, e0194346. [CrossRef]
28. Thorsen, L.; Budde, B.B.; Henrichsen, L.; Martinussen, T.; Jakobsen, M. Cereulide formation by *Bacillus weihenstephanensis* and mesophilic emetic *Bacillus cereus* at temperature abuse depends on pre-incubation conditions. *Int. J. Food Microbiol.* **2009**, *134*, 133–139. [CrossRef]
29. Saleh-Lakha, S.; Leon-Velarde, C.G.; Chen, S.; Lee, S.; Shannon, K.; Fabri, M.; Downing, G.; Keown, B. A Study To Assess the Numbers and Prevalence of *Bacillus cereus* and Its Toxins in Pasteurized Fluid Milk. *J. Food Prot.* **2017**, *80*, 1085–1089. [CrossRef]
30. Park, K.M.; Kim, H.J.; Jeong, M.; Koo, M. Enterotoxin Genes, Antibiotic Susceptibility, and Biofilm Formation of Low-Temperature-Tolerant *Bacillus cereus* Isolated from Green Leaf Lettuce in the Cold Chain. *Foods* **2020**, *9*, 249. [CrossRef] [PubMed]

31. Rouzeau-Szynalski, K.; Stollewerk, K.; Messelhäußer, U.; Ehling-Schulz, M. Why be serious about emetic *Bacillus cereus*: Cereulide production and industrial challenges. *Food Microbiol.* **2020**, *85*, 103279. [CrossRef]
32. Messelhäußer, U.; Frenzel, E.; Blöchinger, C.; Zucker, R.; Kämpf, P.; Ehling-Schulz, M. Emetic *Bacillus cereus* Are More Volatile Than Thought: Recent Foodborne Outbreaks and Prevalence Studies in Bavaria (2007–2013). *BioMed Res. Int.* **2014**, *2014*, 465603. [CrossRef] [PubMed]
33. Warda, A.K.; Siezen, R.J.; Boekhorst, J.; Wells-Bennik, M.H.; de Jong, A.; Kuipers, O.P.; Nierop Groot, M.N.; Abee, T. Linking *Bacillus cereus* Genotypes and Carbohydrate Utilization Capacity. *PLoS ONE* **2016**, *11*, e0156796. [CrossRef]
34. Hendriksen, N.B.; Hansen, B.M. Diagnostic properties of three conventional selective plating media for selection of *Bacillus cereus*, *B. thuringiensis* and *B. weihenstephanensis*. *Folia Microbiol.* **2011**, *56*, 535–539. [CrossRef]
35. Chon, J.W.; Song, K.Y.; Kim, H.; Seo, K.H. Comparison of 3 Selective Media for Enumeration of *Bacillus cereus* in Several Food Matrixes. *J. Food Sci.* **2014**, *79*, M2480–M2484. [CrossRef]
36. Kabir, M.S.; Hsieh, Y.H.; Simpson, S.; Kerdahi, K.; Sulaiman, I.M. Evaluation of Two Standard and Two Chromogenic Selective Media for Optimal Growth and Enumeration of Isolates of 16 Unique *Bacillus* Species. *J. Food Prot.* **2017**, *80*, 952–962. [CrossRef]
37. Chon, J.W.; Seo, K.H. Development of a new chromogenic medium for the enumeration of *Bacillus cereus* in various ready-to-eat foods. *Food Control* **2021**, *128*, 108188. [CrossRef]
38. López, A.C.; Alippi, A.M. Phenotypic and genotypic diversity of *Bacillus cereus* isolates recovered from honey. *Int. J. Food Microbiol.* **2007**, *117*, 175–184. [CrossRef]
39. Chaves, J.Q.; Pires, E.S.; Vivoni, A.M. Genetic diversity, antimicrobial resistance and toxigenic profiles of *Bacillus cereus* isolated from food in Brazil over three decades. *Int. J. Food Microbiol.* **2011**, *147*, 12–16. [CrossRef]
40. Miller, R.A.; Jian, J.; Beno, S.M.; Wiedmann, M.; Kovac, J. Intraclade Variability in Toxin Production and Cytotoxicity of *Bacillus cereus* Group Type Strains and Dairy-Associated Isolates. *Appl. Environ. Microbiol.* **2018**, *84*, e02479-17. [CrossRef] [PubMed]
41. Slamti, L.; Perchat, S.; Gominet, M.; Vilas-Bôas, G.; Fouet, A.; Mock, M.; Sanchis, V.; Chaufaux, J.; Gohar, M.; Lereclus, D. Distinct Mutations in PlcR Explain Why Some Strains of the *Bacillus cereus* Group Are Nonhemolytic. *J. Bacteriol.* **2004**, *186*, 3531–3538. [CrossRef] [PubMed]
42. *ISO 11290*; Microbiology of the Food Chain-Horizontal Method for the Detection and Enumeration of *Listeria monocytogenes* and of *Listeria* spp.—Part 1: Detection Method; International Organization for Standardization: Geneva, Switzerland, 2017.
43. *ISO 6579*; Microbiology of the Food Chain-Horizontal Method for the Detection, Enumeration and Serotyping of Salmonella—Part 1: Detection of *Salmonella* spp.; International Organization for Standardization: Geneva, Switzerland, 2017.
44. Webb, M.D.; Barker, G.C.; Goodburn, K.E.; Peck, M.W. Risk presented to minimally processed chilled foods by psychrotrophic *Bacillus cereus*. *Trends Food Sci. Technol.* **2019**, *93*, 94–105. [CrossRef]
45. Cairo, J.; Gherman, I.; Day, A.; Cook, P.E. *Bacillus cytotoxicus*-A potentially virulent food-associated microbe. *J. Appl. Microbiol.* **2021**, *132*, 31–40. [CrossRef]
46. Heini, N.; Stephan, R.; Ehling-Schulz, M.; Johler, S. Characterization of *Bacillus cereus* group isolates from powdered food products. *Int. J. Food Microbiol.* **2018**, *283*, 59–64. [CrossRef]
47. Porcellato, D.; Narvhus, J.; Skeie, S.B. Detection and quantification of *Bacillus cereus* group in milk by droplet digital PCR. *J. Microbiol. Methods* **2016**, *127*, 1–6. [CrossRef]
48. Ramarao, N.; Tran, S.L.; Marin, M.; Vidic, J. Advanced methods for detection of *Bacillus cereus* and its pathogenic factors. *Sensors* **2020**, *20*, 2667. [CrossRef] [PubMed]
49. Walser, V.; Kranzler, M.; Dawid, C.; Ehling-Schulz, M.; Stark, T.D.; Hofmann, T.F. Distribution of the Emetic Toxin Cereulide in Cow Milk. *Toxins* **2021**, *13*, 528. [CrossRef] [PubMed]
50. Porcellato, D.; Aspholm, M.; Skeie, S.B.; Mellegård, H. Application of a novel amplicon-based sequencing approach reveals the diversity of the *Bacillus cereus* group in stored raw and pasteurized milk. *Food Microbiol.* **2019**, *81*, 32–39. [CrossRef]
51. Bonis, M.; Felten, A.; Pairaud, S.; Dijoux, A.; Maladen, V.; Mallet, L.; Radomski, N.; Duboisset, A.; Arar, C.; Sarda, X.; et al. Comparative phenotypic, genotypic and genomic analyses of *Bacillus thuringiensis* associated with foodborne outbreaks in France. *PLoS ONE* **2021**, *16*, e0246885. [CrossRef] [PubMed]
52. Schwenk, V.; Riegg, J.; Lacroix, M.; Märtlbauer, E.; Jessberger, N. Enteropathogenic potential of *Bacillus thuringiensis* isolates from soil, animals, food and biopesticides. *Foods* **2020**, *9*, 1484. [CrossRef]
53. Raymond, B.; Federici, B.A. In defence of *Bacillus thuringiensis*, the safest and most successful microbial insecticide available to humanity-a response to EFSA. *FEMS Microbiol. Ecol.* **2017**, *93*, fix084. [CrossRef] [PubMed]
54. De Bock, T.; Zhao, X.; Jacxsens, L.; Devlieghere, F.; Rajkovic, A.; Spanoghe, P.; Höfte, M.; Uyttendaele, M. Evaluation of *B. thuringiensis*-based biopesticides in the primary production of fresh produce as a food safety hazard and risk. *Food Control* **2021**, *130*, 108390. [CrossRef]

foods

Review

Intestinal Organoids: New Tools to Comprehend the Virulence of Bacterial Foodborne Pathogens

Mayra Aguirre Garcia [1], Killian Hillion [2], Jean-Michel Cappelier [1], Michel Neunlist [2], Maxime M. Mahe [2,3,4] and Nabila Haddad [1,*]

1 UMR SECALIM, INRAE, Oniris, 44307 Nantes, France; maaguirreg@unal.edu.co (M.A.G.); jean-michel.cappelier@oniris-nantes.fr (J.-M.C.)
2 UMR Inserm 1235-TENS, INSERM, Université de Nantes, Institut des Maladies de l'Appareil Digestif–CHU de Nantes, 44035 Nantes, France; killian.hillion@univ-nantes.fr (K.H.); michel.neunlist@univ-nantes.fr (M.N.); maxime.mahe@cchmc.org (M.M.M.)
3 Department of Pediatric General and Thoracic Surgery, Cincinnati Children's Hospital Medical Center, Cincinnati, OH 45229, USA
4 Department of Pediatrics, University of Cincinnati, Cincinnati, OH 45220, USA
* Correspondence: nabila.haddad@oniris-nantes.fr

Abstract: Foodborne diseases cause high morbidity and mortality worldwide. Understanding the relationships between bacteria and epithelial cells throughout the infection process is essential to setting up preventive and therapeutic solutions. The extensive study of their pathophysiology has mostly been performed on transformed cell cultures that do not fully mirror the complex cell populations, the in vivo architectures, and the genetic profiles of native tissues. Following advances in primary cell culture techniques, organoids have been developed. Such technological breakthroughs have opened a new path in the study of microbial infectious diseases, and thus opened onto new strategies to control foodborne hazards. This review sheds new light on cellular messages from the host–foodborne pathogen crosstalk during in vitro organoid infection by the foodborne pathogenic bacteria with the highest health burden. Finally, future perspectives and current challenges are discussed to provide a better understanding of the potential applications of organoids in the investigation of foodborne infectious diseases.

Keywords: pathogenic mechanism; foodborne bacteria; in vitro cell models; organoids; enteroids

Citation: Aguirre Garcia, M.; Hillion, K.; Cappelier, J.-M.; Neunlist, M.; Mahe, M.M.; Haddad, N. Intestinal Organoids: New Tools to Comprehend the Virulence of Bacterial Foodborne Pathogens. *Foods* **2022**, *11*, 108. https://doi.org/10.3390/foods11010108

Academic Editors: Antonio Afonso Lourenco, Catherine Burgess and Timothy Ells

Received: 15 November 2021
Accepted: 22 December 2021
Published: 1 January 2022

1. Introduction

Foodborne diseases (FBDs) are thought to be a major public health issue that contributes significantly to human morbidity and mortality around the world. The World Health Organization (WHO) estimates that almost one person in 10 falls ill from eating unsafe food every year [1]. Although the European region has the lowest burden in the world, the WHO calculated that more than 23 million people become sick annually because of FBDs [2]. Moreover, foodborne hazards of microbial origin raise a broad number of issues due to their economic burden. The European Food Safety Authority (EFSA) has estimated that the overall economic impact of human salmonellosis in Europe could be as high as EUR 3 billion annually [3]. In addition, antibiotic resistance and increasing food contamination as a consequence of environmental changes and dynamic methods of food production threaten to compound this problem further [4].

The surveillance of FBDs and our ability to tackle the knowledge gaps regarding host–pathogen–environment interactions need to be improved for the better prevention and control of microbial foodborne poisoning. Despite significant results from a large number of studies, their pathophysiology still appears to be poorly characterized, even less so where the pathogen can spread to distant organs and tissues through the blood stream and cause severe complications. One permanent challenge in this area of study is the lack of experimental models to address infection mechanisms and establish a clear picture of FBD biology.

To date, two-dimensional (2D) cultured cell lines have mostly been used, but the reproducibility of the overall physiology remains questionable. Organoids help to overcome the shortcomings of cell line monolayers thanks to their high cell type diversity and closer morphology to native intestinal tissue. They can be used to study the same questions as those addressed with monotypic cell systems, and many more. Organoids may be envisioned as a new tool that holds great promise for addressing novel challenges in the study of foodborne pathogens (FBPs)–host interactions. In this review, we describe the main advances in the field of FBPs relating to the use of organoid model systems and discuss their use for modeling bacterial FBDs, focusing on the foodborne bacteria with the highest disease burden.

2. Moving from Cell Lines to Intestinal Organoids

The oral route is the main entry site of FBPs, and the primary site of infection is the gastrointestinal tract [5]. They generally induce mild to severe enteritis, with widely known symptoms [6]. Because of this common pattern of infection, studies have been mostly focused on what occurs at the intestinal interface. The biology of these diseases remains less explored in other tissues [7], even though FBPs may occasionally spread deeply in the tissues and cause severe complications, permanent disability, and death [8–10].

From a historical perspective of model development and attempts to characterize bacterial FBP pathogenesis, concerns have emerged regarding animal models because bacterial intestinal pathogenesis varies considerably between humans and animals and the occurrence of symptoms in animals remains rare [11]. For example, *Campylobacter jejuni* and *Salmonella enterica*, both considered the main causes of bacterial FBDs worldwide, are mainly responsible for asymptomatic intestinal carriage in livestock [12]. In addition, national and international legislation and regulations restrict the use of animals in scientific procedures. The 3Rs principle (replacement, reduction, and refinement) aims to reduce the number of animals used in experimentation, which has led to the development of alternative methods [13]. In view of this, cell culture models of bacterial interaction with the epithelium have proved valuable for defining bacterium–host interactions [11].

The gold standard in intestinal modelling is based on immortalized cancer-derived cell lines, such as the enterocyte-like Caco-2 cell line. Numerous conclusions have been drawn from infected polarized or unpolarized cell monolayers (Figure 1a), even though it has been widely demonstrated over the last 50 years that these cell systems are outperformed [14]. As they consist of tumor-derived cells, they may not represent the native and healthy human intestine [15]. Several factors are likely to define intestinal homeostasis, and these vary considerably between cancer cell lines and the epithelial cells of native organs [16]. Structurally speaking, cell monolayers do not account for three-dimensional (3D) architecture and the complex cell population of the intestinal epithelium.

In light of these disadvantages, cell coculture systems have been used to mirror the physiology of the human intestine more consistently. For instance, triple or cell coculture models (Figure 1b) have represented mucus-carrying intestinal tissue and basic elements of the innate immune system [17–21]. In parallel, the rotating wall vessel (RWV) facilitated the intestinal cell aggregation and growth in three dimensions (Figure 1c). Three-dimensional spheres resemble the native intestinal epithelium more accurately than monolayers derived from the same cell line [22]. The responses to bacterial pathogens also differ from those observed in 2D cell models [22,23].

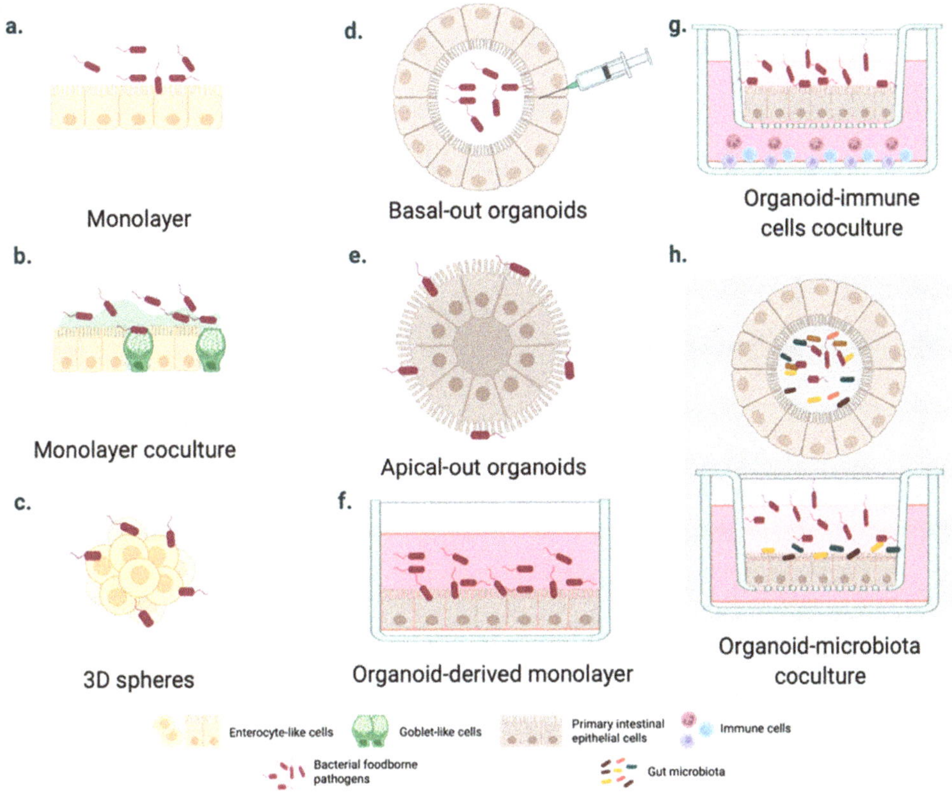

Figure 1. Cell culture systems mimicking intestinal FBD. (**a**–**c**) Intestinal FBD models derived from immortalized cells. (**a**) Polarized homogeneous cell monolayer typically based on immortalized cell lines with an enterocyte-like phenotype (e.g., Caco-2 cell monolayer). (**b**) Heterogeneous cell monolayer coculturing different cell lines to mimic essential intestinal features, such as the mucus-carrying intestinal tissue (e.g., Caco-2 and HT29 co-culture in vitro cell models). (**c**) 3D cell spheres developed from tumor-derived cell lines. (**d**–**f**) Intestinal organoid cultures generated from pluripotent stem cells (PSCs) or adult stem cells (AdSCs). (**d**) Basal-out organoid. The pathogen is generally injected inside the organoid. (**e**) Apical-out organoids might enhance the access of FBP with a high preference for the apical intestinal compartment. (**f**) Organoid-derived monolayers are D cell infection systems, such as the conventional immortalized cell cultures. (**g**–**h**) Coculture of intestinal organoids with immune cells and microbiota. More sophisticated organoid-based cultures, including intestinal epithelium–immune system and epithelium–microbiota interactions during infection.

Owing to the potential of organoids, the number of citations including the term "organoid" has rocketed in the last years. However, there does not seem to be a consensus on a general definition of organoids in the literature. In order to avoid misunderstandings, the recent definition suggested by Fujii and Sato was adopted in this review [24], i.e., "any heterotypic structures that can be reproducibly generated from single cells or cell clusters derived from somatic tissues or pluripotent stem cells, can self-assemble through cell–cell and cell–extracellular matrix (ECM) communications, and have some features of counterpart in vivo tissues" [24]. A further distinction is made according to the type of stem cell used to generate the organoids. While intestinal human organoids can be derived from pluripotent stem cells (PSCs) (including embryonic stem cells (ESCs) and induced pluripotent stem cells (iPSCs)) (Figure 2), adult stem cell (AdSC)-based organoids are initiated from self-renewing tissues, such as the gastrointestinal epithelium (see Figure 2) [25,26]. Two additional terms, enteroids and colonoids, are often used in the context of organoids to refer to the 3D models derived from intestinal and colon adult stem cells that only comprise epithelial cells (Figure 2) [27].

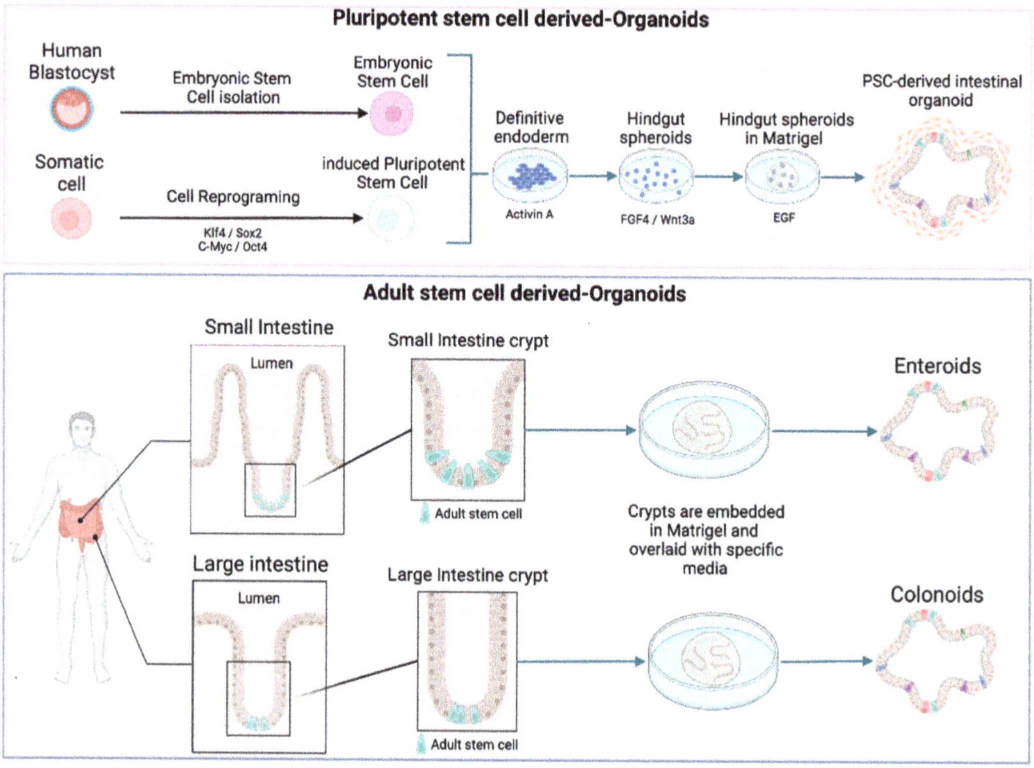

Figure 2. Schematic diagram of intestinal organoid, enteroid, and colonoid generation. Organoids can be derived from pluripotent stem cells (PSCs), including either induced pluripotent stem cells (iPSC) or embryonic stem cells (ESC). Enteroids and colonoids can be grown from the adult stem cells (AdSC) isolated from intestinal crypts.

Contrary to immortalized cancer-derived cell lines, intestinal organoids are characterized by the capacity to generate crypt-like domains with proliferative regions able to differentiate into all of the epithelial cell lineages. They also possess villus-like domains able to maintain cellular polarization toward the tissue. A comparison of 2D versus 3D cell culture systems is provided in Table 1.

Table 1. Comparison of 2D versus 3D cell cultures (as reviewed in [28–30]). The phrase 2D cell culture refers to monolayer epithelial cells (not derived from organoid/enteroid models), whereas 3D cell culture refers to organoid and enteroid models.

Comparison	2D Monolayer Cell Culture	3D Cell Culture
Cell differentiation into enterocyte or goblet cell	✓	✓
Cell differentiation into Paneth cell and enteroendocrine lineages	-	✓
Easily accessible to the apical side of cells	✓	-

Table 1. *Cont.*

Comparison	2D Monolayer Cell Culture	3D Cell Culture
Include immune, nerve, or vascular cells	-	-
Cell polarisation	✓	✓
Formation of cell–cell tight junctions	✓	✓
Development of villus-like and crypt-like structures—three-dimensional architecture	-	✓
Expanded indefinitely	✓ (if derived from tumour cells)	✓
Cryopreservation for long-term storage	✓ (if derived from tumour cells)	✓
Reproducibility	+++	+
Cost	+	+++

Legend: (✓), presence. (-), absence. (+), low. (+++), high.

To mimic the architectural and physiological properties of the in vivo small intestine, the models for foodborne diseases require differentiated crypt-villus structures. Intestinal crypts contain stem cells, which maintain the epithelial progenitor cells pool. Once generated, epithelial cells migrate toward the lumen, and differentiate and die at the tip of the villi. This process leads to a complete regeneration of the intestinal epithelium every 4–5 days [31]. Organoid culture is based on the capacity of the intestinal epithelial stem cells to perpetually divide and produce epithelial progenitor cells. The discovery of *Lgr5* (Leucine-rich repeat-containing G protein-coupled receptor 5) has paved the way for culturing adult stem cells [32]. *Lgr5*$^+$ intestinal stem cells cultured in 3D can undergo multi-lineage differentiation to ultimately form a "mini-gut". In 2009, Sato et al. developed this long-term culture based on crucial signaling pathways, such as the Wnt/β-Catenin pathway and the EGF/EGF receptor (EGFR) with ECM-supported culture [33]. The resulting organoid culture system has been successfully applied to culture other epithelial organs, including stomach, pancreas, colon, and liver organoids [14].

Organoids have been mainly used for the study of cancer and genetic disorders as well as host cell–microorganism interactions [34]. In the organoid–pathogen coculture, several constraints in the mimicking of viral and human host-specific infections have been overcome. Alternatively, organoids generated from genetically modified pluripotent stem cells or from patients harboring mutations of clinical interest have opened a new window onto human infection diseases [35]. Furthermore, these practical and reproducible in vitro models of infection lead to the exploration of additional host–microbe dynamics, e.g., in disseminated infections [7,36,37].

Intestinal organoids usually form structures with budded and branched shapes [38], encapsulating the apical surface and the lumen (Figure 1d) [39]. This makes pathogen delivery inside the organoid interior more challenging from a technical point of view. Even though several studies have employed microinjection (Figure 1e), this is a tedious technique and observations can be disturbed by cellular material accumulating within the luminal side; moreover, cellular material may damage the organoid epithelium [39].

In 2019, Co et al. developed a culture system where organoids could precisely adopt polarity-specific parameters inspired by previous studies of polarity reversal in Madin–Darby canine kidney (MDCK) spheroids [39,40]. The resulting method provided a cell apparatus with an apical-out surface that promoted pathogen inclusion, especially of microbes with a marked preference for interacting with the apical intestinal compartment [39].

Although the study of intestinal epithelial cell (IEC)–pathogen interactions is time and cell consuming [39], most studies have used organoid-derived monolayers on insert/filter membranes (Figure 1f). Two-dimensional cell systems, as with other conventional transmembrane models, provide experimental access to the apical or the basolateral surface [41]. Similarly, monolayers of somatic cells allow adding other nearby intestinal cells to transformed cell lines in coculture to analyze the cellular crosstalk associated with the response to infection (Figure 1g) [42,43]. Although these complex cell systems are still in their infancy, advances have been made in modeling the intestinal microenvironment systems containing macrophages and T-cells (Figure 1g) [42,44] or microbiota (Figure 1h). On a wider scale, hybrid cell cultures could provide insights into the tissue inflammation and carcinogenesis significantly associated with intestinal infections. Table 2 summarizes the main advantages and disadvantages of 3D cell cultures.

Table 2. Main advantages and disadvantages/limitations of 3D cell cultures (as reviewed in [45–49]).

Advantages	Disadvantages
Better mimic endogenous tissues, including organization and spontaneous differentiation of multiple cell types into physiologically relevant 3-D structures, expression and localization of tight junctions, mucus production, polarity, gene expression, cell viability and proliferation, cytokine production	Heterogeneity in size, shape, and viability of organoids within a culture and across different samples, owing to the diversity of individuals and protocols. Protocols for organoid establishment and quality control are not globally standardized.
Contain highly polarized cells that differentiate into the cell lineages of the tissue of origin, i.e., intestinal organoids contain fully mature goblet cells, enterocytes, Paneth cells, and enteroendocrine cells.	Lack of neural innervation, immune cells, vasculature, and amicrobiome → coculture systems with other cell types are not firmly established. Lack of mechanical stress (peristalsis) and luminal and basolateral flow → towards organoid on chip.
Personalization: induced pluripotent stem cells and organoids can be obtained from individuals	Infection experiments: closed system that represents a nonphysiological route for pathogens that infect via the apical/luminal side, i.e., the luminal side is inaccessible without microinjection or disruption of organoid polarization. Microinjection remains a technical challenge.
Genetic engineering: most modern genetic engineering tools can be applied to induced pluripotent stem cells or directly to organoid systems	Relatively costly: organoids cost less than animal models, but they are relatively expensive compared to traditional cell lines (mainly due to medium composition with growth factors and volume required for culturing large numbers of cells).

In the following sections, the main studies related to the use of organoids to decipher the virulence mechanisms of FDPs and the responses of the host cells are discussed.

3. Using Organoids to Explore the Cell and Tissue Tropism of FBPs

Regarding the infection capacity of FBPs, plausible discrepancies can be observed between homogenous cell monolayers and organoids that retain most of the intestinal cell composition and somatic signatures. Early works have shown that bacteria can cause the loss of a tissue's structural integrity in intestinal organoids. Unsurprisingly, a growing body of evidence has assessed this common and fundamental issue. Antibiotic-protection assays coupled to confocal imaging to evaluate changes of the actin network have showed that *Salmonella*-, enterohemorrhagic *Escherichia coli* (EHEC)-, *Listeria monocytogenes*-, or *Shigella*-infected organoids showed intracellular pathogen carriage and damage of intestinal tissue in vitro [39,50–52].

Upon reaching the intestinal epithelium, some pathogens exhibit a higher affinity for regional intestinal segments [53]. Enteroids derived from cells from an anatomical region of the intestine could be a potential starting point for reliably studying segment-specific colonization on an in vitro device, an achievement never attained in whole animal models [54]. VanDussen et al. inoculated various strains of pathogenic *E. coli* to the apical surface of a cell monolayer generated from the dissociation of human intestinal biopsies [41]. *E. coli* EPEC strains preferentially adhered to ileal epithelial cells, whereas *E. coli* EAggEC and EHEC strains instead adhered to rectal epithelial cells. In et al. noted a remarkable difference between the number of EHEC bacteria associated with the apical surface in organoids representing colon and jejunum environments [51]. The authors indicated that the preference of EHEC for these colonoids could be related to the colon-specific differentiation [51]. Each *E. coli* pathotype usually possesses distinct virulence mechanisms to disrupt the host intestinal epithelium. Adherence patterns are one of the key signs generally accepted among *E. coli* pathovars [55]. Rajan et al. mimicked bacterial adhesion using enteroids made from crypts isolated from tissues from four different gut segments. Histopathological comparisons of infected enteroids suggested that *E. coli* EAggEC aggregated in several ways, including those patterns observed in classic in vitro models and new ones, with a remarkable dependency on donor and intestinal segment tropism [56].

Unlike EHEC, *Shigella flexneri* can invade enteroids from the duodenum, ileum, and colon in the same manner [57]. However, these findings substantially contrast with the in vivo shigellosis biology that describes a specificity of *Shigella* to the rectal and colonic mucosae [58]. Thus, other elements of the intestinal microenvironment, such as vasculature, the enteric nervous system, or the resident microbiota contributing human colon infection, were not taken into account with the previous enteroid study [57].

Several studies have showed the preferential attachment of FBP on the apical surface of immortalized cell lines [11,20,59–61]. However, some works have investigated the ability of enterocytes to internalize bacteria for transcellular translocation from the basolateral to the apical compartment. To address this issue, Co et al. developed a reversed polarity apical-out human enteroid model [39]. Thanks to this novel cell culture platform, they were able to compare the binding patterns of *S. enterica* Typhimurium and *L. monocytogenes*. *Salmonella* predominantly invade apical-out enteroids and induce cytoskeletal rearrangement, as described using cancer derived monolayers [62]. Conversely, the Gram-positive *L. monocytogenes* adhered more to the basal-out enteroids. When the author used mixed polarity enteroids, whose polarity had been partially reversed and contained both basal-out and apical-out surfaces, both pathogens preferentially invaded the apical side [39]. Apical-out human enteroids seem to be relevant and accessible models because they highlight the importance of cell polarity to visualize the mechanism of pathogen exit from the epithelium to promote shedding and dissemination. This is particularly true for pathogens that use basolateral receptors for invasion, such as *L. monocytogenes* or *S. flexneri*.

Organoids can be used to model the complex multicellular environment of the intestine. Experimental workflows now finely sum up the interactions of pathogens with highly specialized epithelia cells (i.e., mucus-producing cells, Paneth cells, and microfold (M) cells). This could overcome the limitations of the in vitro cell lines that commonly represent enterocytes [54].

The thick mucus layer is a key component of the physical barrier that protects the gut epithelium from the potential pathogens present in the luminal environment [63]. Transcript-based comparisons using organoids have showed changes in the expression signature of mucin *Muc2*, the major structural component of the intestinal mucus. A study based on fully differentiated enteroids infected with *S. flexneri* indicated the transcriptional upregulation of *Muc2* after apical or basolateral bacterial infection [64]. Similar *Muc2* transcript profiles were observed using the goblet-like cells HT29-MTX infected with *S. flexneri* [64]. While non-motile bacteria, such as *Shigella*, increased the level of *Muc2*, EHEC exposure to human colonoids reduce the thickness of the *Muc2*-positive mucus layer in less than 6 h [51].

The follicle-associated epithelium (FAE) is characterized by the presence of M cells, which constitute a niche for bacteria with an intracellular lifestyle because they naturally internalize foreign particles. M cells are exploited by many different pathogens, including *S. flexneri* [65], *L. monocytogenes* [66], and *S. enterica* Typhimurium [67], as a passage through the intestinal barrier to deeper host tissues [68]. *S. enterica* Typhimurium-infected enteroids derived from human small intestinal crypts confirmed that bacteria could rapidly trigger a transition from FAE enterocytes into M cells via an epithelial-mesenchymal transition (EMT) [69]. Similar findings were reported using cocultures of Caco-2 and Raji-B cells [70]. Stimulation with receptor activator of NF-κB Ligand (RANKL) and tumor necrosis factor alpha (TNF-α) was used to induce M cell differentiation in enteroids [71]. The resulting 3D intestinal in vitro device was used to study *S. flexneri* transcytosis via M cells [64]. The authors confirmed the presence of M cells using glycoprotein 2 immunostaining. *S. flexneri* invaded M cell-containing enteroids more often than it invaded non-stimulated enteroids [64].

FBDs are usually self-limiting and of short duration. Some FBD cases, however, can lead to long-lasting disability. A range of human tissues are currently expandable as organoids, but only a few applications are currently used to explore the interactions of FBPs with tissues or cells once the pathogen has colonized the deeper tissues. Organoids have been used to understand the molecular mechanisms behind the epidemiological association between chronic infection with *Salmonella enterica* and gallbladder carcinoma (GBC) in humans. Scanu et al. developed a murine gallbladder organoid (GBO) genetically predisposed to resemble the analogous *TP53* inactivation in GBC patients. Infected murine cells formed organoids in growth factor-free medium. In addition, they presented polarity loss and large irregular nuclei. These observations indicate a cell transformation driven by *Salmonella* infection [72]. More recent evidence reveals that the human restricted pathogenic serovar Paratyphi A induced DNA damage in human GBO [7]. A detailed analysis of longer-term infected organoids reveals that bacteria could drive the termination of cell replication via the downregulation of the transcriptional programs related to each cell cycle phase (G1/S, S, G2, and G2/M) [7]. Therefore, these studies showed not only a clear *Salmonella* tropism of gallbladder tissue, but also the underlying pathways of the connection between *S. enterica* and cancer.

4. Organoids for Investigating the Host Immune Response Following Foodborne Infection

Studying the interplay between FBPs and the distinct cellular populations in disease ecosystems also requires a large picture of the coordinated factors involved in the host defense mechanisms. Given the fact that the signature of organoids resembles the genetic signature of native intestinal epithelium cells and allows genome editing, organoids have also been used to study host signaling for maintaining a fine balance in the gut environment.

Studies have revealed the global transcriptional changes occurring within organoids during tissue inflammation and host defense. Forbester et al. identified a large spectrum of transcriptional changes by evaluating host–pathogen interactions with *S. enterica* Typhimurium [73]. Six of the most highly upregulated genes in the infected organoids consisted of genes related to the interleukins (ILs) that are essential messengers between immune cells and nonhematopoietic cells [73]. Karve et al. found no significant differences in the gene expression of proteins that are involved in gastrointestinal guarding between commensal *E.coli* and STEC strains. However, inflammatory mediators IL-8 and IL-18 were significantly upregulated upon STEC infection [52]. Organoids have also provided significant clues about host defense against *S. flexneri* infection. Elements of the NF-κB-mediated inflammation, including IL-8, TNF-α and TNFAIP3, were enriched in colonoid monolayers infected by *S. flexneri* [57]. Ranganathan et al. evaluated in more detail the effect of *S. flexneri* infection on IL-8 expression [64]. Enteroid and colonoid monolayers infected with *S. flexneri* secreted IL-8 in a time- and compartment-dependent manner. At the same time, the level of apical IL-8 was significantly higher than the level of basolateral IL-8 at the early phase of *S. flexneri* infection. At 26.5 h post infection, the level of basolateral IL-8 was higher than the level of apical IL-8 in the infected enteroids derived from either segment [64].

Although inflammasomes play diverse roles in innate immunity, their function in the central line of human defense against enteric pathogens has not been dealt with in depth. The big cytoplasmic multiprotein complexes can be activated by bacterial stimuli that unlock the canonical and non-canonical pathways, resulting in the secretion of IL-1β and IL-18 [74]. Moreover, the downstream effectors of inflammasomes are involved in activating signals of pyroptosis, a programmed form of cell death that occurs via IEC shedding [60]. Researchers have attempted to determine the role of each caspase in the defense against *Salmonella* infection. Murine enteroid infection models have showed a specific contribution of caspase-1 (*Casp1*) and caspase-11 (*Casp11*) (the equivalents of caspase-4 and caspase-5 in humans), which induced cellular responses and effector mechanisms. *Casp11−/−* and wild type (WT) enteroid-derived monolayers were much less passive upon *Salmonella* infection compared to *Casp1/11−/−* and *Casp11−/−* enteroid monolayers. This infection profile demonstrates that Casp-1 is sufficient to restrict bacterial invasion. Additional findings suggested that the proinflammatory response could upregulate *Casp-11* expression later in the course of infection, and that caspases acted together against pathogen attack [75]. In a similar fashion, Holly et al. compared the caspase-mediated activities of enteroids from human intestinal epithelium and mouse intestinal epithelium in response to infectious stimulation [50]. The human and murine enteroids responded to the microbe in a specie-dependent manner [50]. Whereas *Casp4*-deficient human enteroids completely stopped IL-18 secretion, the murine equivalent of *Casp4* (*Casp-11*) was found to be important but not essential. Similarly, the contribution of canonical and non-canonical pathways to decreasing the intracellular burden of *S. enterica* Typhimurium was species dependent. While non-canonical pathways play a key role in primary human cells, canonical pathways play a key role in primary mouse cells [50].

Forbester et al. generated organoids from healthy individuals and from a patient harboring a mutation in the IL10RB gene that inactivates the IL-22 receptor [35]. The IL-22 receptor expressed on the basal surface, and the subsequent IL-22 response occurred in organoids derived from healthy cells. In contrast, the IL10RB-defective organoids exhibited a loss of the IL-22 defense function. This highlights the relevance of this method for facilitating studies on phenotypic–genotypic associations. Further results demonstrated the infection-limiting mechanisms and a protective role of IL-22 via phagolysosomal fusion [35].

Beyond the understanding reached with organoids, integrating other cell types critical for intestinal homeostasis appears to be indispensable to mimicking the cellular microenvironment. A reliable model of the crosstalk between immune cells and IEC was created by Noel et al. [42]. The macrophages introduced in the basolateral compartment of a mixed enteroid monolayer system developed the ability to cross the intestinal epithelium without harming the medium upon which they were engrafted [76]. Noel et al. observed the reactions of the human macrophage–enteroid coculture in response to a bacterial stimulus on the apical surface [42]. The number of CFUs in the upward phase of enteroxigenic *E. coli* (ETEC) in the pathogen hybrid coculture was significantly lower than in the macrophage-free enteroids as early as 30 min post-infection [42]. Given that fact, this experiment reflects the successful sensing and bactericidal activity of macrophages. The coordinated work of the intestinal barrier and mucosal immunology to prevent infection of the human gut was also accompanied by lower pro-inflammatory cytokine secretion, including IL-8, IL-6, and IFN-γ [42]. On a wider scale, future studies should deal with mechanistic observations of macrophage transepithelial projections and their contact with enteric pathogens [42]. In the same vein, polymorphonuclear leukocytes (PMN) were added to wells containing organoids, mirroring neutrophil recruitment during EHEC infection on the luminal surface. Images of the control and transcriptional profiles identified PMN cells in the external edge of organoids and the upper production of IL-8, respectively [52], which is known to favor PMN cell attraction. IL-8 is also a key factor in neutrophil recruitment in animal enteric infection model [77]. These results represent an excellent initial step toward increasing the complexity of organoids by including stromal elements.

Incorporating genetic engineering into organoid technology could provide further knowledge on the host factors that influence the functions of the intestinal barrier and intestinal defense mechanisms, and, finally, lead to the development of enteric diseases. For instance, mutated organoids that reflect specific tissue phenotypes have facilitated in-depth experimentation to further analyze infection mechanisms. In 2015, Wilson et al. compared the antimicrobial activity of α-defensins in the epithelial defense against *S. enterica* Typhimurium replication using organoids derived from wild-type and mutated mouse cells for α-defensin production [78]. Comparative assays demonstrated that intraluminal *S. enterica* Typhimurium growth was significantly higher in the deficient genotype model. The intestinal ex vivo system may compensate for the anti-bacterial activity through the expression of human defensin HD5 [78].

In addition to cell host responses to infection, organoid tools can also address infection mechanisms on the bacterial side.

5. Organoids for Studying the Virulence Mechanisms of FBPs

Microorganisms possess a number of interlinked virulence traits that constantly move toward the establishment of infection and which trigger disease and their persistence in the host. The study of pathogen effectors may lead to the development of new rapid diagnostics tools or detection methods, therapeutic drugs, and vaccines to better control foodborne pathogens. Organoids are paving the way for additional and promising investigations of molecular aspects of FBP virulence.

The engineering of genes that encode virulence effectors and host adaptation may well be the keystone to fully understanding the causality between a gene defect and infection developed in organoids.

Interestingly, using enteroids, Geiser et al. attempted to describe the *S. enterica* Typhimurium cycle of infection, and uncovered novelties about the role of known virulence factors [79]. *S. enterica* pathogenesis involves the type three secretion system 1 (TTSS-1), which mediates the translocation of effector proteins into host cells to promote bacterial invasion. According to the authors, TTSS-1 activity and some TTSS-1 effectors (SipA, SopB, SopE, and SopE2) seem to promote *S. enterica* Typhimurium colonization in human enteroids by enabling the bacterial invasion of intestinal epithelial cells. However, flagellar motility does not seem to be required for the efficient bacterial colonization of enteroids; *Salmonella* seems to reach the epithelial surface and invade the intestinal epithelial cells through gravitational sedimentation within enteroids [79].

Intestinal organoids could also be an important tool to shed more light on microbial inter-strain—and even inter-serovar—variation in pathogenicity. For example, infected human ileum-derived organoids were used to evaluate the serovar specificity of disease phenotypes to help analyze the role of the YrbE phospholipid transporter in *S. enterica* Typhi and Typhimurium. Verma et al. established that deletion of the *yrbE* gene induced several changes in *S. enterica* Typhy bacteria, such as the over-expression of flagellin, resulting in uncontrolled motility, elevated IL-8 secretion, and deficient adherence to the organoid of the mutant strain. In contrast, *S. enterica* Typhimurium pathovar did not seem to be affected by the disruption of *yrbE*. These results suggest that YrbE might be involved differently in the pathogenic mechanisms of *S. enterica* serovars, especially in the early steps of infection [80].

A neglected field of study using the overly simplistic 2D models has been the molecular routes likely to be involved in the watery diarrhea that is triggered by the majority of FBPs that colonize the human intestinal epithelium. Based on the advances of culture systems, Tse et al. recreated a colonic environment to evidence the potential enterotoxic effect of extracellular serin protease P (EspP) excreted by EHEC, which displays electronic transport and therefore leads to diarrhea [81]. Measuring changes in active ion movements in human colonoid monolayers, the authors indeed detected a significantly increased transport of colonic electrolytes related to EspP luminal concentrations. Thus, additionally to its protease activity, EspP may be a factor involved in EHEC diarrheic episodes [81]. Broader research should investigate the role of serine protease activity from other enteric infectious agents in organoid-pathogenic phenotypes [82].

A study using organoids derived from intestinal tissue taken from human biopsies revealed novel insights into *S. enterica* Typhi small intestinal mucosa infection. A transmission electron microscopy (TEM) analysis indicated a cytoskeletal change, with microvilli destruction leaving a more accessible surface for pathogen entry and vesicle-contained intracellular bacteria. Secondly, while *S. enterica* Typhimurium invasion predominantly occurred through M cell-facilitated phagocytosis, *S. enterica* Typhi infection mostly progressed via the enterocytes [83].

The characterization of the host cell invasion mechanisms and of the effect of pathogens on intestinal stem cells was studied in *Listeria* organoid models. Co et al. confirmed previous findings that *L. monocytogenes* preferentially binds to basolateral receptors to invade intestinal cells [39]. This bacterium targets sites of cell extrusion, where basolateral proteins are apically exposed, and enters the apical epithelium in human enteroids [39]. Five hours post-infection, *L. monocytogenes* translocated in greater numbers across the distal small intestine epithelial monolayers derived from organoids than they did across the proximal monolayers [84]. In addition, invasion by *L. monocytogenes* altered the morphology of the intestinal organoids, especially the intestinal stem cells, and reduced the budding rate [85]. *L. monocytogenes* modulated organoid proliferation by regulating stem cell niches, which disrupted normal intestinal turnover [85]. In addition, this pathogen affected the expression of *Hes1*, *Math1*, and *Sox9*, and this interfered with the differentiation of intestinal stem cells [85]. Besides investigating the molecular mechanisms associated with the enteritis caused by foodborne pathogens, some works have used organoid/enteroid models to explore the other pathologies induced by these pathogens. For example, *Campylobacter jejuni* is known to be the major cause of bacterial enteritis worldwide. Moreover, *Campylobacter* spp. have been observed in patients with colorectal cancer (CRC), and has been associated with the development of inflammatory bowel disease, a known risk factor of CRC [86–89]. He et al. demonstrated that the human clinical isolate *C. jejuni* 81–176 promotes colorectal tumorigenesis through the action of cytolethal distending toxin (CDT) [90]. The key role of CDT in this process was showed using various models, such as mice (germ free $Apc^{Min/+}$), a non-transformed rat small intestine epithelial cell line (IEC-6), a human colon cancer cell line (HT-29), and cultured enteroids [90]. Cultured enteroids were used to evaluate the effect of *cdtB* on DNA damage in primary intestinal cells. Exposure of enteroids to *C. jejuni* lysates enhanced γH2AX induction (a surrogate marker of DNA damage) compared with the control, while this response was attenuated in enteroids exposed to *C. jejuni* with an inactivated *cdtB* gene [90]. These findings demonstrate that *cdtB* plays an important role in *C. jejuni*-induced DNA damage and cell cycle arrest in vitro.

6. Using Organoids to Investigate the Anti-FBP Activities of Probiotic (-like) Bacterial Strains

Organoids are receiving much attention due to their high resemblance to the physiology of the gastrointestinal environment. They have not showed their full potential yet, and there are still shortcomings when modeling complex environments, such as the intestinal microbiota. However, they provide the initial steps toward a more refined understanding of potential microbe-based therapies, such as probiotics. This fact is consistent with the widespread interest in the development of a robust line of new drugs and innovative pathways to bring solutions to patients suffering from either drug-resistant bacterial infections or—even more critically—infectious diseases with only supportive treatment (i.e., EHEC infections).

The commensal strain *E. coli* Nissle has been used as a probiotic for more than a century, and, more recently, to treat intestinal disorders. However, this strain is highly related to a pathogenic *E. coli* strain isolated from a patient with pyelonephritis [91]. Pradhan and Weiss have used human intestinal organoids to assess the safety and protective effects of the probiotic strain against *E. coli* pathogenic strains [92]. In single-strain infection studies, Nissle did not cause damage to organoids. However, in co-infection experiments, Nissle protected organoids from the EHEC-mediated loss of the epithelial barrier function and EHEC-induced apoptosis [92]. The results also suggest that Nissle can be vulnerable to phages and that lysogens can produce the Shiga toxin, which would limit the usefulness of the probiotic as a therapeutic alternative [92].

Introducing potential probiotic microbes into organoids has recently emerged from disease mimicking based on the crosstalk between microbial components and their microenvironment. In 2020, Lu et al. investigated the use of *Lactobacillus acidophilus*, a recognized probiotic microorganism, to drive protective mechanisms on the gut barrier exposed to *Salmonella* [93]. Pre-treatment with the *L. acidophilus* caused more active mucus secretion, resulting probably from the general IEC response to contact with microorganisms [93]. Furthermore, *L. acidophilus* modulated toll-like receptors (TLRs), which are involved in the hyperplasia and inflammation caused by *Salmonella* infection [93]. In the same way, the ability of five lactic acid bacteria strains to modulate the vitamin D receptor (VDR) pathways and *S. enterica* enteritidis-induced inflammation and infection was evaluated using murine organoids [94]. Some of these strains protected organoids from *Salmonella* inflammation by increasing VDR expression [94]. In addition, VDR deletion in organoids resulted in more severe inflammation and bacterial invasion upon *Salmonella* infection [94].

The well-orchestrated communication between epithelial and non-epithelial cells is essential to decipher the arsenal of infection-related responses set up by the host. In the particular case of foodborne infections, gut immunology, for instance, plays a crucial role in maintaining the host–microbiota interactions, and it is interesting to elucidate the crosstalk between the intestinal epithelium and immune cells.

7. Current Challenges and Future Prospects

In-depth investigation of pathogenic mechanisms. The evolution of cell models towards the design of structures that approximate the real microenvironment to which pathogens are exposed in the gut is still of interest in order to improve the understanding of host–pathogen interaction. For example, *Campylobacter jejuni* is unanimously recognized as the leading cause of bacterial enteritis in the world. Paradoxically, however, despite numerous studies on animal and "traditional" cell models, its pathogenic mechanism has still not been fully described. It seems that the models used so far do not sufficiently reproduce the relationship between the bacteria and intestinal cells. The mechanism of *C. jejuni* translocation is especially controversial and not well understood. Consequently, enteroids are therefore likely to investigate more deeply the transmigration of *C. jejuni* across the intestinal epithelium and to provide new information on intestinal campylobacteriosis. In addition, using intestinal organoids from livestock animals can help to investigate the host specificity of zoonotic bacteria in a one health context [95,96]. In addition, new approaches for improving the accessibility of the pathogen to the apical surface of organoids have been investigated. A robotically articulated microinjection platform showed enhanced performance by transporting a bacterial suspension at a rate of approximately 90 organoids per hour. Nevertheless, the efficiency of the device varied considerably due to great organoid heterogeneity in terms of size, shapes, luminal volumes, and monolayer width [97].

Increasing model complexity to assess interactions of FDP with other organs and the environment. Intestinal organoids are mainly exploited as single-organ systems representing the gut epithelium, lacking for mesenchymal or immune cell populations naturally present in the gut mucosa. In order to better model human disease and to evaluate the role of the mucosal compartment and epithelial–immune cell communication occurring in FPD, cocultures of epithelial organoids with other organ-specific elements are of interest, such as with macrophages and T cells. The cellular diversity gain from organoids can also be exploited by interconnecting multiple organ systems in fluidic systems under dynamic conditions. Organ-on-chip devices that use organoids derived from stem cells can model multi-organ complexity, such as the gut–brain axis or the interaction between the gut and kidney, allowing for the study of infection progression from primary to secondary infection sites. In addition, this "organoids-on-chip" technology can reproduce the mechanical forces to which the enteric pathogens can be exposed in the intestinal environment, such as flow and peristalsis. These mechanical constraints seem essential for infectivity.

Towards personalized medicine in foodborne infectious diseases? One of the most pressing clinical challenges is developing precision medicine in FBP infection. Biobanks can be built using enteroids from different normal or genetically and clinically diverse individuals to facilitate fundamental research, but also to study the effect of pharmacological compounds in a heterogeneous population. Existing human intestinal organoid biobanks derived from healthy and diseased tissues have been established, especially from cancers, but also other diseases, such as inflammatory bowel disease and cystic fibrosis [41,45,98]. Co-clinical trials have already been performed to confirm the usefulness of organoids in drug screening by comparing them with other models (e.g., animal models) and with patients' responses, showing in vitro to in vivo correlations [99–101]. Most applications of organoids for precision medicine are currently related to the screening of anticancer therapeutics. These biobanks can be used for high-throughput screening assays to assess the efficacy and toxicity of drugs in a personalized fashion. The genetic engineering of organoids or patient-derived organoids harboring mutations related to pathogenic bacterial infections may disclose the potential associations between genetic signatures and susceptibility to infectious diseases, and can be used to predict responses to drugs. However, the use of human organoids to fully understand infectious diseases requires the development of technologies that are sufficiently simple for routine use in infectious disease laboratories and adequately robust for use in preclinical studies. The addition of a functional immune system, a complete microbial influence, and the generation of M cells remain to be optimized. Moreover, the generation of standardized protocols and mainstream organoid media will make the model more accessible for laboratories and clinics willing to adopt the model and to provide more accurate data.

8. Conclusions

Over the past decade, organoids have appeared that could act as a human model for studying the virulence of enteric bacterial pathogens. To move closer to in vivo pathophysiological mechanisms, the next stage of disease modelling using organoids will require more complex and robust strategies. Recent evidence has revealed that introducing non-epithelial cells, e.g., microbiota and immune cells [42,97] (Figure 1g,h), and improving pathogen attachment through more refined techniques, such as microinjection techniques, apical phase reversion, or using primary epithelial cell monolayers, may considerably empower the study of interactions of the intestinal ecosystem–pathogen interface using organoids. As the complexity of these model systems increases with cocultures and organ-on-chip systems, new opportunities and challenges arise, and the host–pathogen interaction landscape will benefit from them.

Author Contributions: Original draft preparation, M.A.G., J.-M.C., M.M.M. and N.H.; writing—review and editing, M.A.G., K.H., J.-M.C., M.N., M.M.M. and N.H. All authors have read and agreed to the published version of the manuscript.

Funding: MAG was granted by the MAN-IMAL Master, which is the IDEFI Laureat programme ANR-11-IDFI-0003. This work was part of the CAMPI-ES project funded by MICA department of INRAE institute. This work was supported by #ANR-17-CE14-0021 (M.M.M) and a "New Team" grant (BOGUS to M.M.M) from the Bioregate Regenerative Medicine Cluster University of Nantes and Région Pays de la Loire. We gratefully acknowledge Annie Buchwalter for English editing.

Institutional Review Board Statement: Not applicable.

Informed Consent Statement: Not applicable.

Data Availability Statement: Not applicable.

Conflicts of Interest: The authors declare no conflict of interest. The funders had no role in the design of the study; in the collection, analyses, or interpretation of data; in the writing of the manuscript, or in the decision to publish the results.

References

1. WHO. *WHO Estimates of the Global Burden of Foodborne Diseases*; World Health Organization: Geneva, Switzerland, 2015; p. 254.
2. WHO. *The Burden of Foodborne Diseases in the WHO European Region*; World Health Organization: Copenhagen, Danemark, 2017.

3. EFSA. EFSA Explains Zoonotic Diseases: *Salmonella*. Available online: https://data.europa.eu/doi/10.2805/61217 (accessed on 10 November 2021).
4. Newell, D.G.; Koopmans, M.; Verhoef, L.; Duizer, E.; Aidara-Kane, A.; Sprong, H.; Opsteegh, M.; Langelaar, M.; Jacobson, A.; Scheutz, F.; et al. Food-borne diseases—The challenges of 20years ago still persist while new ones continue to emerge. *Int. J. Food Microbiol.* **2010**, *139*, S3–S15. [CrossRef]
5. Drolia, R.; Bhunia, A.K. Crossing the intestinal barrier via *Listeria* adhesion protein and Internalin A. *Trends Microbiol.* **2019**, *27*, 408–425. [CrossRef]
6. Abebe, E.; Gugsa, G.; Ahmed, M. Review on major food-borne zoonotic bacterial pathogens. *J. Trop. Med.* **2020**, *2020*, 4674235. [CrossRef] [PubMed]
7. Sepe, L.P.; Hartl, K.; Iftekhar, A.; Berger, H.; Kumar, N.; Goosmann, C.; Chopra, S.; Schmidt, S.C.; Gurumurthy, R.K.; Meyer, T.F.; et al. Genotoxic effect of *Salmonella* Paratyphi A infection on human primary gallbladder cells. *mBio* **2020**, *11*, e01911-20. [CrossRef] [PubMed]
8. Mayer, C.L.; Leibowitz, C.S.; Kurosawa, S.; Stearns-Kurosawa, D.J. Shiga toxins and the pathophysiology of hemolytic uremic syndrome in humans and animals. *Toxins* **2012**, *4*, 1261–1287. [CrossRef] [PubMed]
9. McLauchlin, J.; Mitchell, R.T.; Smerdon, W.J.; Jewell, K. *Listeria monocytogenes* and listeriosis: A review of hazard characterisation for use in microbiological risk assessment of foods. *Int. J. Food Microbiol.* **2004**, *92*, 15–33. [CrossRef]
10. Rees, J.H.; Soudain, S.E.; Gregson, N.A.; Hughes, R.A. *Campylobacter jejuni* infection and Guillain-Barre syndrome. *N. Engl. J. Med.* **1995**, *333*, 1374–1379. [CrossRef]
11. Haddad, N.; Marce, C.; Magras, C.; Cappelier, J.M. An overview of methods used to clarify pathogenesis mechanisms of *Campylobacter jejuni*. *J. Food Prot.* **2010**, *73*, 786–802. [CrossRef]
12. EFSA; ECDC. The European Union one health 2018 zoonoses report. *EFSA J.* **2019**, *17*, e05926. [CrossRef]
13. Eskes, C. The usefulness of integrated strategy approaches in replacing animal experimentation. *Ann. Ist. Super Sanita.* **2019**, *55*, 400–404. [CrossRef]
14. Simian, M.; Bissell, M.J. Organoids: A historical perspective of thinking in three dimensions. *J. Cell Biol.* **2017**, *216*, 31–40. [CrossRef]
15. Ross, D.T.; Scherf, U.; Eisen, M.B.; Perou, C.M.; Rees, C.; Spellman, P.; Iyer, V.; Jeffrey, S.S.; Van de Rijn, M.; Waltham, M.; et al. Systematic variation in gene expression patterns in human cancer cell lines. *Nat. Genet.* **2000**, *24*, 227–235. [CrossRef] [PubMed]
16. Langerholc, T.; Maragkoudakis, P.A.; Wollgast, J.; Gradisnik, L.; Cencic, A. Novel and established intestinal cell line models–An indispensable tool in food science and nutrition. *Trends Food Sci. Technol.* **2011**, *22*, S11–S20. [CrossRef] [PubMed]
17. Ahmad, I.; Lamprokostopoulou, A.; Le Guyon, S.; Streck, E.; Barthel, M.; Peters, V.; Hardt, W.-D.; Römling, U. Complex c-di-GMP signaling networks mediate transition between virulence properties and biofilm formation in *Salmonella enterica* serovar Typhimurium. *PLoS ONE* **2011**, *6*, e28351. [CrossRef] [PubMed]
18. Bahrami, B.; Macfarlane, S.; Macfarlane, G.T. Induction of cytokine formation by human intestinal bacteria in gut epithelial cell lines. *J. Appl. Microbiol.* **2011**, *110*, 353–363. [CrossRef]
19. Martinez-Argudo, I.; Jepson, M.A. *Salmonella* translocates across an in vitro M cell model independently of SPI-1 and SPI-2. *Microbiology* **2008**, *154*, 3887–3894. [CrossRef]
20. Rodrigues, R.C.; Pocheron, A.-L.; Cappelier, J.-M.; Tresse, O.; Haddad, N. An adapted in vitro assay to assess *Campylobacter jejuni* interaction with intestinal epithelial cells: Taking into stimulation with TNFα. *J. Microbiol. Met.* **2018**, *149*, 67–72. [CrossRef]
21. Zamora, C.Y.; Ward, E.M.; Kester, J.C.; Kelly Chen, W.L.; Velazquez, J.G.; Griffith, L.G.; Imperiali, B. Application of a gut-immune co-culture system for the study of N-glycan-dependent host-pathogen interactions of *Campylobacter jejuni*. *Glycobiology* **2020**, *30*, 374–381. [CrossRef]
22. Nickerson, C.A.; Goodwin, T.J.; Terlonge, J.; Ott, C.M.; Buchanan, K.L.; Uicker, W.C.; Emami, K.; LeBlanc, C.L.; Ramamurthy, R.; Clarke, M.S.; et al. Three-dimensional tissue assemblies: Novel models for the study of *Salmonella enterica* serovar Typhimurium pathogenesis. *Infect. Immun.* **2001**, *69*, 7106–7120. [CrossRef]
23. Höner zu Bentrup, K.; Ramamurthy, R.; Ott, C.M.; Emami, K.; Nelman-Gonzalez, M.; Wilson, J.W.; Richter, E.G.; Goodwin, T.J.; Alexander, J.S.; Pierson, D.L.; et al. Three-dimensional organotypic models of human colonic epithelium to study the early stages of enteric salmonellosis. *Microb. Infect.* **2006**, *8*, 1813–1825. [CrossRef]
24. Fujii, M.; Sato, T. Somatic cell-derived organoids as prototypes of human epithelial tissues and diseases. *Nat. Mater.* **2020**, *20*, 156–169. [CrossRef]
25. Clevers, H. Modeling development and disease with organoids. *Cell* **2016**, *165*, 1586–1597. [CrossRef]
26. Flatres, C.; Loffet, E.; Neunlist, M.; Mahe, M.M. From human pluripotent stem cells to custom-made intestinal organoids. *Med. Sci.* **2019**, *35*, 549–555. [CrossRef]
27. Mahe, M.M.; Sundaram, N.; Watson, C.L.; Shroyer, N.F.; Helmrath, M.A. Establishment of human epithelial enteroids and colonoids from whole tissue and biopsy. *J. Vis. Exp.* **2015**, *97*, e52483. [CrossRef] [PubMed]
28. Aguilar-Rojas, A.; Olivo-Marin, J.-C.; Guillen, N. Human intestinal models to study interactions between intestine and microbes. *Open Biol.* **2020**, *10*, 200199. [CrossRef]
29. Ho, B.X.; Pek, N.M.Q.; Soh, B.-S. Disease modeling using 3D organoids derived from human induced pluripotent stem cells. *Int. J. Mol. Sci.* **2018**, *19*, 936. [CrossRef]
30. Jensen, C.; Teng, Y. Is it time to start transitioning from 2D to 3D cell culture? *Front. Mol. Biosci.* **2020**, *7*, 33. [CrossRef] [PubMed]

31. Barker, N. Adult intestinal stem cells: Critical drivers of epithelial homeostasis and regeneration. *Nat. Rev. Mol. Cell Biol.* **2014**, *15*, 19–33. [CrossRef] [PubMed]
32. Barker, N.; van Es, J.H.; Kuipers, J.; Kujala, P.; van den Born, M.; Cozijnsen, M.; Haegebarth, A.; Korving, J.; Begthel, H.; Peters, P.J.; et al. Identification of stem cells in small intestine and colon by marker gene Lgr5. *Nature* **2007**, *449*, 1003–1007. [CrossRef] [PubMed]
33. Sato, T.; Vries, R.G.; Snippert, H.J.; van de Wetering, M.; Barker, N.; Stange, D.E.; van Es, J.H.; Abo, A.; Kujala, P.; Peters, P.J.; et al. Single Lgr5 stem cells build crypt-villus structures in vitro without a mesenchymal niche. *Nature* **2009**, *459*, 262–265. [CrossRef]
34. Dutta, D.; Clevers, H. Organoid culture systems to study host–pathogen interactions. *Curr. Op. Immunol.* **2017**, *48*, 15–22. [CrossRef]
35. Forbester, J.L.; Lees, E.A.; Goulding, D.; Forrest, S.; Yeung, A.; Speak, A.; Clare, S.; Coomber, E.L.; Mukhopadhyay, S.; Kraiczy, J.; et al. Interleukin-22 promotes phagolysosomal fusion to induce protection against *Salmonella enterica* Typhimurium in human epithelial cells. *Proc. Natl. Acad. Sci. USA* **2018**, *115*, 10118–10123. [CrossRef]
36. DesRochers, T.M.; Kimmerling, E.P.; Jandhyala, D.M.; El-Jouni, W.; Zhou, J.; Thorpe, C.M.; Leong, J.M.; Kaplan, D.L. Effects of Shiga toxin type 2 on a bioengineered three-dimensional model of human renal tissue. *Infect. Immun.* **2015**, *83*, 28–38. [CrossRef] [PubMed]
37. Marquez, L.B.; Araoz, A.; Repetto, H.A.; Ibarra, F.R.; Silberstein, C. Effects of shiga toxin 2 on cellular regeneration mechanisms in primary and three-dimensional cultures of human renal tubular epithelial cells. *Microb. Pathog.* **2016**, *99*, 87–94. [CrossRef]
38. Rozman, J.; Krajnc, M.; Ziherl, P. Collective cell mechanics of epithelial shells with organoid-like morphologies. *Nat. Commun.* **2020**, *11*, 3805. [CrossRef] [PubMed]
39. Co, J.Y.; Margalef-Català, M.; Li, X.; Mah, A.T.; Kuo, C.J.; Monack, D.M.; Amieva, M.R. Controlling epithelial polarity: A human enteroid model for host-pathogen interactions. *Cell Rep.* **2019**, *26*, 2509–2520.e4. [CrossRef]
40. Wang, A.Z.; Ojakian, G.K.; Nelson, W.J. Steps in the morphogenesis of a polarized epithelium. II. Disassembly and assembly of plasma membrane domains during reversal of epithelial cell polarity in multicellular epithelial (MDCK) cysts. *J. Cell Sci.* **1990**, *95 Pt 1*, 153–165. [CrossRef]
41. VanDussen, K.L.; Marinshaw, J.M.; Shaikh, N.; Miyoshi, H.; Moon, C.; Tarr, P.I.; Ciorba, M.A.; Stappenbeck, T.d.S. Development of an enhanced human gastrointestinal epithelial culture system to facilitate patient-based assays. *Gut* **2015**, *64*, 911–920. [CrossRef]
42. Noel, G.; Baetz, N.W.; Staab, J.F.; Donowitz, M.; Kovbasnjuk, O.; Pasetti, M.F.; Zachos, N.C. A primary human macrophage-enteroid co-culture model to investigate mucosal gut physiology and host-pathogen interactions. *Sci. Rep.* **2017**, *7*, 45270. [CrossRef] [PubMed]
43. Yuki, K.; Cheng, N.; Nakano, M.; Kuo, C.J. Organoid models of tumor immunology. *Trends Immunol.* **2020**, *41*, 652–664. [CrossRef]
44. Rogoz, A.; Reis, B.S.; Karssemeijer, R.A.; Mucida, D. A 3-D enteroid-based model to study T-cell and epithelial cell interaction. *J. Immunol. Methods* **2015**, *421*, 89–95. [CrossRef]
45. Barrila, J.; Crabbé, A.; Yang, J.; Franco, K.; Nydam, S.D.; Forsyth, R.J.; Davis, R.R.; Gangaraju, S.; Ott, C.M.; Coyne, C.B.; et al. Modeling host-pathogen interactions in the context of the microenvironment: Three-dimensional cell culture comes of age. *Infect. Immun.* **2018**, *86*, e00282-18. [CrossRef]
46. Bartfeld, S. Modeling infectious diseases and host-microbe interactions in gastrointestinal organoids. *Dev. Biol.* **2016**, *420*, 262–270. [CrossRef] [PubMed]
47. Han, X.; Mslati, M.A.; Davies, E.G.; Chen, Y.; Allaire, J.M.; Vallance, B.A. Creating a more perfect union: Modeling intestinal bacteria-epithelial interactions using organoids. *Cell. Mol. Gastroenterol. Hepatol.* **2021**, *12*, 769–782. [CrossRef]
48. Kim, J.; Koo, B.-K.; Knoblich, J.A. Human organoids: Model systems for human biology and medicine. *Nat. Rev. Mol. Cell Biol.* **2020**, *21*, 571–584. [CrossRef] [PubMed]
49. Yu, H.; Hasan, N.M.; In, J.G.; Estes, R.K.; Kovbasnjuk, O.; Zachos, N.C.; Donowitz, M. The contributions of human mini-intestines to the study of intestinal physiology and pathophysiology. *Annu. Rev. Physiol.* **2017**, *79*, 291–312. [CrossRef]
50. Holly, M.K.; Han, X.; Zhao, E.J.; Crowley, S.M.; Allaire, J.M.; Knodler, L.A.; Vallance, B.A.; Smith, J.G. *Salmonella enterica* infection of murine and human enteroid-derived monolayers elicits differential activation of epithelial-intrinsic inflammasomes. *Infect. Immun.* **2020**, *88*, e00017-20. [CrossRef]
51. In, J.; Foulke-Abel, J.; Zachos, N.C.; Hansen, A.M.; Kaper, J.B.; Bernstein, H.D.; Halushka, M.; Blutt, S.; Estes, M.K.; Donowitz, M.; et al. Enterohemorrhagic *Escherichia coli* reduce mucus and intermicrovillar bridges in human stem cell-derived colonoids. *Cell. Mol. Gastroenterol. Hepatol.* **2016**, *2*, 48–62.e3. [CrossRef]
52. Karve, S.S.; Pradhan, S.; Ward, D.V.; Weiss, A.A. Intestinal organoids model human responses to infection by commensal and Shiga toxin producing *Escherichia coli*. *PLoS ONE* **2017**, *12*, e0178966. [CrossRef] [PubMed]
53. McCall, L.I.; Siqueira-Neto, J.L.; McKerrow, J.H. Location, location, location: Five facts about tissue tropism and pathogenesis. *PLoS Pathog.* **2016**, *12*, e1005519. [CrossRef]
54. In, J.G.; Foulke-Abel, J.; Estes, M.K.; Zachos, N.C.; Kovbasnjuk, O.; Donowitz, M. Human mini-guts: New insights into intestinal physiology and host–pathogen interactions. *Nat. Rev. Gastroenterol. Hepatol.* **2016**, *13*, 633–642. [CrossRef]
55. Harrington, S.M.; Dudley, E.G.; Nataro, J.P. Pathogenesis of enteroaggregative *Escherichia coli* infection. *FEMS Microbiol. Lett.* **2006**, *254*, 12–18. [CrossRef]

56. Rajan, A.; Vela, L.; Zeng, X.-L.; Yu, X.; Shroyer, N.F.; Blutt, S.E.; Poole, N.M.; Carlin, L.G.; Nataro, J.P.; Estes, M.K.; et al. Novel segment- and host-specific patterns of enteroaggregative *Escherichia coli* adherence to human intestinal enteroids. *mBio* **2018**, *9*, e02419-17. [CrossRef]
57. Koestler, B.J.; Ward, C.M.; Fisher, C.R.; Rajan, A.; Maresso, A.W.; Payne, S.M. Human intestinal enteroids as a model system of *Shigella* pathogenesis. *Infect. Immun.* **2019**, *87*, e00733-18. [CrossRef] [PubMed]
58. Sansonetti, P.J. Rupture, invasion and inflammatory destruction of the intestinal barrier by *Shigella*, making sense of prokaryote-eukaryote cross-talks. *FEMS Microbiol. Rev.* **2001**, *25*, 3–14. [CrossRef]
59. Hamon, M.; Bierne, H.; Cossart, P. *Listeria monocytogenes*: A multifaceted model. *Nat. Rev. Microbiol.* **2006**, *4*, 423–434. [CrossRef]
60. Knodler, L.A.; Crowley, S.M.; Sham, H.P.; Yang, H.; Wrande, M.; Ma, C.; Ernst, R.K.; Steele-Mortimer, O.; Celli, J.; Vallance, B.A. Noncanonical inflammasome activation of Caspase-4/Caspase-11 mediates epithelial defenses against enteric bacterial pathogens. *Cell Host Microb.* **2014**, *16*, 249–256. [CrossRef] [PubMed]
61. Pentecost, M.; Kumaran, J.; Ghosh, P.; Amieva, M.R. *Listeria monocytogenes* Internalin B activates junctional endocytosis to accelerate intestinal invasion. *PLoS Pathog.* **2010**, *6*, e1000900. [CrossRef] [PubMed]
62. Galán, J.E. *Salmonella* interactions with host cells: Type III secretion at work. *Annu. Rev. Cell Dev. Biol.* **2001**, *17*, 53–86. [CrossRef]
63. Johansson, M.E.; Phillipson, M.; Petersson, J.; Velcich, A.; Holm, L.; Hansson, G.C. The inner of the two Muc2 mucin-dependent mucus layers in colon is devoid of bacteria. *Proc. Natl. Acad. Sci. USA* **2008**, *105*, 15064–15069. [CrossRef]
64. Ranganathan, S.; Doucet, M.; Grassel, C.L.; Delaine-Elias, B.; Zachos, N.C.; Barry, E.M. Evaluating *Shigella flexneri* pathogenesis in the human enteroid model. *Infect. Immun.* **2019**, *87*, e00740-18. [CrossRef]
65. Rey, C.; Chang, Y.-Y.; Latour-Lambert, P.; Varet, H.; Proux, C.; Legendre, R.; Coppée, J.-Y.; Enninga, J. Transcytosis subversion by M cell-to-enterocyte spread promotes *Shigella flexneri* and *Listeria monocytogenes* intracellular bacterial dissemination. *PLoS Pathog.* **2020**, *16*, e1008446. [CrossRef]
66. Chiba, S.; Nagai, T.; Hayashi, T.; Baba, Y.; Nagai, S.; Koyasu, S. Listerial invasion protein internalin B promotes entry into ileal Peyer's patches in vivo. *Microbiol. Immunol.* **2011**, *55*, 123–129. [CrossRef]
67. Jones, B.D.; Ghori, N.; Falkow, S. *Salmonella* Typhimurium initiates murine infection by penetrating and destroying the specialized epithelial M cells of the Peyer's patches. *J. Exp. Med.* **1994**, *180*, 15–23. [CrossRef]
68. Mabbott, N.A.; Donaldson, D.S.; Ohno, H.; Williams, I.R.; Mahajan, A. Microfold (M) cells: Important immunosurveillance posts in the intestinal epithelium. *Mucosal Immunol.* **2013**, *6*, 666–677. [CrossRef] [PubMed]
69. Rouch, J.D.; Scott, A.; Lei, N.Y.; Solorzano-Vargas, R.S.; Wang, J.; Hanson, E.M.; Kobayashi, M.; Lewis, M.; Stelzner, M.G.; Dunn, J.C.Y.; et al. Development of functional Microfold (M) cells from intestinal stem cells in primary human enteroids. *PLoS ONE* **2016**, *11*, e0148216. [CrossRef]
70. Tahoun, A.; Mahajan, S.; Paxton, E.; Malterer, G.; Donaldson, D.S.; Wang, D.; Tan, A.; Gillespie, T.L.; O'Shea, M.; Roe, A.J.; et al. *Salmonella* transforms follicle-associated epithelial cells into M cells to promote intestinal invasion. *Cell Host Microb.* **2012**, *12*, 645–656. [CrossRef]
71. Wood, M.B.; Rios, D.; Williams, I.R. TNF-α augments RANKL-dependent intestinal M cell differentiation in enteroid cultures. *Am. J. Physiol. Cell. Physiol.* **2016**, *311*, C498–C507. [CrossRef]
72. Scanu, T.; Spaapen, R.M.; Bakker, J.M.; Pratap, C.B.; Wu, L.-E.; Hofland, I.; Broeks, A.; Shukla, V.K.; Kumar, M.; Janssen, H.; et al. *Salmonella* manipulation of host signaling pathways provokes cellular transformation associated with gallbladder carcinoma. *Cell Host Microb.* **2015**, *17*, 763–774. [CrossRef] [PubMed]
73. Forbester, J.L.; Goulding, D.; Vallier, L.; Hannan, N.; Hale, C.; Pickard, D.; Mukhopadhyay, S.; Dougan, G. Interaction of *Salmonella enterica* serovar Typhimurium with intestinal organoids derived from human induced pluripotent stem cells. *Infect. Immun.* **2015**, *83*, 2926–2934. [CrossRef] [PubMed]
74. Zheng, D.; Liwinski, T.; Elinav, E. Inflammasome activation and regulation: Toward a better understanding of complex mechanisms. *Cell Discov.* **2020**, *6*, 36. [CrossRef]
75. Crowley, S.M.; Han, X.; Allaire, J.M.; Stahl, M.; Rauch, I.; Knodler, L.A.; Vallance, B.A. Intestinal restriction of *Salmonella* Typhimurium requires caspase-1 and caspase-11 epithelial intrinsic inflammasomes. *PLoS Pathog.* **2020**, *16*, e1008498. [CrossRef] [PubMed]
76. Zhang, Y.-G.; Wu, S.; Xia, Y.; Sun, J. *Salmonella*-infected crypt-derived intestinal organoid culture system for host-bacterial interactions. *Physiol. Rep.* **2014**, *2*, e12147. [CrossRef]
77. Singer, M.; Sansonetti, P.J. IL-8 is a key chemokine regulating neutrophil recruitment in a new mouse model of *Shigella*-induced colitis. *J. Immunol.* **2004**, *173*, 4197–4206. [CrossRef] [PubMed]
78. Wilson, S.S.; Tocchi, A.; Holly, M.K.; Parks, W.C.; Smith, J.G. A small intestinal organoid model of non-invasive enteric pathogen–epithelial cell interactions. *Mucosal Immunol.* **2015**, *8*, 352–361. [CrossRef] [PubMed]
79. Geiser, P.; Di Martino, M.L.; Ventayol, P.S.; Eriksson, J.; Sima, E.; Al-Saffar, A.K.; Ahl, D.; Phillipson, M.; Webb, D.-L.; Sundbom, M.; et al. *Salmonella enterica* serovar Typhimurium exploits cycling through epithelial cells to colonize human and murine enteroids. *mBio* **2021**, *12*, e02684-20. [CrossRef]
80. Verma, S.; Prescott, R.A.; Ingano, L.; Nickerson, K.P.; Hill, E.; Faherty, C.S.; Fasano, A.; Senger, S.; Cherayil, B.J. The YrbE phospholipid transporter of *Salmonella enterica* serovar Typhi regulates the expression of flagellin and influences motility, adhesion and induction of epithelial inflammatory responses. *Gut Microb.* **2019**, *11*, 526–538. [CrossRef]

81. Tse, C.; In, J.G.; Yin, J.; Donowitz, M.; Doucet, M.; Foulke-Abel, J.; Ruiz-Perez, F.; Nataro, J.P.; Zachos, N.C.; Kaper, J.B.; et al. Enterohemorrhagic *E. coli* (EHEC)—secreted serine protease EspP stimulates electrogenic ion transport in human colonoid monolayers. *Toxins* **2018**, *10*, 351. [CrossRef]
82. Xue, R.-Y.; Liu, C.; Xiao, Q.-T.; Sun, S.; Zou, Q.-M.; Li, H.-B. HtrA family proteases of bacterial pathogens: Pros and cons for their therapeutic use. *Clin. Microbiol. Infect.* **2021**, *27*, 559–564. [CrossRef]
83. Nickerson, K.P.; Senger, S.; Zhang, Y.; Lima, R.; Patel, S.; Ingano, L.; Flavahan, W.A.; Kumar, D.K.V.; Fraser, C.M.; Faherty, C.S.; et al. *Salmonella* Typhi colonization provokes extensive transcriptional changes aimed at evading host mucosal immune defense during early infection of human intestinal tissue. *EBioMedicine* **2018**, *31*, 92–109. [CrossRef]
84. Roodsant, T.; Navis, M.; Aknouch, I.; Renes, I.B.; van Elburg, R.M.; Pajkrt, D.; Wolthers, K.C.; Schultsz, C.; van der Ark, K.C.H.; Sridhar, A.; et al. A human 2D primary organoid-derived epithelial monolayer model to study host-pathogen interaction in the small intestine. *Front. Cell. Infect. Microbiol.* **2020**, *10*, 272. [CrossRef] [PubMed]
85. Huang, J.; Zhou, C.; Zhou, G.; Li, H.; Ye, K. Effect of *Listeria monocytogenes* on intestinal stem cells in the co-culture model of small intestinal organoids. *Microb. Pathog.* **2021**, *153*, 104776. [CrossRef]
86. Allali, I.; Delgado, S.; Marron, P.I.; Astudillo, A.; Yeh, J.J.; Ghazal, H.; Amzazi, S.; Keku, T.; Azcarate-Peril, M.A. Gut microbiome compositional and functional differences between tumor and non-tumor adjacent tissues from cohorts from the US and Spain. *Gut Microb.* **2015**, *6*, 161–172. [CrossRef] [PubMed]
87. Gradel, K.O.; Nielsen, H.L.; Schønheyder, H.C.; Ejlertsen, T.; Kristensen, B.; Nielsen, H. Increased short- and long-term risk of inflammatory bowel disease after *Salmonella* or *Campylobacter* gastroenteritis. *Gastroenterology* **2009**, *137*, 495–501. [CrossRef]
88. Newman, A.; Lambert, J.R. *Campylobacter jejuni* causing flare-up in inflammatory bowel disease. *Lancet* **1980**, *2*, 919. [CrossRef]
89. Warren, R.L.; Freeman, D.J.; Pleasance, S.; Watson, P.; Moore, R.A.; Cochrane, K.; Allen-Vercoe, E.; Holt, R.A. Co-occurrence of anaerobic bacteria in colorectal carcinomas. *Microbiome* **2013**, *1*, 16. [CrossRef]
90. He, Z.; Gharaibeh, R.Z.; Newsome, R.C.; Pope, J.L.; Dougherty, M.W.; Tomkovich, S.; Pons, B.; Mirey, G.; Vignard, J.; Hendrixson, D.R.; et al. *Campylobacter jejuni* promotes colorectal tumorigenesis through the action of cytolethal distending toxin. *Gut* **2019**, *68*, 289–300. [CrossRef] [PubMed]
91. Vejborg, R.M.; Friis, C.; Hancock, V.; Schembri, M.A.; Klemm, P. A virulent parent with probiotic progeny: Comparative genomics of *Escherichia coli* strains CFT073, Nissle 1917 and ABU 83972. *Mol. Genet. Genom.* **2010**, *283*, 469–484. [CrossRef]
92. Pradhan, S.; Weiss, A.A. Probiotic properties of *Escherichia coli* Nissle in human intestinal organoids. *mBio* **2020**, *11*, e01470-20. [CrossRef] [PubMed]
93. Lu, X.; Xie, S.; Ye, L.; Zhu, L.; Yu, Q. *Lactobacillus* protects against *S.* Typhimurium–induced intestinal inflammation by determining the fate of epithelial proliferation and differentiation. *Mol. Nutr. Food Res.* **2020**, *64*, 1900655. [CrossRef]
94. Lu, R.; Shang, M.; Zhang, Y.G.; Jiao, Y.; Xia, Y.; Garrett, S.; Bakke, D.; Bäuerl, C.; Martinez, G.P.; Kim, C.H.; et al. Lactic acid bacteria isolated from korean kimchi activate the vitamin D receptor-autophagy signaling pathways. *Inflamm. Bowel. Dis.* **2020**, *26*, 1199–1211. [CrossRef]
95. Beaumont, M.; Blanc, F.; Cherbuy, C.; Egidy, G.; Giuffra, E.; Lacroix-Lamandé, E.; Wiedemann, A. Intestinal organoids in farm animals. *Vet. Res.* **2021**, *52*, 33. [CrossRef] [PubMed]
96. Derricott, H.; Luu, L.; Fong, W.Y.; Hartley, C.S.; Johnston, L.J.; Armstrong, S.D.; Randle, N.; Duckworth, C.A.; Campbell, B.J.; Wastling, J.M.; et al. Developing a 3D intestinal epithelium model for livestock species. *Cell Tissue Res.* **2019**, *375*, 409–424. [CrossRef] [PubMed]
97. Williamson, I.A.; Arnold, J.W.; Samsa, L.A.; Gaynor, L.; DiSalvo, M.; Cocchiaro, J.L.; Carroll, I.; Azcarate-Peril, M.A.; Rawls, J.F.; Allbritton, N.L.; et al. A high-throughput organoid microinjection platform to study gastrointestinal microbiota and luminal physiology. *Cell. Mol. Gastroenterol. Hepatol.* **2018**, *6*, 301–319. [CrossRef]
98. Perrone, F.; Zilbauer, M. Biobanking of human gut organoids for translational research. *Exp. Mol. Med.* **2021**, *53*, 1451–1458. [CrossRef]
99. Pasch, C.A.; Favreau, P.F.; Yueh, A.E.; Babiarz, C.P.; Gillette, A.A.; Sharick, J.T.; Karim, M.R.; Nickel, K.P.; DeZeeuw, A.K.; Sprackling, C.M.; et al. Patient-derived cancer organoid cultures to predict sensitivity to chemotherapy and radiation. *Clin. Cancer Res.* **2019**, *25*, 5376–5387. [CrossRef]
100. Vlachogiannis, G.; Hedayat, S.; Vatsiou, A.; Jamin, Y.; Fernández-Mateos, J.; Khan, K.; Lampis, A.; Eason, K.; Huntingford, I.; Burke, R.; et al. Patient-derived organoids model treatment response of metastatic gastrointestinal cancers. *Science* **2018**, *359*, 920–926. [CrossRef] [PubMed]
101. Yan, H.H.N.; Siu, H.C.; Law, S.; Ho, S.L.; Yue, S.S.K.; Tsui, W.Y.; Chan, D.; Chan, A.S.; Ma, S.; Lam, K.O.; et al. A comprehensive human gastric cancer organoid biobank captures tumor subtype heterogeneity and enables therapeutic screening. *Cell. Stem Cell* **2018**, *23*, 882–897. [CrossRef]

foods

MDPI

Article

Resistance-Nodulation-Cell Division (RND) Transporter AcrD Confers Resistance to Egg White in *Salmonella enterica* Serovar Enteritidis

Xiaojie Qin [1], Yanhong Liu [2] and Xianming Shi [3,*]

[1] School of Health Science and Engineering, University of Shanghai for Science and Technology, Shanghai 200093, China; qxj19900709@163.com

[2] Molecular Characterization of Foodborne Pathogens Research Unit, Eastern Regional Research Center, Agricultural Research Service, United States Department of Agriculture, Wyndmoor, PA 19038, USA; yanhong.liu@usda.gov

[3] School of Agriculture and Biology, Shanghai Jiao Tong University, Shanghai 200240, China

* Correspondence: xmshi@sjtu.edu.cn

Abstract: The excellent survival ability of *Salmonella enterica* serovar Enteritidis (*S.* Enteritidis) in egg white leads to outbreaks of salmonellosis frequently associated with eggs and egg products. Our previous proteomic study showed that the expression of multidrug efflux RND transporter AcrD in *S.* Enteritidis was significantly up-regulated (4.06-fold) in response to an egg white environment. In this study, the potential role of AcrD in the resistance of *S.* Enteritidis to egg white was explored by gene deletion, survival ability test, morphological observation, Caco-2 cell adhesion and invasion. It was found that deletion of *acrD* had no apparent effect on the growth of *S.* Enteritidis in Luria-Bertani (LB) broth but resulted in a significant ($p < 0.05$) decrease in resistance of *S.* Enteritidis to egg white and a small number of cell lysis. Compared to the wild type, a 2-log population reduction was noticed in the Δ*acrD* mutant with different initial concentrations after incubation with egg white for 3 days. Furthermore, no significant difference ($p > 0.05$) in the adhesion and invasion was found between the wild type and Δ*acrD* mutant in LB broth and egg white, but the invasion ability of the Δ*acrD* mutant in egg white was significantly ($p < 0.05$) lower than that in LB broth. This indicates that *acrD* is involved in virulence in *Salmonella*. Taken together, these results reveal the importance of AcrD on the resistance of *S.* Enteritidis to egg white.

Keywords: *Salmonella* Enteritidis; egg white; AcrD; stress resistance; cell invasion

Citation: Qin, X.; Liu, Y.; Shi, X. Resistance-Nodulation-Cell Division (RND) Transporter AcrD Confers Resistance to Egg White in *Salmonella enterica* Serovar Enteritidis. *Foods* **2022**, *11*, 90. https://doi.org/10.3390/foods11010090

Academic Editors: Catherine Burgess, Antonio Afonso Lourenco and Timothy Ells

Received: 22 November 2021
Accepted: 29 December 2021
Published: 30 December 2021

1. Introduction

Eggs are an important and integral part of the human diet. These are consumed all over the world and possess natural physical and chemical defenses to prevent the contamination of microorganisms [1]. Egg white, as a chemical barrier, is generally a hostile environment for bacterial survival and growth because of its unfavorable conditions, such as alkaline pH, nutritional limitations and antibacterial molecules [2,3]. However, the risk of *Salmonella* contamination is a serious threat to human health as well as egg production and processing. In particular, *Salmonella enterica* serovar Enteritidis (*S.* Enteritidis) represents the predominant serotype that is involved in food-borne diseases due to the consumption of eggs and egg products. More importantly, *S.* Enteritidis presents an exceptional ability to survive in egg white in contrast to other *Salmonella* serotypes [4–6].

It is important to understand the resistance mechanisms of *S.* Enteritidis to egg white. Previous workers have revealed key information through the use of molecular biological techniques such as site-directed mutagenesis, transposon-mediated insertional mutagenesis, in vivo expression and DNA arrays at the transcriptional level, which may be helpful in explaining the underlying survival mechanism in this foodborne pathogen [7–9]. While genes identified in those studies were mainly involved in iron transport, biotin synthesis, energy metabolism, cell envelope maintenance, DNA synthesis and repair, motility and pathogenicity. Furthermore, genes such as outer membrane channel-related gene *tolC* [10], DNA repair-related gene *yafD* [7] and stress response-related genes *uspAB* [11] were identified by mutagenic analysis and were considered as main players for the survival of *S.* Enteritidis in egg white. On the other hand, according to our previous study, the transcriptomic and proteomic profiles of *S.* Enteritidis exposed to egg white were analyzed using RNA-Seq and iTRAQ analysis to reveal potential important metabolic pathways, such as stress response, iron acquisition, amino acid and biotin synthesis, transport and regulation [6,12]. A highly up-regulated expression (4.06-fold) of the stress response related protein AcrD was found in *S.* Enteritidis in response to whole egg white at the protein level [12].

Gram-negative bacteria such as *Salmonella* usually have multidrug efflux transporters, which have been found to recognize and excrete various structurally unrelated compounds from the cell. Among the multidrug efflux pumps, members of the RND (Resistance-Nodulation-cell Division) family appear to be the most effective efflux systems in those bacteria. *Salmonella* has five RND-type efflux systems: AcrAB, AcrAD, AcrEF, MdtABC and MdsABC [13]. The RND transporter AcrD has a unique biological role, which can remove antimicrobial compounds, such as aminoglycosides, from the bacterial cell. Inactivation of *acrD* resulted in changes in the expression of 403 genes involved in basic metabolism, stress responses and virulence [14]. Furthermore, the deletion of *acrD* led to a significant reduction in biofilm formation and down-regulated expression of key biofilm formation-related proteins encoded by *csgBD* [15]. Previously, the deletion of *acrD* resulted in an increased sensitivity to antibiotics, dyes and detergents in *S.* Typhimurium [16,17]. Furthermore, AcrD also contributes to copper and zinc resistance in *Salmonella* [18]. Previous works have demonstrated that *Salmonella* usually infects the human host through the ingestion of contaminated food products. This bacterium is able to resist the adverse environment of the gastrointestinal tract and then adhere, colonize and invade host intestinal epithelial cells, leading to human infections and diseases [19,20]. To our knowledge, the role of *acrD* in *S.* Enteritidis resistance to antibacterial egg white is not yet clear.

Hence, this study aimed to uncover the role of *acrD* in the resistance and virulence of *S.* Enteritidis to egg white by gene expression analysis, gene deletion, survival ability test, cellular morphology analysis, Caco-2 cell adhesion and invasion assays. These results will provide new information to help elucidate the resistance mechanisms of *S.* Enteritidis to egg white.

2. Materials and Methods

2.1. Bacterial Strains

S. Enteritidis strain SJTUF10978, isolated from chicken wings, was used as the wild-type (WT) strain in this study. *Escherichia coli* DH5α and *Salmonella* MRL0026 were utilized as reference strains for cell adhesion and invasion assays. These strains were stored at −80 °C in LB (Luria-Bertani) broth, including 50% (*v*/*v*) glycerol. All strains were propagated overnight at 37 °C on LB agar before experiments.

2.2. Caco-2 Cell Culture Preparation

The human colon adenocarcinoma cell line Caco-2, obtained from Shanghai Fuheng Biotechnology Co., Ltd. (Shanghai, China) (FH0029), were routinely maintained in DMEM (Dulbecco's Modified Eagle's Medium, Gibco, Pittsburgh, PA, USA) medium containing 1% non-essential amino acids (Coolaber, Beijing, China), 10% fetal bovine serum (Fuheng biology, Shanghai, China), 100 U/mL penicillin (Hyclone, Shanghai, China) and 100 μg/mL streptomycin (Hyclone, Shanghai, China) at 37 °C with 5% CO_2. Meanwhile, cells were sub-cultured every 2–3 days and used between passages 5 and 10.

2.3. Egg White and Its Filtrate Preparation

SPF (Specific Pathogen Free) eggs used in this study were purchased from Boehringer Ingelheim Vital Biotechnology Co., Ltd. (Beijing, China). Fifty eggs were stored in a 37 °C incubator with 65% RH (Relative Humidity) for 5 days as required for every independent biological repeat. Egg white was collected by homogenization and centrifugation as previously described [12]. Egg white filtrate (FEW, less than 10 kDa) was acquired by centrifugation using ultrafiltration tubes with the cut-off limit of 30 kDa and 10 kDa according to our previous method [21].

2.4. Gene Expression Analysis

Total RNA from log-phase cells of S. Enteritidis in whole egg white and LB broth was extracted using Trizol reagent (Invitrogen, Carlsbad, CA, USA) following the manufacturer's instructions. The RNA concentrations were determined using a NanoDrop 2000c spectrophotometer (Thermo scientific, South Logan, UT, USA), and the quality of RNA was evaluated using 1% agarose gel electrophoresis. Furthermore, genomic DNA treatment and cDNA synthesis was conducted using the PrimeScript RT reagent kit following the manufacturer's instructions (TaKaRa, Dalian, China). The gene expression of *acrD* was tested by RT-qPCR analysis (Eppendorf, Hamburg, Germany) as previously described [12], using primer pair of *acrD*-F (5′-ACGCAACAGCAGACCC-3′) and *acrD*-R (5′-GCCCAGACCGCTAATT-3′). The relative expression of *acrD* in S. Enteritidis was calculated by the comparative cycle threshold method [22]. For data normalization, 16S rRNA was utilized as a reference gene.

2.5. Construction of acrD Deletion Mutant Strain

In-frame deletion of *acrD* was generated based on the previously described homologous recombination knockout method [23]. The primers used are shown in Table 1. In addition, strains and plasmids utilized for the deletion are listed in Table 2. Firstly, the fragment of homologous arms (i.e., upper arm and lower arm) was amplified from the genomic DNA of wild-type S. Enteritidis SJTUF10978 by overlap extension PCR. Secondly, this fragment was cloned into the pMD_{19}-T plasmid carrying an ampicillin resistance gene to produce pMD_{19}-*ΔacrD*. The correct pMD_{19}-*ΔacrD* plasmid was digested with *Sac* I and *Xba* I and then ligated into the pRE112 plasmid carrying a chloramphenicol resistance gene and a sucrose-sensitive gene. Then, the obtaining pRE112-*ΔacrD* plasmid was imported into *E. coli* SM10 λpir using $CaCl_2$ transformation method. The recombinant plasmid was then extracted and transformed into the S. Enteritidis wild-type strain by electroporation to obtain a single-crossover strain. The resulting strain was induced by 8%(*w/v*) sucrose to finish a second crossover. Finally, suspected colonies were chosen and confirmed by DNA sequencing and PCR analysis to acquire the *acrD* deletion mutant (*ΔacrD*).

Table 1. Primers used for Δ*acrD* mutant construction.

Primer	Sequence (5′ to 3′)
acrD-F1	GCTCTAGACTCTACGCCGCTGCTGA (*Xba* I)
acrD-R1	**GGCCGGGAGCTAAAGGGGAACCTCGTGTTT**
acrD-F2	**TTCCCCTTTAGCTCCCGGCCAGCCTGATAC**
acrD-R2	CGAGCTCGGCGACGAATAAGTTGCTGTG (*Sac* I)

The 20-bp overlap sequences for amplification of the fragments of homologous arms is shown in bold. Restriction enzyme sites are underlined.

Table 2. Strains and plasmids used in this study.

Strains or Plasmids	Relevant Characteristics	Reference or Source
S. Enteritidis SJTUF10978	Wild-type strain	Lab stock
ΔacrD	*acrD* deletion mutant of *S.* Enteritidis SJTUF10978	This study
E. coli DH5α	Host for cloning	Lab stock
E. coli SM10 (*λpir*)	*thi thr*-1 *leu*6 *pro*A2 *his*-4 *arg* E2 *lac*Y1 *gal*K2, *ara*14*xyl*5 *sup*E44, *λpir*	[24]
pMD$_{19}$-T	Cloning vector, Amp^r	TaKaRa, China
pMD$_{19}$-*ΔacrD*	pMD$_{19}$-T containing a 3113 bp *acrD* deletion PCR product	This study
pRE112	pGP704 suicide plasmid, *pir* dependent, *ori*T *ori*V *sac*B, Cm^r	[25]
pRE112-*ΔacrD*	pRE112 containing a 3113 bp *acrD* deletion PCR product	This study

2.6. Measurement of Bacterial Growth

Overnight cultures of *S.* Enteritidis wild-type and *ΔacrD* strains in LB broth were collected by centrifugation and diluted to the cell density of $OD_{600} \approx 0.1$. Then, cultures were incubated at 37 °C with continuous shaking at 200 rpm. The bacterial growth curve was measured at regular time (1 h) intervals by a Bioscreen C Analyzer (OY Growth Curves, Finland).

2.7. Survival Ability of *S.* Enteritidis Strains in Egg White and Its Filtrate

The survival of *S.* Enteritidis wild-type and *ΔacrD* strains in egg white and its filtrate was measured according to our previously described method [12]. Bacteria (1 mL) at logarithmic phase were collected, washed twice using sterile PBS (Phosphate-Buffered Saline, 1.8 mM KH_2PO_4, 10 mM Na_2HPO_4, 2.7 mM KCl, 137 mM NaCl, pH 7.2) and suspended in PBS. The bacterial suspension was adjusted to approximately 1×10^7 CFU/mL and 1×10^4 CFU/mL by dilution in PBS. Then in a 96-well microplate, 20 μL aliquots of the bacterial suspensions were inoculated into 180 μL of egg white and 180 μL of its filtrate, respectively. It was mixed to give a final concentration of 1×10^6 CFU/mL and 1×10^3 CFU/mL, respectively. The above mixtures were incubated at 37 °C for 24 h. Viable bacteria after incubation were enumerated by plating 100 μL of the treated cell suspensions on LB agar and incubated at 37 °C for 24 h.

2.8. Scanning Electron Microscopy (SEM) Analysis

The cell morphology of *S.* Enteritidis wild-type and *ΔacrD* strains in LB broth, egg white and egg white filtrate at 37 °C for 1 day was observed using a Sirion 200 SEM (FEI Company, Hillsboro, OR, USA) as previously described [21].

2.9. Adhesion and Invasion Assays

The adhesive and invasive ability of *Salmonella* strains and *E. coli* DH5α were investigated according to a previous method [26] with some modifications. The 48-h, 80% confluent Caco-2 monolayers were sub-cultured and placed into a 12-well plate at a density of approximately 1×10^5 cells/well. Bacterial strains were inoculated in LB broth overnight at 37 °C. Bacterial cells were recovered by centrifugation at 13,800× *g* for 5 min, washed twice using DEME medium and suspended in DEME medium to a final concentration of 10^7 CFU/mL. Then, bacterial suspensions and Caco-2 cells were mixed at a ratio of 100:1 and then incubated for 1 h at 37 °C in a 5% CO_2 incubator. The unattached bacteria after incubation were removed after incubation by washing with PBS. 1% Triton X-100 was added to release the attached bacteria at 37 °C for 5 min. Then, the suspensions were serially diluted, and 20 μL of each dilution was plated on LB agar and then incubated at 37 °C for 24 h. Counted colonies were recorded as the total adhesive bacterial population. The adhesion rate of bacteria was represented as the ratio of the number of adhesive bacteria compared to that of initial inoculated bacteria.

Similarly, in the invasion test, infected Caco-2 cells were incubated in a DMEM medium containing 1% penicillin and streptomycin for 1 h at 37 °C to kill extracellular bacteria. Serial dilutions of the lysates were plated on LB agar to enumerate invading bacterial populations. The invasion rate of bacteria to Caco-2 cells was represented as the ratio of the number of invading bacteria compared to that of cell-adhesion bacteria.

2.10. Statistical Analysis

Three independent experiments were conducted in all assays, and each treatment was carried out in triplicate. Data were evaluated via one-way ANOVA using SAS software. Duncan's multiple range test ($p < 0.05$, $p < 0.001$) was used to identify the difference in survival, cell adhesion and invasion ability between wild-type *S.* Enteritidis and the mutant.

3. Results and Discussion

3.1. Expression of acrD in S. Enteritidis in Response to Whole Egg White and Construction of acrD Mutant

In previous studies, up-regulated expression of *acrD* at the mRNA level has been demonstrated in *S.* Enteritidis in response to low concentrations of egg white, e.g., 10% egg white and 80% egg white [6,9]. To test whether *acrD* was up-regulated in the whole egg white (i.e., 100% egg white), the expression of *acrD* in *S.* Enteritidis exposed to the whole egg white was further analyzed using RT-qPCR in this study. As shown in Figure 1A, the expression of *acrD* was significantly ($p < 0.001$) up-regulated (16.09-fold) in whole egg white compared with that in LB broth at the mRNA level. This gene expression data at the mRNA level was consistent with the proteomic data, which also showed up-regulated expression of AcrD in *S.* Enteritidis exposed to whole egg white at the protein level [12]. Hence, these results suggest that *acrD* may play a potential role in the resistance of *S.* Enteritidis to egg white.

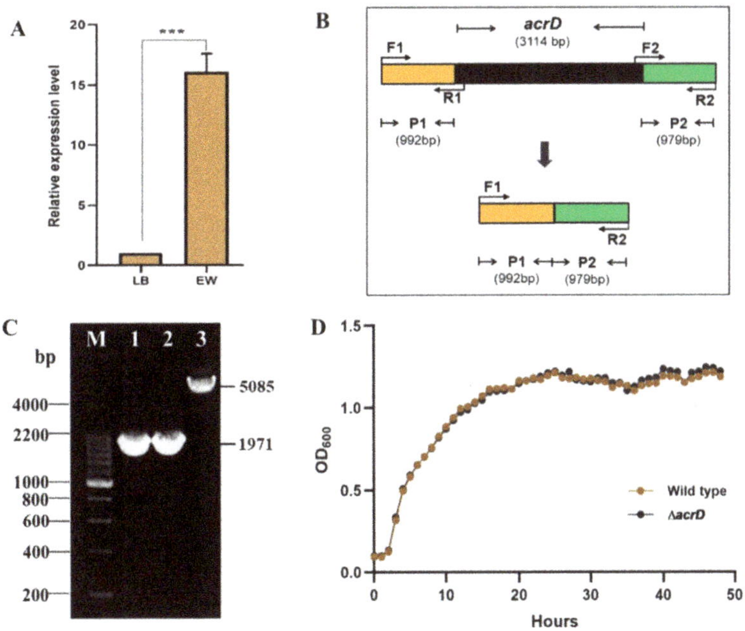

Figure 1. (**A**) Relative expression level of *acrD* of S. Enteritidis in whole egg white (EW) compared to that of Luria-Bertani (LB) broth at the mRNA level. Vertical bars show standard deviation. Asterisk indicates statistical differences according to Duncan's multiple range test at $p < 0.001$ (***) level. (**B**) The in-frame deletion of *acrD*. P1: upstream fragment, P2: downstream fragment. (**C**) Confirmation of the successful construction of *ΔacrD* mutant by PCR using F1 and R2 primers. M: DNA marker, 1/2: *ΔacrD* mutant, 3: wild type. (**D**) The growth curve of S. Enteritidis wild type and *ΔacrD* mutant in LB broth.

Salmonella can resist adverse conditions through gene expression regulation. For example, two universal stress-related genes, *uspA* and *uspB*, of *S.* Enteritidis 147^{str} are highly expressed in egg white, and a decreased colonization ability was observed to the magnum and isthmus of the oviduct when these genes were deleted [11]. The promoter of out membrane channel gene *tolC* was activated by egg white at 42 °C, and mutagenic analysis showed that *tolC* had an important role in *S.* Enteritidis survival in egg white [10]. Furthermore, the specific gene SEN1393 results in higher survivability of *S.* Enteritidis in egg white [27]. Although the *acrD* gene was expressed under different environmental stress conditions (e.g., antibiotics, detergents, metal) and seemed to contribute to stress resistance [15,16,18,20], there is no concrete evidence on the functional role of *acrD* in *S.* Enteritidis under egg white stress.

To better understand the role of *acrD* (encoding 1037 amino acids, a multidrug efflux RND transporter) in the resistance of *S.* Enteritidis to egg white, this gene (3114 bp) was deleted successfully in *S.* Enteritidis strain SJTUF10978 to obtain a *ΔacrD* deletion mutant (Figure 1B,C). The in-frame deletion mutant was confirmed by PCR and DNA sequencing (Figure 1B,C). Moreover, similar growth patterns of wild-type and *ΔacrD* strains in LB broth were found at 37 °C (Figure 1D), indicating that *acrD* is not required for *S.* Enteritidis growth in LB broth.

3.2. *Survival Study of S. Enteritidis ΔacrD Mutant in Egg White and Its Filtrate*

To explore the role of *acrD* in the survival of *S.* Enteritidis in egg white, the wild type and *ΔacrD* mutant were exposed to egg white and surviving bacteria were enumerated via plate counts on LB agar. As shown in Figure 2, the survival ability of *ΔacrD* mutant was significantly lower than that of the wild type in egg white ($p < 0.05$) (Figure 2A,B). More importantly, a 2-log population reduction was observed for the *ΔacrD* mutant after incubation in egg white for 3 days with different initial concentrations (i.e., 10^3 and 10^6 CFU/mL) (Figure 2A,B). These results demonstrate that *acrD* confers resistance to egg white in *S.* Enteritidis.

Generally, egg white filtrate (FEW, less than 10 kDa) has been used as a food matrix to reveal the antibacterial activity of egg white proteins [10,28]. Therefore, to explore the role of *acrD* in the resistance of *S.* Enteritidis to antibacterial egg white proteins in the present study, the wild-type strain and its *ΔacrD* mutant were exposed to egg white (containing various types of proteins) and egg white filtrate without the main antibacterial proteins. As shown in Figure 2C,D, the loss of *acrD* had no significant ($p > 0.05$) effect on the resistance of *S.* Enteritidis in egg white filtrate regardless of the initial cell concentrations, indicating that *acrD* plays a critical role in *S.* Enteritidis resistance to egg white proteins.

The RND transporter AcrD has a unique biological role in multidrug resistance, which can remove antimicrobial drugs such as aminoglycosides from the bacterial cell, and the RND family requires interaction with outer membrane channel TolC to function [13,29]. In addition, a previous study has demonstrated that RND transporters, such as AcrD, are necessary for the secretion of enterobactin, which chelates iron to enable bacterial growth under iron-limiting conditions [30]. Meanwhile, TolC has an important role in the resistance of *S.* Enteritidis to egg white ovotransferrin at 42 °C [10]. These results suggest that AcrD may contribute to *Salmonella* iron homeostasis to resist ovotransferrin in egg white. Antibacterial component experiments are needed to further confirm this hypothesis.

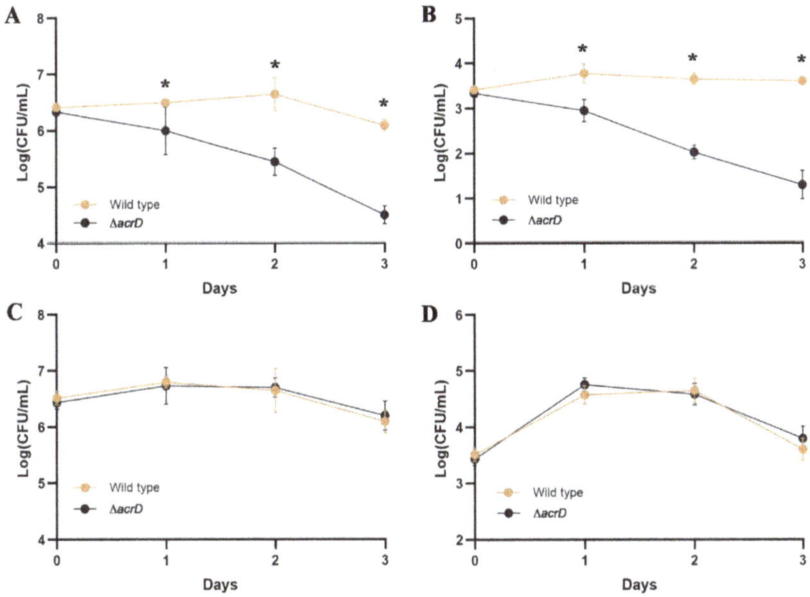

Figure 2. Survival ability of *S.* Enteritidis wild type and *ΔacrD* mutant in whole egg white (**A**,**B**) and its filtrate (**C**,**D**). Survival of *S.* Enteritidis in whole egg white with initial concentrations of 10^6 CFU/mL (**A**) and 10^3 CFU/mL (**B**). Survival of *S.* Enteritidis in egg white filtrate with initial concentrations of 10^6 CFU/mL (**C**) and 10^3 CFU/mL (**D**). Three independent experiments were performed, and the results of representative experiments were shown. Error bars indicate the standard deviation of three replicates. Asterisks (*) indicate significant differences in the survival ability between wild type and *ΔacrD* mutants ($p < 0.05$).

3.3. Cellular Morphology of S. *Enteritidis in Egg White under SEM*

The cellular morphology of *S.* Enteritidis strains (wild type and *ΔacrD* mutant) in LB broth, egg white and egg white filtrate at 37 °C for 1 day was observed by SEM. As shown in Figure 3, no significant morphological change was found between WT and *ΔacrD* in LB broth. However, a small number of *ΔacrD* cells exposed to egg white were lysed, compared with that of the WT. In contrast, there was no apparent morphological difference between WT and *ΔacrD* in egg white filtrate.

It has been commonly suggested that antibacterial proteins and peptides (e.g., lysozyme, ovotransferrin, defensins) are the main antibacterial factors of egg white that prevent bacterial growth, and the bacterial cell membrane is the main target of these antibacterial components [3,31]. For example, the bactericidal mechanisms of egg white lysozyme are mainly involved in hydrolyzing the β-1,4 glycosidic bonds of bacterial peptidoglycan, whereas the peptidoglycan layer is a key shape determining factor of the bacterial cell membrane [32,33]. Cationic peptides produced by the degradation of lysozyme and ovotransferrin, as well as other antibacterial peptides from egg white such as β-defensins, could interact electrostatically with negative charges on the outer membrane (e.g., anionic phospholipids, lipopolysaccharides and lipoteichoic acid) of bacterial cells, leading to bacterial death due to the leakage of substances [34]. Hence, in combination with the results of survival ability of the *S.* Enteritidis WT and its *ΔacrD* mutant in egg white filtrate, we speculated that damaged cells of *ΔacrD* are caused by antibacterial components in egg white.

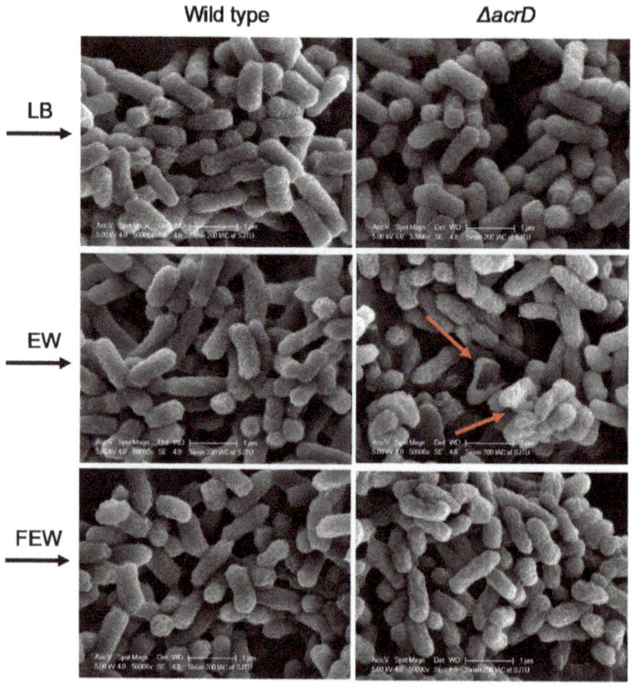

Figure 3. SEM micrographs of S. Enteritidis wild type and *ΔacrD* mutants in LB broth, egg white (EW) and egg white filtrate (FEW) at 37 °C for 1 day. Red arrows highlight clear examples of morphology changes. Magnification = 50,000×, bar marker = 1 μm.

3.4. The Adhesion and Invasion Ability of S. Enteritidis to Caco-2 Cells

Cell adhesive and invasive ability are usually used to evaluate the potential virulence of bacteria [31,35,36]. In this study, the adhesion and invasion abilities of Caco-2 cells by the S. Enteritidis wild type and its *ΔacrD* mutant were further investigated in LB broth and egg white. Non-adherent/invasive *E. coli* DH5α and the highly invasive *Salmonella* MRL0026 strain were used as negative and positive controls, respectively. The results showed that no significant difference ($p > 0.05$) in the adhesion rate was observed between the wild type and *ΔacrD* mutant in LB broth or egg white (Figure 4), indicating that the loss of *acrD* had no significant effect on the adhesion ability of S. Enteritidis. Similarly, the invasion rate of the *ΔacrD* mutant was basically consistent with that of the wild type in LB and egg white ($p > 0.05$). However, the invasion rate of the *ΔacrD* mutant in egg white (4.54%) was significantly ($p < 0.05$) lower than that in LB broth (9.30%) (Figure 4). These results indicate that the invasion ability of S. Enteritidis was influenced by egg white for cells lacking *acrD*.

Previous studies have confirmed that AcrD is related to the virulence of bacteria. For example, the infected ability of *Salmonella* was significantly reduced in its ability to infect INT 407 cells when either AcrD, AcrB or AcrF were missing [37]. Inactivation of *acrD* resulted in changes in the expression of some virulence-related genes [14]. Although no significant differences ($p > 0.05$) in the adhesion and invasion rates between the wild-type and *ΔacrD* mutant in LB broth or egg white were found in this study, there was a significant difference ($p < 0.05$) between the invasion rate of the *ΔacrD* mutant when in egg white versus LB broth (Figure 4). Combined with the other authors' findings, the results of this study indicated that *acrD* is involved in virulence in *Salmonella* in response to egg white; however, the extent of this role requires further investigation.

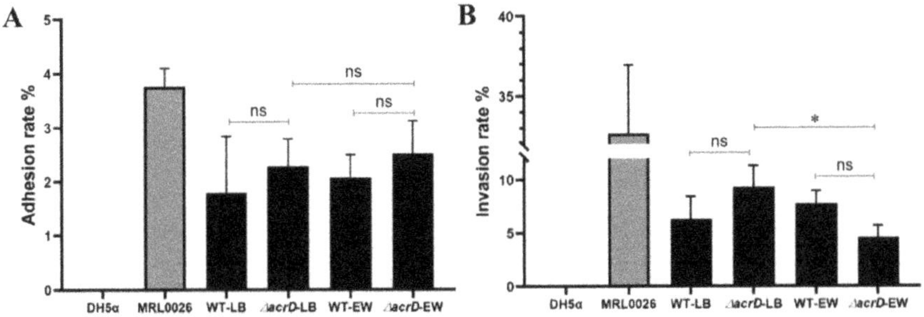

Figure 4. Adhesion (**A**) and invasion (**B**) ability of Caco-2 cells by *S.* Enteritidis wild type (WT), *ΔacrD* mutant and control strains in egg white (EW) and LB broth at 37 °C. Non-adherent/invasive *E. coli* DH5α and highly invasive *Salmonella* MRL0026 were used as negative and positive controls, respectively. Mean values and standard deviations were calculated from three replicates. Asterisk (*) indicates that there is a significant difference between the invasion rate of *ΔacrD* in egg white and that in LB broth ($p < 0.05$). ns, no significant difference.

4. Conclusions

This study revealed that AcrD conferred resistance to egg white in *S.* Enteritidis strain SJTUF10978 by analyzing an *acrD* deletion mutant. Meanwhile, this protein appears to be involved in virulence in *S.* Enteritidis in response to egg white. These findings broaden the understanding of the RND protein related to efflux pumps that mediates the resistance of *Salmonella* in egg white. Collectively, this study provides some novel insights into the resistance mechanism of *S.* Enteritidis to egg white.

Author Contributions: Conceptualization, X.Q.; methodology, X.Q.; validation, X.Q.; formal analysis, X.Q.; investigation, X.Q.; writing—original draft preparation, X.Q.; writing—review and editing, X.Q., Y.L. and X.S.; visualization, X.Q.; supervision, X.Q. and X.S.; project administration, X.Q.; funding acquisition, X.Q. All authors have read and agreed to the published version of the manuscript.

Funding: This work was supported by the National Natural Science Foundation of China (Grant No. 32102111), the Project funded by the China Postdoctoral Science Foundation (2021M702194) and the Shanghai Post-Doctoral Excellence Program (2020338).

Data Availability Statement: All data related to the research are presented in the article.

Acknowledgments: We thank Qiyun Zhuo and Xinxin Lu for the Caco-2 cell culture preparation and Muhammad Zohaib Aslam for his constructive suggestions on the manuscript.

Conflicts of Interest: The authors declare no conflict of interest.

References

1. Li, Y.; Yang, X.R.; Zhang, H.N.; Jia, H.Y.; Liu, X.G.; Yu, B.; Zeng, Y.C.; Zhang, Y.; Pei, X.Y.; Yang, D.J. Prevalence and antimicrobial susceptibility of *Salmonella* in the commercial eggs in China. *Int. J. Food Microbiol.* **2020**, *325*, 108623. [CrossRef]
2. Kang, H.; Loui, C.; Clavijo, R.I.; Riley, L.W.; Lu, S. Survival characteristics of *Salmonella enterica* serovar Enteritidis in chicken egg albumen. *Epidemiol. Infect.* **2006**, *134*, 967–976. [CrossRef]
3. Baron, F.; Nau, F.; Guérin-Dubiard, C.; Bonnassie, S.; Gautier, M.; Andrews, S.C.; Jan, S. Egg white versus *Salmonella* Enteritidis! A harsh medium meets a resilient pathogen. *Food Microbiol.* **2016**, *53*, 82–93. [CrossRef] [PubMed]
4. Clavijo, R.I.; Loui, C.; Andersen, G.L.; Riley, L.W.; Lu, S. Identification of genes associated with survival of *Salmonella enterica* serovar Enteritidis in chicken egg albumen. *Appl. Environ. Microbiol.* **2006**, *72*, 1055–1064. [CrossRef] [PubMed]
5. De Vylder, J.; Raspoet, R.; Dewulf, J.; Haesebrouck, F.; Ducatelle, R.; Van Immerseel, F. *Salmonella* Enteritidis is superior in egg white survival compared with other *Salmonella* serotypes. *Poult. Sci.* **2013**, *92*, 842–845. [CrossRef]
6. Huang, X.Z.; Zhou, X.J.; Jia, B.; Li, N.; Jia, J.Y.; He, M.; He, Y.C.; Qin, X.J.; Shi, C.L.; Shi, X.M. Transcriptional sequencing uncovers survival mechanisms of *Salmonella enterica* Serovar Enteritidis in antibacterial egg white. *mSphere* **2019**, *4*, e00700-18. [CrossRef] [PubMed]
7. Lu, S.W.; Killoran, P.B.; Riley, L.W. Association of *Salmonella enterica* Serovar Enteritidis YafD with resistance to chicken egg albumen. *Infect. Immun.* **2003**, *71*, 6734–6741. [CrossRef]

8. Gantois, I.; Ducatelle, R.; Pasmans, F.; Haesebrouck, F.; Van Immerseel, F. *Salmonella enterica* serovar Enteritidis genes induced during oviduct colonization and egg contamination in laying hens. *Appl. Environ. Microbiol.* **2008**, *74*, 6616–6622. [CrossRef]
9. Baron, F.; Bonnassie, S.; Alabdeh, M.; Cochet, M.; Nau, F.; Guérin-Dubiard, C.; Gautier, M.; Andrews, S.C.; Jan, S. Global gene-expression analysis of the response of *Salmonella* Enteritidis to egg white exposure reveals multiple egg white-imposed stress responses. *Front. Microbiol.* **2017**, *8*, 829. [CrossRef]
10. Raspoet, R.; Eeckhaut, V.; Vermeulen, K.; De Smet, L.; Wen, Y.; Nishino, K.; Haesebrouck, F.; Ducatelle, R.; Devreese, B.; Van Immerseel, F. The *Salmonella* Enteritidis TolC outer membrane channel is essential for egg white survival. *Poult. Sci.* **2019**, *98*, 2281–2289. [CrossRef]
11. Raspoet, R.; Gantois, I.; Devloo, R.; Martel, A.; Haesebrouck, F.; Pasmans, F.; Ducatelle, R.; Van Immerseel, F. *Salmonella* Enteritidis universal stress protein (usp) gene expression is stimulated by egg white and supports oviduct colonization and egg contamination in laying hens. *Vet. Microbiol.* **2011**, *153*, 186–190. [CrossRef] [PubMed]
12. Qin, X.J.; He, S.K.; Zhou, X.J.; Cheng, X.; Huang, X.Z.; Wang, Y.Y.; Wang, S.Y.; Cui, Y.; Shi, C.L.; Shi, X.M. Quantitative proteomics reveals the crucial role of YbgC for *Salmonella enterica* serovar Enteritidis survival in egg white. *Int. J. Food Microbiol.* **2019**, *289*, 115–126. [CrossRef] [PubMed]
13. Blair, J.M.A.; Richmond, G.E.; Piddock, L.J.V. Multidrug efflux pumps in Gram-negative bacteria and their role in antibiotic resistance. *Future Microbiol.* **2014**, *9*, 1165–1177. [CrossRef] [PubMed]
14. Buckner, M.M.C.; Blair, J.M.A.; Ragione, R.M.L.; Newcombe, J.; Dwyer, D.J.; Lvens, A.; Piddock, L.J.V. Beyond antimicrobial resistance: Evidence for a distinct role of the AcrD efflux pump in *Salmonella* biology. *mBio* **2016**, *7*, e01916-16. [CrossRef] [PubMed]
15. Stephanie, B.; Ekanayaka, A.S.; Piddock, L.J.; Webber, M.A. Loss of or inhibition of all multidrug resistance efflux pumps of *Salmonella enterica* serovar Typhimurium results in impaired ability to form a biofilm. *J. Antimicrob. Chemother.* **2012**, *67*, 2409–2417. [CrossRef]
16. Eaves, D.J.; Ricci, V.; Piddock, L.J.V. Expression of *acrB*, *acrF*, *acrD*, *marA*, and *soxS* in *Salmonella enterica* Serovar Typhimurium: Role in multiple antibiotic resistance. *Antimicrob. Agents Chemother.* **2004**, *48*, 1145–1150. [CrossRef] [PubMed]
17. Nishino, K.; Latifi, T.; Groisman, E.A. Virulence and drug resistance roles of multidrug efflux systems of *Salmonella enterica* serovar Typhimurium. *Mol. Microbiol.* **2006**, *59*, 126–141. [CrossRef]
18. Nishino, K.; Nikaido, E.; Yamaguchi, A. Regulation of multidrug efflux systems involved in multidrug and metal resistance of *Salmonella enterica* Serovar Typhimurium. *J. Bacteriol.* **2007**, *189*, 9066–9075. [CrossRef] [PubMed]
19. Álvarez-Ordóñez, A.; Begley, M.; Prieto, M.; Messens, W.; López, M.; Bernardo, A.; Hill, C. *Salmonella* spp. survival strategies within the host gastrointestinal tract. *Microbiology* **2011**, *157*, 3268–3281. [CrossRef]
20. Budak, G.; Ozsoy, O.E.; Son, Y.A.; Can, T.; Tuncbag, N. Reconstruction of the temporal signaling network in *Salmonella*-infected human cells. *Front. Microbiol.* **2015**, *6*, 730. [CrossRef]
21. Qin, X.J.; Dong, R.; He, S.K.; Zhou, X.J.; Zhang, Z.F.; Cui, Y.; Shi, C.L.; Liu, Y.H.; Shi, X.M. Characterization of the role of *ybgC* in lysozyme resistance of *Salmonella* Enteritidis. *Food Control* **2020**, *109*, 106732. [CrossRef]
22. Livak, K.J.; Schmittgen, T.D. Analysis of relative gene expression data using real-time quantitative PCR and the $2^{-\Delta\Delta CT}$ method. *Methods* **2001**, *25*, 402–408. [CrossRef]
23. Zhou, X.J.; Liu, B.; Shi, C.L.; Shi, X.M. Mutation of a *Salmonella* serogroup-C1-specific gene abrogates O7-antigen biosynthesis and triggers NaCl-dependent motility deficiency. *PLoS ONE* **2014**, *9*, e106708. [CrossRef] [PubMed]
24. Rubirés, X.; Saigi, F.; Piqué, N.; Climent, N.; Merino, S.; Alberti, S.; Tomas, J.M.; Regué, M. A gene (*wbbL*) from Serratia marcescens N28b (O4) complements the rfb-50 mutation of *Escherichia coli* K-12 derivatives. *J. Bacteriol.* **1997**, *179*, 7581–7586. [CrossRef] [PubMed]
25. Edward, R.A.; Keller, L.H.; Schifferli, D.M. Improved allelic exchange vectors and their use to analyze 987P fimbria gene expression. *Gene* **1998**, *207*, 149–157. [CrossRef]
26. Chakroun, I.; Cordero, H.; Mahdhi, A.; Morcillo, P.; Fedhila, K.; Cuesta, A.; Bakhrouf, A.; Mahdouani, K.; Esteban, M.Á. Adhesion, invasion, cytotoxic effect and cytokine production in response to atypical *Salmonella* Typhimurium infection. *Microb. Pathog.* **2017**, *106*, 40–49. [CrossRef]
27. Liu, B.; Hou, W.W.; Li, K.; Chen, Q.; Liu, Y.X.; Yue, T.L. Specific gene SEN 1393 contributes to higher survivability of *Salmonella* Enteritidis in egg white by regulating sulfate assimilation pathway. *Int. J. Food Microbiol.* **2019**, *337*, 108927. [CrossRef] [PubMed]
28. Alabdeh, M.; Lechevalier, V.; Nau, F.; Gautier, M.; Cochet, M.; Gonnet, F.; Jan, S.; Baron, F. Role of incubation conditions and protein fraction on the antimicrobial activity of egg white against *Salmonella* Enteritidis and *Escherichia coli*. *J. Food Protect.* **2011**, *74*, 24–31. [CrossRef]
29. Zhang, C.Z.; Chen, P.X.; Yang, L.; Li, W.; Chang, M.X.; Jiang, H.X. Coordinated expression of acrAB-tolC and eight other functional efflux pumps through activating *ramA* and *marA* in *Salmonella enterica* serovar Typhimurium. *Microb. Drug Resist.* **2018**, *24*, 120–125. [CrossRef]
30. Horiyama, T.; Nishino, K. AcrB, AcrD, and MdtABC multidrug efflux systems are involved in enterobactin export in *Escherichia coli*. *PLoS ONE* **2014**, *9*, e108642. [CrossRef]
31. McWhorter, A.R.; Chousalkar, K.K. Comparative phenotypic and genotypic virulence of *Salmonella* strains isolated from Australian layer farms. *Front. Microbiol.* **2015**, *6*, 12. [CrossRef] [PubMed]

32. Mine, Y.; Ma, F.; Lauriau, S. Antimicrobial peptides released by enzymatic hydrolysis of hen egg white lysozyme. *J. Agric. Food Chem.* **2004**, *52*, 1088–1094. [CrossRef]
33. Rojas, E.R.; Billings, G.; Odermatt, P.D.; Auer, G.K.; Zhu, L.; Miguel, A.; Chang, F.; Weibel, D.B.; Theriot, J.A.; Huang, K.C. The outer membrane is an essential load-bearing element in Gram-negative bacteria. *Nature* **2018**, *559*, 617–621. [CrossRef] [PubMed]
34. Hervé-Grépinet, V.; Réhault-Godbert, S.; Labas, L.; Magallon, T.; Derache, C.; Lavergne, M.; Gautron, J.; Lalmanach, A.C.; Nys, Y. Purification and characterization of avian β-defensin 11, an antimicrobial peptide of the hen egg. *Antimicrob. Agents Chemother.* **2010**, *54*, 4401–4408. [CrossRef] [PubMed]
35. Liu, Y.H.; Yoo, B.B.; Hwang, C.; Suo, Y.J.; Sheen, S.; Khosravi, P.; Huang, L.H. LMOf2365_0442 encoding for a fructose specific PTS permease IIA may be required for virulence in *L. monocytogenes* strain F2365. *Front. Microbiol.* **2017**, *8*, 1611. [CrossRef]
36. Pradhan, D.; Negi, V.D. Stress-induced adaptations in *Salmonella*: A ground for shaping its pathogenesis. *Microbiol. Res.* **2019**, *229*, 126311. [CrossRef]
37. Blair, J.M.A.; Smith, H.E.; Vito, R.; Lawler, A.J.; Thompson, L.J.; Piddock, L.J.V. Expression of homologous RND efflux pump genes is dependent upon AcrB expression: Implications for efflux and virulence inhibitor design. *J. Antimicrob. Chemother.* **2015**, *70*, 424–431. [CrossRef]

foods

MDPI

Article

Antibiotic Resistant *Enterobacteriaceae* in Milk Alternatives

Winnie Mukuna [1], Abdullah Ibn Mafiz [1], Bharat Pokharel [1], Aniume Tobenna [1] and Agnes Kilonzo-Nthenge [2,*]

[1] Department of Agriculture and Environmental Sciences, Tennessee State University, 3500 John A. Merritt Boulevard, Nashville, TN 37209, USA; wmukuna@tnstate.edu (W.M.); amafiz@tnstate.edu (A.I.M.); bpokhare@tnstate.edu (B.P.); taniume@tnstate.edu (A.T.)

[2] Department of Human Sciences, Tennessee State University, 3500 John A. Merritt Boulevard, Nashville, TN 37209, USA

* Correspondence: akilonzontheng@tnstate.edu; Tel.: +1-(615)-963-5437; Fax: +1-(615)-963-5557

Abstract: The consumption of non-dairy milk is on the rise due to health benefits. Although there is increasing inclination towards milk alternatives (MA), there is limited data on antibiotic resistant bacteria in these substitutes. The aim of this study was to investigate antimicrobial resistance of bacteria isolated from MA. A total of 138 extracts from almonds ($n = 63$), cashew nuts ($n = 36$), and soybeans ($n = 39$) were analyzed for *Enterobacteriaceae*. The identification of the bacteria was based on biochemical and PCR methods. Antibiotic sensitivity was determined by using the Kirby-Bauer disk diffusion technique. Overall, 31% (43 of 138) of extracts were positive for *Enterobacteriaceae*. Ten bacterial species were identified, of which *Enterobacter cloacae* (42.7%) and *Enterobacter cancerogenus* (35.4%) were the most predominant species ($p < 0.05$). Antibiotic resistance was exhibited to vancomycin (88.3%), novobiocin (83.8%), erythromycin (81.1%), which was significantly higher ($p < 0.05$) than in tetracycline (59.5%), cefpodoxime (30.6%), and nalidixic acid (6.3%). There was no resistance displayed to kanamycin and imipenem. ERY-NOV-VAN-TET and ERY-NOV-CEP-VAN-TET were the most common resistant patterns displayed by *Enterobacter cloacae*. The findings of this study suggest that MAs, though considered healthy, may be a reservoir of multidrug resistant opportunist pathogens.

Keywords: multidrug-resistant bacteria; milk alternatives; food safety

Citation: Mukuna, W.; Mafiz, A.I.; Pokharel, B.; Tobenna, A.; Kilonzo-Nthenge, A. Antibiotic Resistant *Enterobacteriaceae* in Milk Alternatives. *Foods* **2021**, *10*, 3070. https://doi.org/10.3390/foods10123070

Academic Editor: Karl R. Matthews

Received: 29 September 2021
Accepted: 8 December 2021
Published: 10 December 2021

1. Introduction

Milk is considered a superior source of micro- and macro-nutrients compared to milk alternatives (MA) [1]. However, its association with increased risks of cardiovascular diseases, diabetes, cancer, and as a principal vehicle for transmission of foodborne pathogens continues to make it unfavorable. Generally, cow milk is frequently consumed and dominates global milk production [2], accounting for 85% of the world's production, followed by buffalo milk at 11%, goat (2.3%), sheep (1.4%), and camel (0.2%) [3]. However, due to the current changes in lifestyles towards a healthier diet, there has been an increasing trend in the consumption of MA [4]. The U.S. market for MA is increasing and has reached an annual sales volume of $1.8 billion [4]. The increased market growth is attributed to the consumers' preference for vegan diets, increasing instances of lactose intolerance, and a growing demand for fortified non-dairy food and beverages [5–7]. Generally, consumers' perception is that MA are healthier than milk [8]. Milk alternatives are becoming increasingly popular; however, they are characterized by low protein content, and poor bioavailability of minerals and vitamins [9]. With the increasing demand for these MA, different plants with varying functional attributes are being explored as bases for primary materials for processing [10]. Soymilk, which originated from Asia [11], is the most globally consumed MA while almond milk is the most prevalently used, solely based on sales volume [12]. Other available MA are sourced from cashew nuts, hemp, coconut, rice, etc. [10]. The majority of non-dairy consumers purchase their MA from grocery stores, though a sector of the population make these milk substitutes at home by using raw nuts or seeds. Hence, home-made milk alternative might potentially be contaminated if food safety is not practiced during preparation and storage. Although MA are an intensifying trend, the usage of the term "milk" to mean plant-based substitutes to milk is debatable and is protected by legislation in several countries [8].

There are abundant MA in the market, such as almond, cashew, soy, rice, hazelnut, and oat milk [13,14]. Nuts and seeds, the primary raw materials for milk alternatives, may come into direct contact with soil and be contaminated with pathogenic bacteria at pre- or post-harvest period [15]. It is usually thought that, due to less moisture content, nuts, seeds, and grains are less susceptible to microbial contamination [16]. Regrettably, this attribute does not exempt nuts and seeds from contamination with foodborne pathogens. For instance, *Salmonella* serovar has been detected in almonds, pecans, and peanuts [15,17,18], *E. coli* O157:H7 in walnuts [19], and *Listeria* spp. in peanuts, almonds, cashews, and hazelnuts [20]. Moreover, *Pseudomonas* spp., *Clostridium* spp., and *Klebsiella* spp. have been detected in other nuts [21]. These bacteria and others that are prevalent in raw nuts and seeds belong to the family *Enterobacteriaceae*, the most prevalent human opportunistic pathogens [22].

The increasing frequency of antimicrobial resistant bacteria is a global threat [23]. Accordingly, it is important to study the presence of antimicrobial resistant *Enterobacteriaceae* in MA, especially with the continuous increase trend in consumption. Antimicrobial resistant bacteria cause illnesses that have high morbidity and mortality [24], one of the greatest health challenges in the 21st century [25]. Around 99,000 individuals die every year in the USA owing to drug-resistant infections [26]. Antimicrobial-resistant *Enterobacteriaceae* in milk and milk products has been reported in numerous studies [27]. Just like milk, milk substitutes can also be potential vehicles for transmission of antimicrobial resistant foodborne pathogens to consumers. Antimicrobial resistant pathogens originating from raw nuts or seeds might be transferred to MA during preparation at processing facilities or at home. To the best of our knowledge, there has been limited enquiry of the possible occurrence of antimicrobial resistant bacteria in raw MA. Consequently, this study aims to investigate the presence of opportunist *Enterobacteriaceae* in MA and their resistance to antibiotics used both in human and animal medicine.

2. Materials and Methods

2.1. Sample Collection and Preparation

Raw nuts (almond, cashew) and soybeans were randomly purchased from 3 local stores in Davidson County, Tennessee, depending on availability. The preparation of almonds, cashew nuts, and soybean extracts involved schematic steps as displayed in the flowchart (Figure 1). Briefly, in duplicates, 5 g of each sample (almonds, cashews, and soybeans) were sorted from unwanted materials (damaged, split seeds or nuts), followed by soaking separately overnight in 45 mL sterile distilled water at room temperature. Next, in duplicates, each sample was disintegrated in a laboratory blender (Waring Division, Dynamics Corporation, New Hartford, CT, USA) for 3 min at high speed. The resulting slurry was filtered through a cheesecloth (Farberware, Fairfield, CA, USA) to attain milk extracts which were then placed in sterile capped containers. A total of 138 extracts (almond nuts = 63, cashew nuts = 36, and soybeans = 39) were analyzed for *Enterobacteriaceae* and AMR by using biochemical and molecular tests.

Enrichment of Milk and Bacterial Identification

One ml of nuts and seed extracts was enriched in 9 mL *Enterobacteriaceae* enrichment (EE) broth Mossel enrichment (BD, Sparks, MD) and incubated at 37 °C for 24 h. From each enriched sample, 1 μL was streaked onto violet red bile agar (Oxoid, Basingstoke, and Hants, UK) and incubated for 18–24 h at 37 °C. Red to dark purple colonies surrounded by red-purple halos were identified as presumptive *Enterobacteriacea*. *Enterobacteriacea* isolates and further characterized by using oxidase and API 20E (bioMerieux, Hazelwood, MO, USA) tests. Three colonies per plate were selected for API biochemical testing. Due to the role played by *Klebsiella pneumoniae* and *ronobacter sakazakii* as opportunist pathogens in clinical settings, isolates above the 90% confidence interval were stored at −80 °C for further testing.

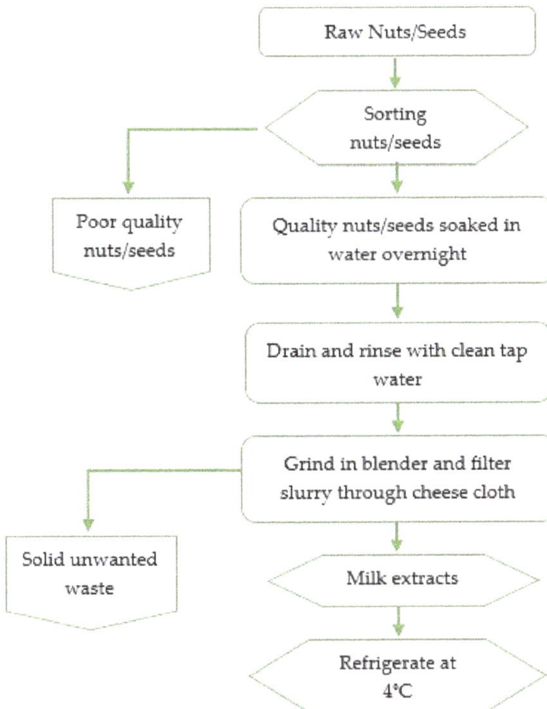

Figure 1. A schematic diagram for extracting milk from almonds, cashews, and soybeans.

2.2. DNA Extraction and Confirmation of Klebsiella and Cronobacter Sakazakii

Biochemically identified *K. pneumoniae* and *C. sakazakii* isolates from almond and cashew extracts, respectively, were further confirmed by PCR. DNA was extracted from overnight cultures ($\leq 2 \times 10^9$ cells) using the PureLink Genomic DNA Mini Kit (Invitrogen, Carlsbad, CA, USA). DNA concentrations and integrity were determined using a NanoDrop 2000 (Thermo Scientific, Pittsburgh, PA, USA) and agarose gel electrophoresis, respectively. Oligonucleotide primer pairs were synthesized (Operon Technologies, Huntsville, AL, USA) and used to amplify genes of interest. The sequences of the primer pair used for targeting *C. sakazakii* target gene *omp*A (469 bp) was 3′-GGATTTAACCGTGAACTTTTCC-5′ and 5′-CGCCAGCGATGTTAGAAGA-3′ [28,29]. Each reaction mixture (20 µL) contained 4 µL DNA template, 1 µL of each primer (×2), 10 µL master mix, 2 µL RNase free water and, 2 µL coral load (supplied with the kit). *C. muytjensii* (ATCC 51329) was used as a positive control for the detection and identification methods. Reaction conditions for PCR were initial denaturation at 95 °C for 5 min, 30 cycles of denaturation at 95 °C for 30 s, annealing at 55 °C for 1 min, extension at 72 °C for 10 min, and final extension at 72 °C for 10 min.

A Multiplex PCR plus kit (Qiagen, Hillden, Germany) was used to amplify *K. pneumoniae* and *Klebsiella* spp primers in a single reaction [30]. Primer pair 5′-CAA CGG TGT GGT TAC TGA CG-3′ and 5′-TCT ACG AAG TGG CCG TTT TC-3′ targeted gene *rpo*B (108 bp) in *K. pneumoniae* isolates as described by [30], and 5′- CGC GTA CTA TAC GCC ATG AAC GTA-3′ and 5′-ACC GTT GAT CAC TTC GGT CAG G-3′ targeted gene *gyr*A (441bp) in *Klebsiella* spp. [31]. Each 50 µL reaction mixture contained 25 µL of master mix, 5 µL of 10 × primer mix (2.5 µM each primer), 100 ng DNA template, 5 µL Q-solution, 5 µL Coral Load dye, and 10 µL RNase free water. Reaction conditions for PCR were: initial denaturation at 95 °C for 5 min, 35 cycles of denaturation at 95 °C for 30 s, annealing at 60 °C for 90 s, extension at 72 °C for 90 s, and final extension at 68 °C for 10 min. *K. pneumoniae* (ATCC 49131) and *Salmonella typhimurium* (ATCC 13311) were used as positive and negative control, respectively. A nexus gradient Thermal Cycler (Eppendorf, Hauppauge, New York) was used for all amplifications. PCR products were electrophoresed in agarose gel stained with 0.1 µg/mL of ethidium bromide (Sigma-Aldrich, Madrid, Spain) and photographed under UV light.

2.3. Antibiotic Resistant Profiles

For all identified *Enterobacteriaceae* isolates ($n = 110$), the characterization of the strain resistance/susceptibility profiles was carried out as recommended by the Clinical and Laboratory Standards Institute guidelines [32]. The antimicrobial susceptibility test was conducted on isolates that were identified at ≤90 confidence interval by API 20E system. Antimicrobial disks ($n = 8$), with strength in parentheses were: vancomycin (VAN; 30 µg), novobiocin (NOVO; 30 µg), erythromycin (ERY; 15 µg), tetracycline (TET; 30 µg), cefpodoxime (CEF; 10 µg), kanamycin (KAN; 10 µg), nalidixic acid (NAL; 30 µg), and imipenem (IPM; 30 µg). The results were interpreted as susceptible, intermediate, and resistant based on the Clinical and Laboratory Standards Institute recommendations [32]. *Escherichia coli* ATCC 25922 and *Staphylococcus aureus* ATCC 25923 were used as control strains. Reference standard bacterial strains were verified simultaneously with controls.

2.4. Statistical Analysis

The bacterial data were expressed as percentages and analyzed using Microsoft Excel 2016 (Microsoft Corp., Redmond, WA, USA). Chi-square tests were used to measure the significance of difference in the incidence of *Enterobacteriaceae* and antimicrobial resistance. Data were analyzed using SPSS v. 25.0 (IBM SPSS, Chicago, IL, USA). p values of less than 0.05 were considered statistically significant.

3. Results and Discussion

3.1. Enterobacteriaceae in Nuts and Seeds Extract

Overall, 31% (43 of 138) of extracts were positive for *Enterobacteriaceae*. Specifically, *Enterobacteriaceae* isolation rates were 33.3% (21/63), 30.5% (11/36), and 28.2% (11/39) of almond, cashew, and soybean extracts, respectively (data not shown). *Enterobacteriaceae* offers valuable information on the hygienic conditions during food preparation or post-process contamination [33]. Overall, 79.7% (110 out of 138) *Enterobacteriaceae* isolates were identified from almond, cashew, and soybean milk extracts (Table 1).

Table 1. Presence (%) of *Enterobacteriaceae* in Nut and Seed Extracts.

Bacterial Species	Total Isolates (N = 110)	No. (%) of ENT Isolates in Extracts		
		Almond Milk ($n = 56$)	Cashew Milk ($n = 28$)	Soy Milk ($n = 26$)
Enterobacter cancerogenus	39 (35.4) [a]	22 (39.28) [a]	5 (17.9) [b]	12 (46.2) [a]
Enterobacter cloacae	47 (42.7) [a]	21 (37.5) [a]	15 (17.9) [a]	11 (42.3) [a]
Klebsiella pneumoniae spp. *ozaenae*	5 (4.5) [bc]	5 (8.9) [b]	0 (0) [c]	0 (0) [c]
Pantoea spp. 3	8 (7.3) [b]	8 (14.2) [b]	0 (0) [c]	0 (0) [c]
Chryseomonas luteola	1 (0.9) [c]	0 (0) [c]	1 (3.6) [b,c]	0 (0) [bc]
Citrobacter youngae	1 (0.9) [c]	0 (0) [c]	1 (3.6) [b,c]	0 (0) [bc]
Cronobacter sakazakii	3 (2.7) [b,c]	0 (0) [c]	3 (10.7) [b]	0 (0) [bc]
Klebsiella pneumoniae spp. *pneumoniae*	3 (2.7) [b,c]	0 (0) [c]	3 (10.7) [b]	0 (0) [bc]
Escherichia Vulneris	2 (1.8) [b,c]	0 (0) [c]	0 (0) [c]	2 (7.7) [c]
Rahnella aquatilis	1 (0.9) [c]	0 (0) [c]	0 (0) [c]	1 (3.8) [c]

N: Total number of *Enterobacteriaceae* isolates. n: Total number of *Enterobacteriaceae* isolates from various extracts. [a–c] Mean percentages in the same column followed by different letters are significantly different ($p < 0.05$).

In 2020, almond milk (MILKLAB and Blue Dimond Almond Breeze Chocolate Almond Milk) recalls were reported in Australia due to contamination with *Pseudomonas* [34]. These recalls support our results that MA can be contaminated with pathogenic bacteria. Our findings also suggest that MA may be contaminated with harmful microorganism. Pathogens such as *Salmonella* serovar., *Listeria* spp., *E. coli* spp., *Campylobacter* spp., *Brucella* spp. or *Shigella* spp. [35–37] have been associated with milk.

According to our findings, a total of 10 different commensal and pathogenic genera of *Enterobacteriaceae* were identified with the most common strain being *Enterobacter cloacae* at 42.7% (47 of 110), which was not significantly different ($p > 0.05$) from *Enterobacter cancerogenus* at 35.4% (39 of 110). *E. cloacae* is a commensal microorganism found in human and animal guts and widely found in food, soil, and water [38]. Although *E. cloacae* is not a common foodborne pathogen, its presence in MA is a concern as it is a widely known nosocomial pathogen and the third most prevalent acquired bacteria causing illness in hospital after *E. coli* and *K. pneumoniae* [39].

Our results indicate that clinically significant *C. sakazakii* accounted for 2.7% (3 out of 110) of the identified isolates. *C. sakazakii* was only detected in cashew extracts and was confirmed through amplification of the *Omp*A gene (469 bp) (Figure 2).

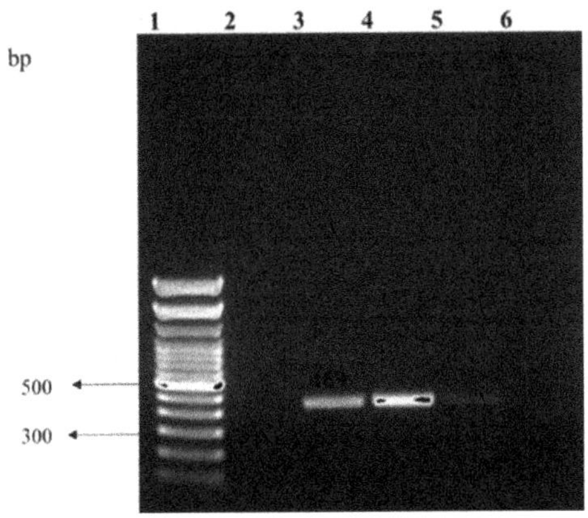

Figure 2. Represents PCR amplification of the *omp*A gene in *Cronobacter sakazakii*, Lane 1: 1 kb ladder; lane 2: negative control; lane 3: positive control; lane 4–5: *C. sakazakii isolates*.

Earlier findings showed that *Omp*A is a determinant that causes *C. sakazakii* invasion of brain microvascular endothelial cells in vitro, and possibly contributes to pathogenesis of neonatal meningitis [40]. *Cronobacter* spp. is an emerging pathogen and a major concern, especially to hypersensitive clusters of the population including children and the elderly [41,42]. *C. sakazakii* is also considered as an evolving opportunistic pathogen [43] that has been detected in milk, and powdered infant milk among other sources [44]. Although there is no data on the incidence of *C. sakazakii* in MA, nuts and seeds are important raw materials in these substitutes which might be contaminated with pathogenic bacteria at any point during production, harvest, storage, and transportation [45]. At production and harvesting stages, pathogenic bacteria might transfer from the soils onto the nuts/seeds when they are in contact with the ground. One possible scenario is during almond harvesting as was the case of *Salmonella* in almonds grown in California [15].

K. pneumoniae spp. *ozaenae* (4.5%) and *K. pneumoniae* spp. *pneumoniae* (2.7%) in the current study were also isolated from almond and cashew extracts, respectively. As these two bacteria are emerging pathogens of concern, they were confirmed by multiplex PCR through amplification of *rpo*B (108 bp) and *gyr*A (441 bp) genes for *K. pneumoniae* and *Klebsiella* spp., respectively (Figure 3).

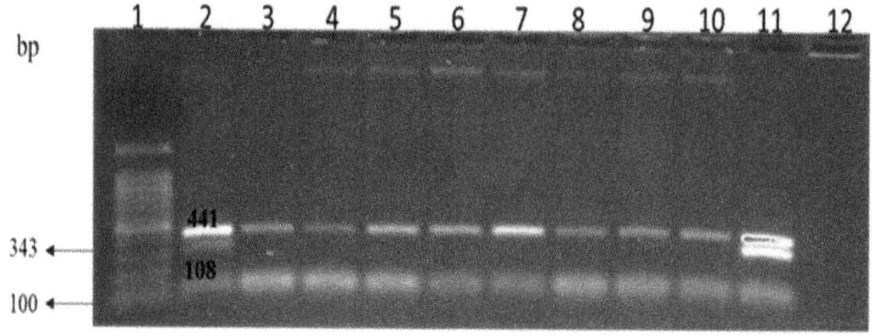

Figure 3. Multiplex PCR amplification of *gyr*A and *rpo*B genes in *K. pneumoniae* and *Klebsiella* spp. Lane 1: 1 kb ladder; lane 2 & 11: positive control; lane 3–10: *K. pneumoniae* and *Klebsiella* spp.; lane 12: negative control.

During harvesting, almond trees are shaken to release the nuts and might stay on the ground for up to 2 weeks before collection [46]. Through this period, bacteria in the soil may be transferred to the hulls which might infiltrate to the kernel as has been demonstrated in *Salmonella* on almonds and pecans [18]. *Klebsiella* spp. have recently become significant pathogens in nosocomial infections [47] such as urinary tract infection, bacteremia, pneumonia, sepsis, and meningitis [48]. With the increased trend in adoption of MA and with some consumers making their nut and seed extracts at home, they may also be at risk of nosocomial infections originating from contaminated and unpasteurized extracts. To avoid potential infections from *Klebsiella* spp., consumers should be encouraged to adhere to food safety practices or drink pasteurized MA.

Other bacteria in the *Enterobacteriaceae* family were also identified in the current study (Table 1). Our findings present *E. vulneris* (1.8%) in soybean extracts. It is possible that the soybeans used in this study were contaminated with *E. vulneris* through soil or water that was used at preharvest or post-harvest. Jain et al. [49] hypothesized that an infant infected with gastroenteritis may have been infected by contaminated formula or water that was used to reconstitute the formula. *Escherichia vulneris* has previously been recovered from water, soil, human beings, and animals [50]. *Rahnella aquatilis* (0.9%) was another *Enterobacteriaceae* isolated from soybean milk extract in our study. *Rahnella aquatilis* is considered a primary and opportunistic pathogen that has been associated with diarrhea and endocarditis [51]. Milk alternatives may be extracted from nuts and seeds that may directly touch the soil during pre- or post-harvest [15]. Hence, restricted precautions must be taken during planting and harvesting of nuts and seeds and processing of MA. Additionally, nuts and seeds should be stored in dry facilities that are protected from rain and ground water, insects and pests, and that have optimal temperatures that avert microbial growth [45].

3.2. Antimicrobial Drug Resistance in Enterobacteriaceae

Detailed presentation of antimicrobial resistant *Enterobacteriaceae* species from MA extracts is shown in Table 2. In the present study, *Enterobacteriaceae* resistance in isolated bacteria was higher ($p < 0.05$) in vancomycin (90.0%), novobiocin (83.7%), and erythromycin (80.9%) than in tetracycline (60.0%), cefpodoxime (31.8%), and nalidixic acid (6.4%). The majority of *Enterobacteriaceae* in our study are opportunistic pathogens that cause nosocomial infections; hence their antimicrobial resistance might lead to impediments in treating infected individuals [52]. Our findings agree with a previous study that documented *C. sakazakii* resistance to both erythromycin and tetracycline [53]. Occurrence of antibiotic resistant *C. sakazakii* in nut and seed extracts is a concern because antibiotic therapy is a chosen path to avert *Cronobacter* infection in humans [54]. Resistance to erythromycin, tetracycline, vancomycin, and novobiocin was also exhibited by *K. pneumoniae* isolates in nut and seed extracts in our study, hence a concern, since *K. pneumoniae* is a significant multidrug-resistant (MDR) pathogen that causes hospital infections leading to high morbidity and death [55], one of the most severe challenges in clinical practice.

Table 2. Resistant Antibiotics Profile and *Enterobacteriaceae* Nut and Seed Extracts.

Antibiotics (µg)	No. (%) of *Enterobacteriaceae* Resistant to Antimicrobial Agents									No. (%) of Total Resistant (* R) Isolates
	Almond milk (*n* = 56)			Cashew Milk (*n* = 28)			Soy Milk (*n* = 26)			
	R	I	S	R	I	S	R	I	S	
Erythromycin (15)	48 (85.7) ^{b}	8 (14.3) ^{b}	0 (0) ^{e}	28 (100) ^{a}	0 (0) ^{c}	0 (0) ^{d}	13 (50) ^{b}	2 (7.7) ^{bc}	11 (42.3) ^{b}	89 (80.9) ^{a}
Novobiocin (30)	51 (91.1) ^{b}	1 (1.8) ^{d}	4 (7.1) ^{d}	27 (96.4) ^{a}	13.6) ^{a}	0 (0) ^{d}	14 (53.9) ^{b}	12 (46.1) ^{a}	0 (0) ^{d}	92 (83.7) ^{a}
Cefpodoxime (10)	13 (23.2) ^{d}	24 (42.9) ^{a}	19 (33.9) ^{c}	5 (17.9) ^{c}	8 (28.6) ^{b}	15 (53.6) ^{b}	17 (65.4) ^{ab}	5 (19.2) ^{b}	4 (15.4) ^{c}	35 (31.8) ^{c}
NalidixicAcid (30)	5 (8.9) ^{e}	2 (3.6) ^{cd}	49 (87.5) ^{b}	2 (7.1) ^{cd}	0 (0) ^{c}	26 (92.9) ^{a}	0 (0) ^{c}	2 (7.7) ^{b}	24 (92.3) ^{a}	7 (6.4) ^{d}
Imipenem (30)	0 (0) ^{f}	0 (0) ^{d}	56 (100) ^{a}	0 (0) ^{d}	0 (0) ^{c}	28 (100) ^{a}	0 (0) ^{c}	0 (0) ^{c}	26 (100) ^{a}	0 (0) ^{e}
Kanamycin (10)	0 (0) ^{f}	6 (10.7) ^{c}	50 (89.3) ^{b}	0 (0) ^{c}	2 (7.1) ^{c}	26 (92.9) ^{a}	0 (0) ^{c}	2 (7.7) ^{bc}	24 (92.3) ^{a}	0 (0) ^{e}
Vancomycin (30)	56 (100) ^{a}	0 (0) ^{d}	0 (0) ^{e}	28 (100) ^{a}	0 (0) ^{c}	0 (0) ^{d}	15 (57.7) ^{ab}	1 (3.8) ^{c}	10 (38.5) ^{bc}	99 (90.0) ^{a}
Tetracycline (30)	31 (55.4) ^{c}	17 (30.4) ^{a}	8 (14.3) ^{d}	14 (50) ^{b}	8 (28.6) ^{a}	6 (21.4) ^{c}	21 (80.8) ^{a}	5 (19.2) ^{b}	0 (0) ^{d}	66 (60.0) ^{b}

R = Resistant, I = Intermediate, S = Susceptible (CLSI, 2018). * R = Total number of resistant isolates from all extracts (µg). *n*: Total number of *Enterobacteriacea* isolates from various extracts. $^{a–f}$ Mean percentages in the same column followed by different letters are significantly different ($p < 0.05$).

According to Zhou et al. [56], *Klebsiella*-resistant strains have increased more quickly than those of any other bacteria in the past decade. The consumption of both MA and milk may result in foodborne illnesses if not controlled [57]. According to our study, antibiotic resistant *E. vulneris* was detected in MA. Our data is supported by previous studies [58] which displayed multiple antibiotic resistant *E. coli* strains in milk.

The absence of resistance among all *Enterobacteriaceae* strains to kanamycin was also noted in the current study. Additionally, *Enterobacteriaceae* strains in this study did not display resistance to imipenem which agrees with previous findings [59]. Although no imipenem resistance was indicated in our findings, carbapenems have been used to treat numerous *Enterobacteriaceae* infections, hence there has been a rapid development in their resistance to the same. The rapid spread of carbapenem resistant *Enterobacteriaceae* (CRE) in the community is a national epidemiologic concern, since *Enterobacteriaceae* are common causes of nosocomial and community infections.

A total of seven multidrug-resistance patterns were observed in *Enterobacteriaceae* in this study (Table 3). Out of 110 *Enterobacteriaceae* isolates, 87 (79.1%) from nuts and seed extracts were multidrug-resistant. According to Nguyen et al. [60], an MDR isolate displays resistance to three or more classes of antibiotic. Overall, the most common resistance pattern (ERY-NOV-VAN-TET) in our study was exhibited in by *Citrobacter youngae* (1), *E. cancerogenus* (7), *E. cloacae* (18), *E. vulneris* (1) *Pantoea* spp. 3 (3), and *Rahnella aquatilis* (1) (number of isolates in parenthesis). Forty-four (44) *E. cloacae* and 28 *E. cancerogenus* isolates recovered from nuts and seed extracts were multidrug-resistant (MDR). ERY-NOV-VAN-TET was the most significant ($p < 0.05$) multidrug resistance pattern among *E. cloacae* isolates. *E. cloacae* and *E. cancerogenus* presented a common resistance pattern: ERY-NOV-CEP-NAL-VAN-TET, which was resistant to six out of eight antibiotics.

Table 3. Antibiotic Resistance Patterns of *Enterobacteriaceae* in Nuts and Seed Extracts.

Bacterial Species [A]	No of Isolates	Resistance Profile [B]
Chryseomonas luteola	1 [d]	ERY-NOV-CEP-VAN-TET
Citrobacter youngae	1 [d]	ERY-NOV-VAN-TET
Enterobacter Cancerogenus	8 [b,c]	CEP-TET
	1 [d]	CEP-VAN-TET
	1 [d]	ERY-CEP-TET
	4 [c,d]	ERY-NOV-CEP-NAL-VAN-TET
	3 [c,d]	ERY-NOV-CEP-VAN-TET
	10 [b]	ERY-NOV-VAN
	7 [b,c]	ERY-NOV-VAN-TET
	3 [c,d]	ERY-VAN
	2 [c,d]	NOV-CEP-TET
Enterobacter cloacae	3 [c,d]	ERY-NOV-CEP-NAL-VAN-TET
	12 [a,b]	ERY-NOV-CEP-VAN-TET
	11 [a,b]	ERY-NOV-VAN
	18 [a]	ERY-NOV-VAN-TET
	2 [c,d]	NOV-VAN
	1 [d]	VAN
Cronobacter sakazakii	2 [c,d]	ERY-NOV-VAN
	1 [d]	ERY-VAN
Escherichia vulneris	1 [d]	ERY-NOV-VAN-TET
	1 [d]	VAN
Klebsiella pneumoniae spp. *ozaenae*	5 [c,d]	NOV-VAN
Klebsiella pneumoniae spp. *pneumoniae*	3 [c,d]	ERY-NOV-VAN
Pantoea spp. 3	3 [c,d]	ERY-NOV-VAN
	3 [c,d]	ERY-NOV-VAN-TET
	1 [d]	ERY-VAN
	1 [d]	VAN
Rahnella aquatilis	1 [d]	ERY-NOV-VAN-TET

[A] Bacterial species isolated from milk extracts (MA). [B] Antibiotic resistance patterns against eight antibiotics: vancomycin (VAN), novobiocin (NOVO), erythromycin (ERY), tetracycline (TET), cefpodoxime (CEF), kanamycin (KAN), nalidixic acid (NAL), and imipenem (IPM). [a–d] Number of isolates in the same column followed by different letters are significantly different ($p < 0.05$).

4. Conclusions

Processed MA and milk food safety can be improved by implementation of high sanitary standards that reduce risk of contamination. Milk contamination with micro-organisms can occur before harvest, during milking or postharvest, and in storage. Similarly, MA may be contaminated by use of pathogen tinted nuts or seeds before harvest, during collection, and processing, and in storage. With the increased trend in adoption of MA, consumers may also be at risk of infection with AMR bacteria from ingestion of unpasteurized MA. Therefore, it is imperative that consumers should be educated on safe milk handling practices. Although many consumers are aware of foodborne illnesses, they have limited knowledge of food storage, time, and temperature abuse that may increase bacterial growth.

Although MA are considered healthy, our data suggest that they are reservoirs of antibiotic resistant *Enterobacteriaceae*. Consumers should be aware of the impending risks of ingesting unpasteurized milk substitutes in their homes, which can harbor AMR bacteria that can pose serious health risks.

Author Contributions: Conceptualization, W.M. and A.K.-N.; methodology, W.M., A.K.-N. and B.P.; validation, W.M.; formal analysis, W.M. and A.K.-N.; investigation, W.M. and A.K.-N.; resources, A.K.-N.; writing—original draft preparation, W.M.; writing—review and editing, W.M., A.K.-N., A.I.M. and A.T.; visualization, W.M. and A.K.-N.; supervision, A.K.-N.; project administration, A.K.-N.; and funding acquisition, A.K.-N. All authors have read and agreed to the published version of the manuscript.

Funding: This research was funded by United States Department of Agriculture/National Institute of Food and Agriculture. Grant No. Accession No: 1014615; Project: TENX-1813-FS.

Acknowledgments: We are grateful for the PCR technical suggestions by Boniface Kimathi and Collins Khwatenge.

Conflicts of Interest: The authors declare no conflict of interest. The funders had no role in the design of the study; in the collection, analyses, or interpretation of data; in the writing of the manuscript, or in the decision to publish the results.

References

1. Rice, B.H.; Quann, E.E.; Miller, G.D. Meeting and exceeding dairy recommendations: Effects of dairy consumption on nutrient intakes and risk of chronic disease. *Nutr. Rev.* **2013**, *71*, 209–223. [CrossRef] [PubMed]
2. Claeys, W.; Verraes, C.; Cardoen, S.; De Block, J.; Huyghebaert, A.; Raes, K.; Dewettinck, K.; Herman, L. Consumption of raw or heated milk from different species: An evaluation of the nutritional and potential health benefits. *Food Control* **2014**, *42*, 188–201. [CrossRef]
3. Gerosa, S.; Skoet, J. *Milk Availability: Trends in Production and Demand and Medium-Term Outlook*; FAO, Agricultural Development Economics Division: Rome, Italy, 2012.
4. Mintel. Added Value in Dairy Drinks, Milk and Cream. Available online: https://reports.mintel.com/display/792665/ (accessed on 2 September 2021).
5. Mäkinen, O.E.; Wanhalinna, V.; Zannini, E.; Arendt, E.K. Foods for special dietary needs: Non-dairy plant-based milk substitutes and fermented dairy-type products. *Crit. Rev. Food Sci. Nutr.* **2016**, *56*, 339–349. [CrossRef]
6. Stall, S.; Adams, G. Can almond milk be called milk? *J. Ren. Nutr.* **2017**, *27*, e15–e17. [CrossRef]
7. Fiocchi, A.; Schünemann, H.J.; Brozek, J.; Restani, P.; Beyer, K.; Troncone, R.; Martelli, A.; Terracciano, L.; Bahna, S.L.; Rancé, F. Diagnosis and rationale for action against cow's milk allergy (DRACMA): A summary report. *J. Allergy Clin. Immunol.* **2010**, *126*, 1119–1128.e1112. [CrossRef]
8. McCarthy, K.; Parker, M.; Ameerally, A.; Drake, S.; Drake, M. Drivers of choice for fluid milk versus plant-based alternatives: What are consumer perceptions of fluid milk? *J. Dairy Sci.* **2017**, *100*, 6125–6138. [CrossRef] [PubMed]
9. Aydar, E.F.; Tutuncu, S.; Ozcelik, B. Plant-based milk substitutes: Bioactive compounds, conventional and novel processes, bioavailability studies, and health effects. *J. Funct. Foods* **2020**, *70*, 103975.
10. Sethi, S. *Bread, Wine, Chocolate: The Slow Loss of Foods We Love*; HarperCollins: New York, NY, USA, 2015.
11. Afroz, M.; Anjum, W.; Islam, N.; Kobir, A.; Hossain, K.; Sayed, A. Preparation of soymilk using different methods. *J. Food Nutr. Sci.* **2016**, *4*, 11–17. [CrossRef]
12. Watson, E. Almond Breeze Almond Milk only Contains 2% Almonds, Claims False Advertising Lawsuit. 2015. Available online: https://www.foodnavigator-usa.com/Article/2015/07/22/ (accessed on 20 October 2021).
13. Vanga, S.K.; Raghavan, V. How well do plant based alternatives fare nutritionally compared to cow's milk? *J. Food Sci. Technol.* **2018**, *55*, 10–20. [CrossRef] [PubMed]
14. Ziarati, P.; Shirkhan, F.; Mostafidi, M.; Zahedi, M.T. An overview of the heavy metal contamination in milk and dairy products. *Acta Sci. Pharm. Sci.* **2018**, *2*, 1–14.
15. Danyluk, M.D.; Jones, T.M.; Abd, S.J.; Schlitt-Dittrich, F.; Jacobs, M.; Harris, L.J. Prevalence and amounts of *Salmonella* found on raw California almonds. *J. Food Prot.* **2007**, *70*, 820–827. [CrossRef]
16. Willis, C.; Little, C.L.; Sagoo, S.; de Pinna, E.; Threlfall, J. Assessment of the microbiological safety of edible dried seeds from retail premises in the United Kingdom with a focus on *Salmonella* spp. *Food Microbiol.* **2009**, *26*, 847–852. [CrossRef]
17. Calhoun, S.; Post, L.; Warren, B.; Thompson, S.; Bontempo, A.R. Prevalence and concentration of *Salmonella* on raw shelled peanuts in the United States. *J. Food Prot.* **2013**, *76*, 575–579. [CrossRef] [PubMed]
18. Brar, P.K.; Strawn, L.K.; Danyluk, M.D. Prevalence, level, and types of *Salmonella* isolated from North American in-shell pecans over four harvest years. *J. Food Prot.* **2016**, *79*, 352–360. [CrossRef] [PubMed]
19. Davidson, G.R.; Frelka, J.C.; Yang, M.; Jones, T.M.; Harris, L.J. Prevalence of *Escherichia coli O157: H7* and *Salmonella* on inshell California walnuts. *J. Food Prot.* **2015**, *78*, 1547–1553. [CrossRef] [PubMed]
20. Eglezos, S. The bacteriological quality of retail-level peanut, almond, cashew, hazelnut, Brazil, and mixed nut kernels produced in two Australian nut-processing facilities over a period of 3 years. *Foodborne Pathog. Dis.* **2010**, *7*, 863–866. [CrossRef] [PubMed]
21. Al-Moghazy, M.; Boveri, S.; Pulvirenti, A. Microbiological safety in pistachios and pistachio containing products. *Food Control* **2014**, *36*, 88–93. [CrossRef]

22. Liu, S.; Kilonzo-Nthenge, A. Prevalence of multidrug-resistant bacteria from US-grown and imported fresh produce retailed in chain supermarkets and ethnic stores of Davidson County, Tennessee. *J. Food Prot.* **2017**, *80*, 506–514. [CrossRef]
23. Jee, Y.; Carlson, J.; Rafai, E.; Musonda, K.; Huong, T.T.G.; Daza, P.; Sattayawuthipong, W.; Yoon, T. Antimicrobial resistance: A threat to global health. *Lancet Infect. Dis.* **2018**, *18*, 939–940. [CrossRef]
24. Ventola, C.L. The antibiotic resistance crisis: Part 1: Causes and threats. *Pharm. Ther.* **2015**, *40*, 277.
25. Centers for Disease Control and Prevention, Office of Infectious Disease Antibiotic Resistance Threats in the United States. 2013. Available online: http://www.cdc.gov/drugresistance/threat-report-2013 (accessed on 5 September 2021).
26. Golkar, Z.; Bagasra, O.; Pace, D.G. Bacteriophage therapy: A potential solution for the antibiotic resistance crisis. *J. Infect. Dev. Ctries* **2014**, *8*, 129–136. [CrossRef] [PubMed]
27. Ababu, A.; Endashaw, D.; Fesseha, H. Isolation and Antimicrobial Susceptibility Profile of *Escherichia coli O157: H7* from Raw Milk of Dairy Cattle in Holeta District, Central Ethiopia. *Int. J. Microbiol.* **2020**, *2020*, 1–8. [CrossRef]
28. Zhou, Y.; Wu, Q.; Xu, X.; Yang, X.; Ye, Y.; Zhang, J. Development of an immobilization and detection method of *Enterobacter sakazakii* from powdered infant formula. *Food Microbiol.* **2008**, *25*, 648–652. [CrossRef]
29. Kilonzo-Nthenge, A.; Rotich, E.; Godwin, S.; Nahashon, S.; Chen, F. Prevalence and antimicrobial resistance of *Cronobacter sakazakii* isolated from domestic kitchens in middle Tennessee, United States. *J. Food Prot.* **2012**, *75*, 1512–1517. [CrossRef] [PubMed]
30. Chander, Y.; Ramakrishnan, M.; Jindal, N.; Hanson, K.; Goyal, S.M. Differentiation of *Klebsiella pneumoniae* and *K. oxytoca* by multiplex polymerase chain reaction. *Int. J. Appl. Res. Vet. Med.* **2011**, *9*, 138.
31. Brisse, S.; Verhoef, J. Phylogenetic diversity of Klebsiella pneumoniae and *Klebsiella oxytoca* clinical isolates revealed by randomly amplified polymorphic DNA, gyrA and parC genes sequencing and automated ribotyping. *Int. J. Syst. Evol. Microbiol.* **2001**, *51*, 915–924. [CrossRef]
32. Clinical and Laboratory Standards Institute (CLSI). *Performance Standards for Antimicrobial Susceptibility Testing*, 31st ed.; CLSI Supplement M100S; CLSI: Wayne, PA, USA, 2018.
33. Heperkan, D.; Dalkilic-Kaya, G.; Juneja, V.K. *Cronobacter sakazakii* in baby foods and baby food ingredients of dairy origin and microbiological profile of positive samples. *LWT* **2017**, *75*, 402–407. [CrossRef]
34. Food Authority. MILKLAB and Blue Diamond Almond Milk. 2020. Available online: https://www.foodauthority.nsw.gov.au/news/recalls/milklab-and-blue-diamond-almond-milk. (accessed on 20 October 2021).
35. Langer, A.J.; Ayers, T.; Grass, J.; Lynch, M.; Angulo, F.J.; Mahon, B.E. Nonpasteurized dairy products, disease outbreaks, and state laws—United States, 1993–2006. *Emerg. Infect. Dis.* **2012**, *18*, 385. [CrossRef]
36. Allocati, N.; Masulli, M.; Alexeyev, M.F.; Di Ilio, C. *Escherichia coli* in Europe: An overview. *Int. J. Environ. Res. Public Health* **2013**, *10*, 6235–6254. [CrossRef]
37. Vandera, E.; Lianou, A.; Kakouri, A.; Feng, J.; Koukkou, A.-I.; Samelis, J. Enhanced control of *Listeria monocytogenes* by *Enterococcus faecium* KE82, a multiple enterocin–producing strain, in different milk environments. *J. Food Prot.* **2017**, *80*, 74–85. [CrossRef]
38. Davin-Regli, A. *Enterobacter aerogenes* and *Enterobacter cloacae*; versatile bacterial pathogens confronting antibiotic treatment. *Front. Microbiol.* **2015**, *6*, 392. [CrossRef] [PubMed]
39. Jarlier, V.; INVS. Surveillance of Multidrug Resistant Bacteria in French Healthcare Facilities BMR-Raisin Network Données. Saint-Maurice: Institut de Veille Sanitaire. 2014. Available online: http://www.invs.sante.fr (accessed on 3 September 2021).
40. Mohan Nair, M.K.; Venkitanarayanan, K.; Silbart, L.K.; Kim, K.S. Outer membrane protein A (OmpA) of *Cronobacter sakazakii* binds fibronectin and contributes to invasion of human brain microvascular endothelial cells. *Foodborne Pathog. Dis.* **2009**, *6*, 495–501. [CrossRef] [PubMed]
41. Food and Agriculture Organization (FAO); World Health Organization (WHO). *Enterobacter sa-kazakii* (Cronobacter spp.) in Powdered Follow-Up Formulae. In *Microbiological Risk Assessment*; Series No. 15; FAO: Rome, Italy, 2008.
42. Hunter, C.J.; Bean, J. Cronobacter: An emerging opportunistic pathogen associated with neonatal meningitis, sepsis and necrotizing enterocolitis. *J. Perinatol.* **2013**, *33*, 581–585. [CrossRef]
43. Yan, Q.; Condell, O.; Power, K.; Butler, F.; Tall, B.; Fanning, S. Cronobacter species (formerly known as *Enterobacter sakazakii*) in powdered infant formula: A review of our current understanding of the biology of this bacterium. *J. Appl. Microbiol.* **2012**, *113*, 1–15. [CrossRef]
44. Mullane, N.; Iversen, C.; Healy, B.; Walsh, C.; Whyte, P.; Wall, P.; Quinn, T.; Fanning, S. *Enterobacter sakazakii* an emerging bacterial pathogen with implications for infant health. *Minerva Pediatrica* **2007**, *59*, 137–148.
45. Brar, P.K.; Danyluk, M.D. Nuts and grains: Microbiology and preharvest contamination risks. *Preharvest Food Saf.* **2018**, 105–121. [CrossRef]
46. Kader, A.A.; Thompson, J.F. *Postharvest Handling Systems: Tree Nuts*; University California Publication: Davis, CA, USA, 2002; pp. 399–406.
47. Nordmann, P.; Cuzon, G.; Naas, T. The real threat of *Klebsiella pneumoniae* carbapenemase-producing bacteria. *Lancet Infect. Dis.* **2009**, *9*, 228–236. [CrossRef]
48. Zarkotou, O.; Pournaras, S.; Tselioti, P.; Dragoumanos, V.; Pitiriga, V.; Ranellou, K.; Prekates, A.; Themeli-Digalaki, K.; Tsakris, A. Predictors of mortality in patients with bloodstream infections caused by KPC-producing *Klebsiella pneumoniae* and impact of appropriate antimicrobial treatment. *Clin. Microbiol. Infect.* **2011**, *17*, 1798–1803. [CrossRef]
49. Jain, S.; Nagarjuna, D.; Gaind, R.; Chopra, S.; Debata, P.; Dawar, R.; Sardana, R.; Yadav, M. *Escherichia vulneris*: An unusual cause of complicated diarrhoea and sepsis in an infant. A case report and review of literature. *Microbe Infect.* **2016**, *13*, 83–86. [CrossRef]

50. Le Querler, L.; Donnio, P.; Poisson, M.; Rouzet-Gras, S.; Avril, J. Isolation of *Escherichia vulneris* from drinking water. *Ann. Biol. Clin.* **1997**, *55*, 33–35.
51. Matsukura, H.; Katayama, K.; Kitano, N.; Kobayashi, K.; Kanegane, C.; Higuchi, A.; Kyotani, S. Infective endocarditis caused by an unusual gram-negative rod, *Rahnella aquatilis*. *Pediatrica Cardiol.* **1996**, *17*, 108–111. [CrossRef]
52. Aghapour, Z.; Gholizadeh, P.; Ganbarov, K.; Bialvaei, A.Z.; Mahmood, S.S.; Tanomand, A.; Yousefi, M.; Asgharzadeh, M.; Yousefi, B.; Kafil, H.S. Molecular mechanisms related to colistin resistance in *Enterobacteriaceae*. *Infect. Drug Resist.* **2019**, *12*, 965. [CrossRef] [PubMed]
53. Hochel, I.; Růžičková, H.; Krásný, L.; Demnerová, K. Occurrence of *Cronobacter* spp. in retail foods. *J. Appl. Microbiol.* **2012**, *112*, 1257–1265. [CrossRef]
54. Depardieu, F.; Podglajen, I.; Leclercq, R.; Collatz, E.; Courvalin, P. Modes and modulations of antibiotic resistance gene expression. *Clin. Microbiol. Rev.* **2007**, *20*, 79–114. [CrossRef] [PubMed]
55. Navon-Venezia, S.; Kondratyeva, K.; Carattoli, A. *Klebsiella pneumoniae*: A major worldwide source and shuttle for antibiotic resistance. *FEMS Microbiol. Rev.* **2017**, *41*, 252–275. [CrossRef]
56. Zhou, X.-Y.; Ye, X.-G.; He, L.-T.; Zhang, S.-R.; Wang, R.-L.; Zhou, J.; He, Z.-S. In vitro characterization and inhibition of the interaction between ciprofloxacin and berberine against multidrug-resistant *Klebsiella pneumoniae*. *J. Antibiot.* **2016**, *69*, 741–746. [CrossRef]
57. Melini, F.; Melini, V.; Luziatelli, F.; Ruzzi, M. Raw and heat-treated milk: From public health risks to nutritional quality. *Beverages* **2017**, *3*, 54. [CrossRef]
58. Sudda, M.M.; Mtenga, A.B.; Kusiluka, L.J.; Kassim, N. Prevalence and Antibiotic Susceptibility of *Escherichia coli* and *Salmonella* spp. isolated from milk of zero grazed cows in Arusha City. *Afr. J. Microbiol. Res.* **2016**, *10*, 1944–1951.
59. Ombarak, R.A.; Hinenoya, A.; Elbagory, A.-R.M.; Yamasaki, S. Prevalence and molecular characterization of antimicrobial resistance in *Escherichia coli* isolated from raw milk and raw milk cheese in Egypt. *J. Food Prot.* **2018**, *81*, 226–232. [CrossRef]
60. Nguyen, D.P.; Nguyen, T.A.D.; Le, T.H.; Tran, N.M.D.; Ngo, T.P.; Dang, V.C.; Kawai, T.; Kanki, M.; Kawahara, R.; Jinnai, M. Dissemination of extended-spectrum β-lactamase-and AmpC β-lactamase-producing *Escherichia coli* within the food distribution system of Ho Chi Minh City, Vietnam. *Biomed. Res. Int.* **2016**, *2016*, 8182096. [CrossRef]

foods

MDPI

Review

The Prevalence and Epidemiology of *Salmonella* in Retail Raw Poultry Meat in China: A Systematic Review and Meta-Analysis

Tianmei Sun [1], Yangtai Liu [1], Xiaojie Qin [1], Zafeiro Aspridou [2], Jiaming Zheng [1], Xiang Wang [1], Zhuosi Li [1] and Qingli Dong [1,*]

[1] School of Health Science and Engineering, University of Shanghai for Science and Technology, Shanghai 200093, China; usststm@163.com (T.S.); usstlyt@163.com (Y.L.); qxj19900709@163.com (X.Q.); usstzjm@163.com (J.Z.); xiang.wang@usst.edu.cn (X.W.); lizhuosi624@163.com (Z.L.)

[2] Laboratory of Food Microbiology and Hygiene, Department of Food Science and Technology, School of Agriculture, Faculty of Agriculture, Forestry and Natural Environment, Aristotle University of Thessaloniki, 54124 Thessaloniki, Greece; zasprido@agro.auth.gr

* Correspondence: qdong@usst.edu.cn

Abstract: Foodborne disease caused by *Salmonella* is an important public health concern worldwide. Animal-based food, especially poultry meat, is the main source of human salmonellosis. The objective of this study was to evaluate the prevalence and epidemiology of *Salmonella* contamination in raw poultry meat commercialized in China. Following the principle of systematic review, 98 sets of prevalence data were extracted from 74 publications conducted in 21 Chinese provincial regions. The random-effect model was constructed for subgrouping analysis by meat category, preservation type, and geographical location. The prevalence levels differed from high to low among raw poultry meat, including chicken, 26.4% (95% CI: 22.4–30.8%); pigeon, 22.6% (95% CI: 18.2–27.8%); duck, 10.1% (95% CI: 5.3–18.2%); and other poultry meat, 15.4% (95% CI: 12.0–19.5%). Prevalence data on the preservation type revealed that chilled poultry meat might be more likely to experience cross-contamination than non-chilled poultry meat in China. The distribution map of *Salmonella* for raw poultry meat showed that a higher prevalence level was found in the Shaanxi, Henan, Sichuan, and Beijing regions. All subgroups possessed high amounts of heterogeneity ($I^2 > 75\%$). The scientific data regarding the differences in prevalence levels between meat category, preservation method, and geographical region sources might be useful to improve specific interventions to effectively control the incidence of *Salmonella* in poultry meat.

Keywords: foodborne pathogen; salmonellosis; chicken; antibiotic resistance; microbial contamination; food safety

Citation: Sun, T.; Liu, Y.; Qin, X.; Aspridou, Z.; Zheng, J.; Wang, X.; Li, Z.; Dong, Q. The Prevalence and Epidemiology of *Salmonella* in Retail Raw Poultry Meat in China: A Systematic Review and Meta-Analysis. *Foods* **2021**, *10*, 2757. https://doi.org/10.3390/foods10112757

Academic Editors: Antonio Afonso Lourenco, Catherine Burgess and Timothy Ells

Received: 12 October 2021
Accepted: 9 November 2021
Published: 10 November 2021

1. Introduction

Salmonella, one of the most important foodborne pathogens in the world, is frequently implicated in foodborne disease outbreaks. It is estimated that *Salmonella* is responsible for approximately 1.3 billion cases of salmonellosis worldwide each year [1]. China has a high incidence of salmonellosis [2]. It was found that approximately 70% to 80% of foodborne diseases are caused by *Salmonella* in China [3]. Epidemiological studies have suggested that foods of animal origin, especially poultry and poultry products, are common vehicles of *Salmonella* transmission to human beings [4–6].

The monitoring and tracking of *Salmonella* in poultry meat and the establishment of efficient surveillance systems are the basis for effective public health protection and food safety management. In Europe, a baseline survey was conducted to estimate the prevalence of *Salmonella* and *Campylobacter* on broiler carcasses in 2008 [7]. In the USA, the United States Department of Agriculture Food Safety Inspection Service (USDA/FSIS) has established a verification program to inspect raw poultry products for the presence of *Salmonella* and *Campylobacter* [8]. In China, numerous studies investigated retail chicken meat for *Salmonella* contamination, which showed that up to 50% of retail chicken samples were contaminated with *Salmonella* [9,10]. However, limited information is available concerning *Salmonella* contamination of other poultry meats, such as duck, goose, and pigeon. Furthermore, these previous studies only included samples from one or a few regions from the whole territory of China. Given the variations in data availability and quality observed in the European Union [11], it is expected that the prevalence and contamination level will be different among the various regions of China. As such, there is a lack of comprehensive data on *Salmonella* contamination in poultry meat at the retail level in the whole region of China.

Meta-analysis is concerned with the statistical summary of a large number of results from multiple individual studies on a specific research question [12]. With meta-analysis, it may be possible to explain the sources of heterogeneity and differences among the findings of the primary research [13]. At present, the amount of data generated by food safety research is growing increasingly. In the field of food safety, meta-analysis is a valuable tool to deal with a broad range of research interests, such as disease incidence, epidemiology and prevalence of microorganisms, effect of pre- and post-harvest interventions, consumer practices, etc. [14,15]. Thus, meta-analysis results are an important part of quantitative microbial risk assessments (QMRAs), as they can provide more accurate data for risk assessment models than estimates based on a single study or expert opinion.

According to the Food and Agriculture Organization (FAO) [16], China's poultry production is second only to the USA, and its consumption is increasing steadily. Recent poultry-related systematic reviews and meta-analyses were conducted to estimate the prevalence of *Salmonella* in poultry samples from Europe and North America [5,17–20]. To our knowledge, there is a lack of meta-analysis studies to estimate the pooled prevalence of *Salmonella* in retail raw poultry meat in mainland China. The current study attempted to generate pooled prevalence data based on existing publications from China using the meta-analytical approach. The objective of this study was to quantify *Salmonella* prevalence in Chinese retail poultry meat, to analyze the differences in *Salmonella* prevalence among sub-categories, and to evaluate the levels of the heterogeneity of the published prevalence data.

2. Materials and Methods

2.1. Search Strategy and Selection Criteria

Two databases were systematically searched, including the Web of Science (WoS) database and the China National Knowledge Infrastructure (CNKI) database. The following search strategy was carried out for collecting potentially relevant publications from the WoS database: ("prevalence" OR "incidence" OR "occurrence" OR "quality" OR "contamination" OR "survey" OR "sampling" OR "character*" OR "quanti*" OR "epidemiol*" OR "isolate" OR "enumerate*") AND "Salmonella" AND ("chicken" OR "broiler" OR "duck" OR "goose" OR "turkey" OR "poultry" OR "meat") AND ("China" OR "Chinese") AND ("1950:2019"). Another search format for the CNKI database was: "Salmonella" AND ("contamination" OR "monitoring" OR "checking out" OR "inspection" OR "detection" OR "isolation" OR "epidemiological") AND ("chicken" OR "duck" OR "goose" OR "poultry" OR "meat") AND ("1979:2019"); all the terms were used in Chinese. To avoid missing any additional data, we conducted a complementary literature search on the reference list of relevant publications.

The PRISMA statement (Preferred Reporting Items for Systematic Reviews and Meta-Analyses, http://www.prisma-statement.org/ accessed on 16 November 2020) was employed for reporting the screening process. After removing duplicate records, all the publications were checked against a set of exclusion criteria. A study was excluded if (1) it was published as a conference abstract or was not a research paper (review); (2) it was not relevant, such as studies focusing on the detection method, predictive modeling, or hurdle technology; (3) it was a duplicate report; (4) the poultry samples were imported products; (5) the meat category was not clearly indicated; (6) incomplete data on the prevalence and concentration of *Salmonella* on poultry were reported; (7) the samples were not limited to the retail stage; and (8) the sample size was lower than 50.

2.2. Data Extraction

Data for *Salmonella* prevalence on raw poultry were extracted from the studies identified through the systematic review of the literature independently by a single reviewer and validated by a second reviewer. The following data were extracted from each eligible study of records: author, publication year, survey year, meat category, preservation type, sampling location (i.e., provincial, region), detection method, sample size, positive sample number, identified *Salmonella* serovars, the antimicrobial resistance rate of *Salmonella* to each antibiotic. The poultry meat was further categorized into 'Chicken', 'Duck', 'Pigeon', 'Goose', and 'Other'. The preservation type category was subdivided into 'Ambient', 'Chilled', 'Frozen', and 'Unknown'.

2.3. Meta-Analysis and Statistical Analyses

The meta-analysis and forest plot generation of this review were performed using R language (Version 3.4.3, http://www.R-project.org/ accessed on 6 March 2021) with the 'meta' package. For further subgroup analysis, data were grouped by the meat category, preservation type, and sampling location. Due to the fact that the sampling methods and experimental methodologies of the primary studies were not identical, the description of the heterogeneity (or variability) is critical in a meta-analysis [19,21]. As stated by Gonzales-Barron [13], a fixed-effect model may be unsuitable for application in the meta-analysis of the variability of food research. Thus, all eligible information in our study was pooled and analyzed on the basis of a random-effects model [22]. Cochran's Q test and *I*-squared index (I^2) were used for evaluating heterogeneity among studies [23]. The statistical significance for heterogeneity using Cochran's Q test was defined for $p < 0.10$, and the degree of heterogeneity using I^2 was defined as low, moderate, and high when I^2 values (as percentages) were around 25%, 50%, and 75%, respectively [21]. The statistical map was generated based on the Chinese standard geographical map (downloaded at http://bzdt.ch.mnr.gov.cn, accessed on 1 April 2021).

3. Results

3.1. Characteristic of Literature and Datasets

The process for the selection of eligible articles is depicted in Figure 1. A total of 1000 publications were initially identified from the two selected databases. After removing duplicates and manual screening based on the specified criteria, 74 publications (29 in English and 45 in Chinese) of independent studies of *Salmonella* in retail poultry (before 2020) in China were finally included in our systematic review. Following full-text quality checking, a total of 98 sets of prevalence data of *Salmonella* in poultry meat were retrieved. The data encompassed a total of 21,824 samples (5837 positives) from 21 Chinese provinces, major municipalities, and autonomous regions. Due to the limitations of the included information, the origin of retail poultry meat is unknown (e.g., farm household or industry). Most samples belonged to the 'Chicken' category (n = 15,246), followed by the 'Duck' category (n = 794), 'Pigeon' category (n = 292), and 'Other' category (n = 5492). For qualitative or quantitative analysis of *Salmonella*, the pre-enrichment culture and identification process mainly referred to the Chinese national standard GB 4789.4 (versions 2003, 2008, 2010, and 2016) and a few studies (7 out of 74) deployed ISO 6579 or the Most Probable Number (MPN) method.

A total of 42 studies (20 in Chinese and 22 in English) reported the serotype analysis information of *Salmonella* isolates. As shown in Figure 2a, according to the serotyping of 3104 *Salmonella* isolates recovered from 13,119 poultry samples, the three most commonly isolated serovars were *S.* Enteritidis (32.9%), *S.* Indiana (10.0%), and *S.* Typhimurium (9.1%), followed by *S.* Agona (5.0%), *S.* Derby (4.8%), *S.* Kentucky (3.2%), *S.* Corvallis (2.5%), *S.* Shubra (2.2%), *S.* Rissen (1.5%), and *S.* Infantis (1.4%). All serovars were mostly isolated from chicken.

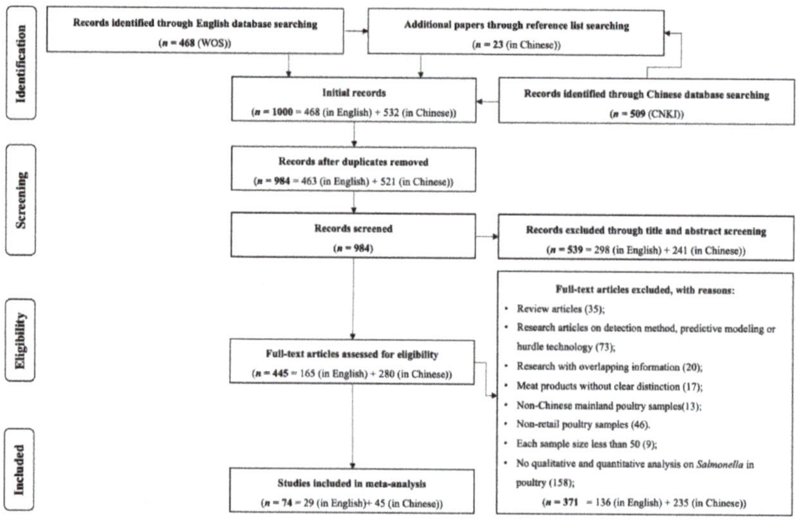

Figure 1. The flowchart of the literature searching and collecting.

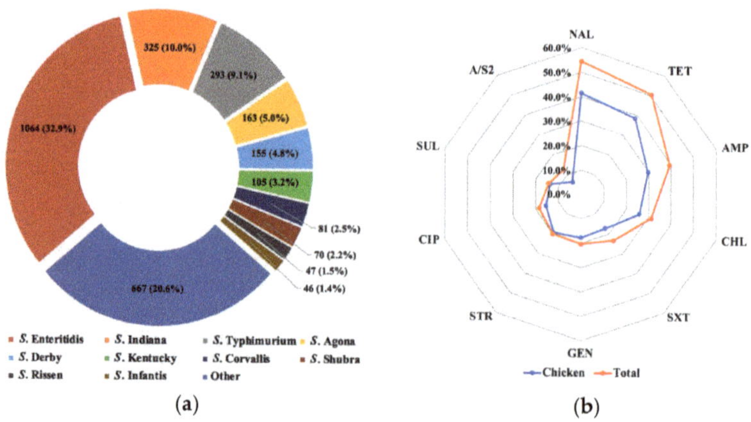

Figure 2. Serovar distribution (**a**) and antimicrobial resistance (**b**) of *Salmonella* strains isolated from Chinese retail raw poultry meat.

Antimicrobial susceptibility of *Salmonella* isolates was evaluated in a total of 27 studies (10 in Chinese and 17 in English). The antibiotic resistance rate was evaluated by dividing the number of resistant *Salmonella* isolates by the number of total *Salmonella* isolates (presented as percentage), when resistant *Salmonella* strains were present. Among the 2249 *Salmonella* isolates from 8920 poultry samples, the results for antimicrobial resistance rates of *Salmonella* are depicted in Figure 2b. The resistance most commonly detected was to nalidixic acid (54.6%), followed by tetracycline (50.6%), ampicillin (39.5%), chloramphenicol (31.4%), trimethoprim/sulfamethoxazole (23.3%), gentamicin (20.3%), streptomycin (20.1%), ciprofloxacin (18.3%), sulfisoxazole (14.2%), and ampicillin/sulbactam (12.4%). Chicken-derived isolates were the majority, and the levels of resistance of them to the above ten antibiotics were basically consistent with the total *Salmonella* isolates.

3.2. *Salmonella Prevalence in Different Poultry Meat Product Types*

The meta-analysis results on the prevalence and heterogeneity of *Salmonella* in poultry meat by poultry type are presented in Table 1. Overall, the pooled prevalence of *Salmonella* in raw poultry meat was 23.0% (95% CI: 19.8–26.8%), with heterogeneity (as indicated by the inverse variance index) as high as 97.0%. Among the different poultry meat categories, chicken presented the highest mean pooled prevalence (26.4%, 95% CI: 22.4–30.8%), followed by pigeon (22.6%, 95% CI: 18.2–27.8%) and duck (10.1%, 95% CI: 5.3–18.2%). Heterogeneity values were relatively low for the prevalence levels reported for pigeon (0%) and duck (87.9%), which may be due to the small number of related studies. In addition, due to the limited information in the included literature on whether the poultry samples were whole carcasses or anatomical pieces (legs, wings, etc.), it was impossible to ascertain the relationship between *Salmonella* prevalence and whole carcasses or anatomical pieces.

Table 1. Meta-analysis results for mean prevalence of *Salmonella* in poultry by meat type based on the included reports.

Meat Category	Total	Positive	Pooled Prevalence (95% CI) [a]	τ^{2} [b]	I^{2} [c]
Raw poultry overall (random effects)	21,824	5837	23.0% (19.8–26.6%)	0.8953	97.0%
Chicken	15,246	4716	26.4% (22.4–30.8%)	0.8821	96.9%
Duck	794	83	10.1% (5.3–18.2%)	0.7475	87.9%
Pigeon	292	66	22.6% (18.2–27.8%)	0.0000	0.0%
Other	5492	972	15.4% (12.0–19.5%)	0.2419	93.1%

[a] 95% CI: 95% confidence interval; [b] τ^2: between-study variance; [c] I^2: inverse variance index.

3.3. *Salmonella Prevalence in Different Geographical Regions*

Due to the inherent limitations of literature retrieval in meta-analyses, the prevalence data of *Salmonella* covered 21 provinces, major municipalities, and autonomous regions in China, occupying a land area of 6,558,251 km^2 (approximately two-thirds of the total). The pooled prevalence estimates of *Salmonella* in poultry meat, to be presented as follows, cannot be generalized to other Chinese regions. The range of *Salmonella* prevalence levels found in raw poultry meat for those regions (indicating low (<15%), medium (≥15% and ≤30%), and high level (>30%) are shown in Figure 3. The highest prevalence level of *Salmonella* in raw poultry meat was reported in Shaanxi (44.3%, 95% CI: 29.9–59.7%), followed by Henan (35.3%, 95% CI: 21.2–52.5%), Sichuan (35.0%, 95% CI: 26.4–44.7%), and Beijing (31.1%, 95% CI:16.5–50.8%).

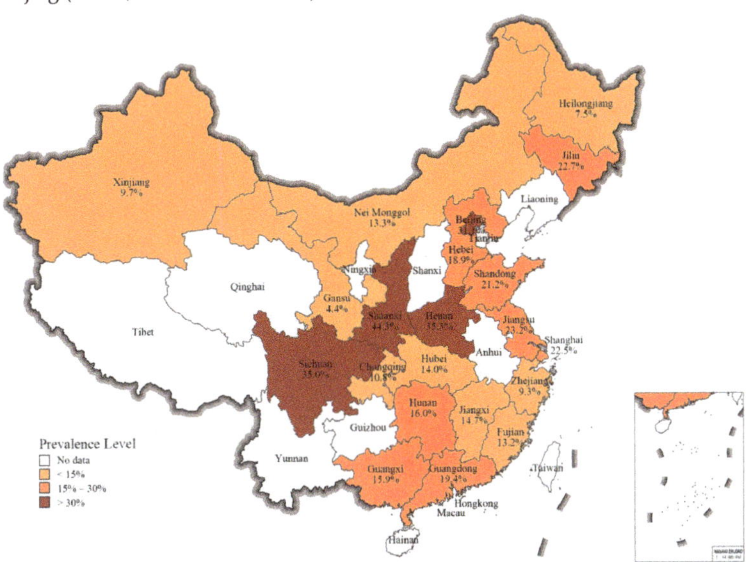

Figure 3. The pooled prevalence of *Salmonella* in raw poultry meat from 21 Chinses provincial regions based on the included reports.

3.4. *Salmonella* Prevalence under Different Preservation Types

The results of the meta-analysis on the prevalence and heterogeneity of *Salmonella* by preservation type are shown in Table 2. The present study recorded 2825 ambient poultry meat samples, 2066 chilled poultry meat samples, and 2173 frozen poultry meat samples, and their pooled prevalence of *Salmonella* were 17.2% (95% CI: 6.6–37.8%), 42.1% (95% CI: 33.7–51.0%), and 25.3% (95% CI: 17.3–35.4%), respectively. Notably, *Salmonella* prevalence in chilled poultry meat was statistically higher than that of frozen poultry meat and ambient poultry meat. Among the included publications, the preservation method was unknown for more than half of the samples (14,760/21,824). In this fraction of the poultry meat samples, the pooled *Salmonella* prevalence was 21.3% (95% CI: 17.9–25.1%). A high heterogeneity was observed among each group.

Table 2. Meta-analysis results for mean prevalence of *Salmonella* in poultry by preservation type based on the included reports.

Preservation Type	Total	Positive	Pooled Prevalence (95% CI) [a]	τ^2 [b]	I^2 [c]
Raw poultry overall (random-effects)	21,824	5837	23.0% (19.8%–26.6%)	0.8953	97.0%
Ambient	2825	649	17.2% (6.6%–37.8%)	2.3137	99.1%
Chilled	2066	974	42.1% (33.7%–51.0%)	0.2596	92.1%
Frozen	2173	535	25.3% (17.3%–35.4%)	0.6518	94.8%
Unknown	14,760	3679	21.3% (17.9%–25.1%)	0.7795	96.3%

[a] 95% CI: 95% confidence interval; [b] τ^2: between-study variance; [c] I^2: inverse variance index.

4. Discussion

Microbiological foodborne hazards have attracted the attention of the food safety management system in China [24]. The Chinese Food Safety Law implemented in 2019 has legally clarified the roles and duties of the national food safety surveillance system for foodborne pathogens in foods [25]. In 2010, this national food safety surveillance system covered all 31 provinces, major municipalities, and autonomous regions in China, to support early detection, diagnosis, and management of foodborne pathogens [26]. Since then, a downward trend is apparent from the publicly available reports on the incidence of foodborne pathogens in foods [27]. However, reducing the incidence of foodborne diseases is a constant topic of concern for the Chinese government as well as the public.

This meta-analysis review demonstrated the widespread prevalence of *Salmonella* in retail poultry meat in China. The contaminated retail poultry may become an issue of concern because the products can be in direct contact and be used by consumers. Although raw meat generally receives a certain lethal treatment (e.g., conventional cooking, microwaving, etc.) before consumption, cross-contamination incidents and undercooking are still the greatest risks in consumers' kitchens [28,29]. In the present study, the pooled prevalence of *Salmonella* in raw poultry meat in China was 23.0%, which is significantly higher than that reported in retail poultry from the European Union (7.1%) [5] and Africa (13.9%) [30]. Thus, raw poultry meat in retail may be an important source of human salmonellosis in China.

According to the prediction by the Organization for Economic Co-operation Development and the Food and Agricultural Organization (OECD-FAO) [31], poultry meat will continue to be the primary driver of meat production growth over the next ten years. Low production costs, a short production cycle, high feed conversion ratios, and low product prices have contributed to making poultry the meat of choice for both producers and consumers. Regarding the different poultry meat categories, chicken is the greatest concern as it bears the highest pooled prevalence of *Salmonella*. The high prevalence of *Salmonella* in raw chicken samples in our study suggests that chicken may be the main vehicle of transmission for *Salmonella* in China. In China's meat consumption structure, chicken takes the largest proportion in poultry meat consumption and is on the rise, becoming the second-largest meat product after pork [32]. Similarly, in Chinese poultry farming operations, densities are generally higher for chickens, while they are considerably lower for ducks and geese (111.2, 27.4, and 6.7 thousand per km^2 maximum, respectively) [33]. Thus, in response to potential public health pressures, more effective intervention strategies during processing should be implemented to control the quality and safety of chicken products.

In terms of geographical distribution, the occurrence of *Salmonella* in retail raw poultry meat is common in China. The pooled prevalence of *Salmonella* in poultry meat samples is the highest in Shaanxi, followed by Henan, Sichuan, and Beijing. There is no known scientific rationale for the observed geographical differences in the prevalence levels of *Salmonella*. From the spatial distribution of poultry animals in China, chickens are most ubiquitous, with high densities across much of eastern China, particularly the Yellow River Basin. Duck densities are highest in southeastern China and the Sichuan Basin [33]. Notably, farm practices can affect the prevalence of *Salmonella* in the final product [34]. Moreover, because the cold chain coverage of agricultural products in China is still much lower (20.0%) than that in developed countries (90.0%) [35], the supply of poultry meat in China's market mainly depends on the centralized distribution of producing regions. This may be the main reason for the high prevalence levels of *Salmonella* in retail poultry meat across several regions of China. In addition, some potential reasons may be related to the differences in the retail environments [36], economic conditions [37], and market supervision [38] between these regions.

In the current study, *Salmonella* prevalence on chilled poultry meat was significantly higher than that on the poultry meat held at both ambient and frozen temperatures. The results showed that preservation methods of poultry meat may be a potential factor indicating cross-contamination at the retail level in China. Chilling is the most commonly utilized processing intervention to control *Salmonella* growth in the poultry meat production chain. Chilled poultry meat is usually kept at a low temperature by maintaining a monitored chill chain through portioning, packaging, transport, and retail storage [39,40]. In China, immersion chilling is employed more frequently. However, once a sample is contaminated with *Salmonella* during the immersion process, the contamination may spread among the whole batch of carcasses, leading to an increase in the prevalence of pathogens on finished products [41]. Consumers generally believe that freshly slaughtered meat has the advantages of higher nutritional value and superior taste [42]. Therefore, Chinese consumers have a preference for ambient meat (60% market share) over chilled meat (25% market share) or frozen meat (15% market share) [42]. Compared with the chilled poultry meat, fresh poultry meat purchased on the market can often be slaughtered, stripped, and eviscerated within 20 min [43] and may be less likely to experience cross-contamination. However, prevalence estimates are not sufficient to assess the probability and severity of illness to which people may be exposed. In a QMRA, implementation of quantitative exposure assessment depends on the concentration data of pathogens in food samples [44,45]. There is a general lack of quantitative data pertaining to *Salmonella* loads in food because most surveillance studies focus on the detection on a presence/absence basis. Therefore, viable cell numbers are often not known because most culture-based standard methods involve enrichment, while molecular methods (aside from RT-qPCR) do not assess viability. According to a few quantitative data on *Salmonella* in poultry meat, the average concentrations of *Salmonella* in the ambient stored samples are higher than that in the chilled samples [46,47]. Therefore, we speculate that although the pooled prevalence of *Salmonella* in freshly slaughtered poultry meat is low, its concentration levels are high, which may pose a greater risk to consumers.

The serotyping results of *Salmonella* isolates obtained from poultry meat in the current study revealed that *S.* Enteritidis, *S.* Indiana, and *S.* Typhimurium were the predominant serovars in poultry meat. The results of previous studies focusing on only one or several cities are consistent with the current nationwide data, indicating that *S.* Enteritidis, *S.* Indiana, and *S.* Typhimurium may be the main serotypes in poultry meat throughout China [6,46,48]. A global epidemiological meta-analysis of *Salmonella* serovars in animal-based foods indicated that *S.* Enteritidis was the most prevalent in Asia, Latin America, Europe, and Africa, while *S.* Typhimurium presented a global distribution [49]. There have been reports of *S.* Indiana in retail raw poultry meat in China since 2009, and this serotype appeared relatively late [50]. In particular, *S.* Enteritidis is most commonly associated with chickens and eggs and has a much smaller relationship with other food animal species [51]. What is more, *Salmonella* serovars Enteritidis, Typhimurium, and Indiana are also reported as the most common serotypes associated with human infections and outbreaks [52,53]. Thus, the high prevalence of these *Salmonella* serotypes in poultry meat indicates a significant risk to consumers. The dominant serotypes of *Salmonella* in food will change over time [54], which reminds us that the monitoring of the emergence and prevalence of different serotypes of *Salmonella* is essential for the better control of salmonellosis.

Nowadays, antimicrobial resistance is becoming an urgent threat and challenge to humans and the public. In the current study, more than half of *Salmonella* isolates were antimicrobial resistant. *Salmonella* isolates recovered from retail poultry meat showed a high frequency of resistance to nalidixic acid, tetracycline, ampicillin, chloramphenicol, trimethoprim/sulfamethoxazole, gentamicin, streptomycin, ciprofloxacin, sulfisoxazole, and ampicillin/sulbactam. Among them, whether in poultry meat or chicken, the highest rates of antimicrobial resistance were observed for nalidixic acid. Nalidixic acid is one of the most widely used antibacterial agents in feed additives and veterinary drugs worldwide. The uncontrolled use of quinolone in China will cause the emergence and increasing prevalence of antimicrobial-resistant *Salmonella*, complicating the treatment of *Salmonella* infections in humans and animals [10,55]. Resistance to tetracycline was the second most frequently observed, with tetracycline and ciprofloxacin also being front-line antibiotics for the treatment of salmonellosis [6]. However, *Salmonella* isolates in the current study were relatively susceptible to ciprofloxacin, a finding that is similar to a previous study in Iran [56]. Unfortunately, in several studies, *Salmonella* strains isolated from food, animals, and humans have been found to show multidrug-resistant (MDR) properties [57,58]. Furthermore, *S.* Indiana isolates with a high detection rate had been found to have high MDR levels [50]. The existence of MDR *Salmonella* isolates poses a major risk to public health, and food safety risk managers should continue to monitor their significant increase in resistance and implement further legislation on the prudent use of antimicrobials.

5. Conclusions

This study systematically reviewed the prevalence and epidemiology of *Salmonella* in retail raw poultry meat in China before 2020. *Salmonella* was more prevalent among chicken samples, especially chilled ones. Among the Chinese provincial regions, Shaanxi, Henan, Sichuan, and Beijing were high-risk areas for *Salmonella* contamination in poultry meat. The recovered *Salmonella* isolates belonged to multiple serovars. *S.* Enteritidis was the most commonly identified serovar in retail raw poultry meat in China. Meanwhile, poultry-derived *Salmonella* isolates showed a high prevalence of antimicrobial resistance, which represents a threat to human health. However, the qualitative sampling data of *Salmonella* accounts for the majority in the published reports on retail raw poultry meat across China. The scarcity of quantitative data on the contamination levels of *Salmonella* on poultry meat indicated the importance of future studies focusing on this topic and making possible quantitative microbial risk assessment studies.

The sampling conditions and laboratory methods of primary studies varied, limiting direct comparability between analyses. High levels of heterogeneity were found for the pooled prevalence of *Salmonella* for most sub-categories. It is concluded that future work should pay more attention to the synchronization of nationwide data and the collection of systematic sub-categories data. The baseline information on the prevalence, concentrations, serotypes, and antimicrobial resistance of *Salmonella* in various meat products from all provincial regions can be used not only to determine the severity of microbial contamination but also to serve as a point of reference for monitoring changes that occur over time. These data will be of great use in the development of effective risk management strategies in the future.

Author Contributions: Conceptualization, T.S. and Y.L.; methodology, T.S. and Y.L.; software, T.S. and Y.L.; validation, X.Q., X.W. and Z.L.; formal analysis, T.S.; investigation, T.S.; resources, Y.L. and Q.D.; data curation, T.S.; writing—original draft preparation, T.S.; writing—review and editing, Z.A., Y.L. and Q.D.; visualization, J.Z.; supervision, Q.D.; project administration, Q.D.; funding acquisition, Q.D. All authors have read and agreed to the published version of the manuscript.

Funding: This work was supported by the Shanghai Agriculture Applied Technology Development Program of China (Grant No. X2021-02-08-00-12-F00782). The partial contribution of the project DiTECT-861915 funded by H2020 was also acknowledged.

Institutional Review Board Statement: Not applicable.

Informed Consent Statement: Not applicable.

Data Availability Statement: All data related to the research are presented in the article.

Conflicts of Interest: The authors declare no conflict of interest.

References

1. Desai, P.T.; Porwollik, S.; Long, F.; Cheng, P.; Wollam, A.; Clifton, S.W.; Weinstock, G.M.; McClelland, M. Evolutionary genomics of *Salmonella* enterica subspecies. *mBio* **2013**, *4*, e00579-12:1–e00579-12:12. [CrossRef]
2. Zhu, J.H.; Wang, Y.R.; Song, X.Y.; Cui, S.H.; Xu, H.B.; Yang, B.W.; Huang, J.L.; Liu, G.H.; Chen, Q.; Zhou, G.; et al. Prevalence and quantification of *Salmonella* contamination in raw chicken carcasses at the retail in China. *Food Cont.* **2014**, *44*, 198–202. [CrossRef]
3. Cai, Y.Q.; Tao, J.; Jiao, Y.; Fei, X.; Zhou, L.; Wang, Y.; Zheng, H.J.; Pan, Z.M.; Jiao, X.N. Phenotypic characteristics and genotypic correlation between *Salmonella* isolates from a slaughterhouse and retail markets in Yangzhou, China. *Int. J. Food Microbiol.* **2016**, *222*, 56–64. [CrossRef] [PubMed]
4. Shao, D.; Shi, Z.; Wei, J.; Ma, Z. A brief review of foodborne zoonoses in China. *Epidemiol. Infect.* **2011**, *139*, 1497–1504. [CrossRef] [PubMed]
5. Gonçalves-Tenório, A.; Silva, B.N.; Rodrigues, V.; Cadavez, V.; Gonzales-Barron, U. Prevalence of pathogens in poultry meat: A meta-analysis of European published surveys. *Foods* **2018**, *7*, 69. [CrossRef] [PubMed]
6. Yang, X.J.; Huang, J.H.; Zhang, Y.X.; Liu, S.R.; Chen, L.; Xiao, C.; Zeng, H.Y.; Wei, X.H.; Gu, Q.H.; Li, Y.; et al. Prevalence, abundance, serovars and antimicrobial resistance of *Salmonella* isolated from retail raw poultry meat in China. *Sci. Total Environ.* **2020**, *713*, 136385:1–136385:9. [CrossRef]
7. Pires, S.M.; Knegt, L.D.; Hald, T. Estimation of the relative contribution of different food and animal sources to human *Salmonella* infections in the European Union. *EFSA Supporting Publ.* **2011**, *8*, 184E:1–184E:80. [CrossRef]
8. USDA/FSIS. Salmonella and Campylobacter Verification Program for Raw Poultry Products. 2021. Available online: https://www.fsis.usda.gov/policy/fsis-directives/10250.1 (accessed on 3 February 2021).
9. Yang, B.W.; Xi, M.L.; Wang, X.; Cui, S.H.; Yue, T.L.; Hao, H.S.; Wang, Y.; Cui, Y.; Alali, W.Q.; Meng, J.H.; et al. Prevalence of *Salmonella* on raw poultry at retail markets in China. *J. Food Prot.* **2011**, *74*, 1724–1728. [CrossRef]
10. Wang, Y.R.; Chen, Q.; Cui, S.H.; Xu, X.; Zhu, J.H.; Luo, H.P.; Wang, D.; Li, F.Q. Enumeration and characterization of *Salmonella* isolates from retail chicken carcasses in Beijing, China. *Foodborne Pathog. Dis.* **2014**, *11*, 126–132. [CrossRef]
11. EFSA Panel on Biological Hazards (BIOHAZ). Analysis of the baseline survey on the prevalence of *Campylobacter* in broiler batches and of *Campylobacter* and *Salmonella* on broiler carcasses, in the EU, 2008-Part B: Analysis of factors associated with *Salmonella* contamination of broiler carcasses. *EFSA J.* **2011**, *8*, 1522. [CrossRef]
12. Glass, G.V. Primary, secondary, and meta-analysis of research. *Educ. Res.* **1976**, *5*, 3–8. [CrossRef]
13. Gonzales-Barron, U.; Butler, F. The use of meta-analytical tools in risk assessment for food safety. *Food Microbiol.* **2011**, *28*, 823–827. [CrossRef] [PubMed]
14. Gonzales-Barron, U.; Cadavez, V.; Sheridan, J.J.; Butler, F. Modelling the effect of chilling on the occurrence of *Salmonella* on pig carcasses at study, abattoir and batch levels by meta-analysis. *Int. J. Food Microbiol.* **2013**, *163*, 101–113. [CrossRef]
15. Cufaoglu, G.; Ambarcioglu, P.; Ayaz, N.D. Meta-analysis of the prevalence of *Listeria* spp. and antibiotic resistant *L. monocytogenes* isolates from foods in Turkey. *LWT-Food Sci. Technol.* **2021**, *144*, 111210:1–111210:6. [CrossRef]
16. FAO. Gateway to Poultry Production and Products. 2021. Available online: http://www.fao.org/poultry-production-products/production/en/ (accessed on 2 August 2021).
17. Young, I.; Rajic, A.; Wilhelm, B.J.; Waddell, L.; Parker, S.; McEwen, S.A. Comparison of the prevalence of bacterial enteropathogens, potentially zoonotic bacteria and bacterial resistance to antimicrobials in organic and conventional poultry, swine and beef production: A systematic review and meta-analysis. *Epidemiol. Infect.* **2009**, *137*, 1217–1232. [CrossRef]
18. Kerr, A.K.; Farrar, A.M.; Waddell, L.A.; Wilkins, W.; Wilhelm, B.J.; Bucher, O.; Wills, R.W.; Bailey, R.H.; Varga, C.; McEwen, S.A.; et al. A systematic review-meta-analysis and meta-regression on the effect of selected competitive exclusion products on *Salmonella* spp. prevalence and concentration in broiler chickens. *Prev. Vet. Med.* **2013**, *111*, 112–125. [CrossRef]

19. Xavier, C.; Gonzales-Barron, U.; Paula, V.; Estevinho, L.; Cadavez, V. Meta-analysis of the incidence of foodborne pathogens in Portuguese meats and their products. *Food Res. Int.* **2014**, *55*, 311–323. [CrossRef]
20. Golden, C.E.; Mishra, A. Prevalence of *Salmonella* and *Campylobacter* spp. in alternative and conventionally produced chicken in the United States: A systematic review and meta-analysis. *J. Food Prot.* **2020**, *83*, 1181–1197. [CrossRef]
21. Higgins, J.P.T.; Thompson, S.G.; Deeks, J.J.; Altman, D.G. Measuring inconsistency in meta-analyses. *Br. Med. J.* **2003**, *327*, 557–560. [CrossRef]
22. Martinez-Rios, V.; Dalgaard, P. Prevalence of *Listeria monocytogenes* in European cheeses: A systematic review and meta-analysis. *Food Cont.* **2018**, *84*, 205–214. [CrossRef]
23. Higgins, J.P.T.; Thompsonm, S.G. Quantifying heterogeneity in a meta-analysis. *Stat. Med.* **2002**, *21*, 1539–1558. [CrossRef]
24. Dong, Q.L.; Barker, G.C.; Gorris, L.G.M.; Tian, M.S.; Song, X.Y.; Malakar, P.K. Status and future of quantitative microbiological risk assessment in China. *Trends Food Sci. Technol.* **2015**, *42*, 70–80. [CrossRef]
25. Pei, X.Y.; Li, N.; Guo, Y.C.; Liu, X.M.; Yan, L.; Li, Y.; Yang, S.R.; Hu, J.; Zhu, J.H.; Yang, D.Y. Microbiological food safety surveillance in China. *Int. J. Environ. Res. Public Health* **2015**, *12*, 10662–10670. [CrossRef]
26. Wu, Y.N.; Chen, J.S. Food safety monitoring and surveillance in China: Past, present and future. *Food Cont.* **2018**, *90*, 429–439. [CrossRef]
27. Liu, Y.T.; Sun, W.X.; Sun, T.M.; Gorris, L.G.M.; Wang, X.; Liu, B.L.; Dong, Q.L. The prevalence of *Listeria monocytogenes* in meat products in China: A systematic literature review and novel meta-analysis approach. *Int. J. Food Microbiol.* **2020**, *312*, 108358:1–108358:9. [CrossRef]
28. Luber, P. Cross-contamination versus undercooking of poultry meat or eggs-which risks need to be managed first? *Int. J. food microbiol.* **2009**, *134*, 21–28. [CrossRef]
29. Moretro, T.; Nguyen-The, C.; Didier, P.; Maitre, I.; Izso, T.; Skuland, S.E.; Cardoso, M.J.; Ferreira, V.B.; Teixeira, P. Consumer practices and prevalence of *Campylobacter*, *Salmonella* and norovirus in kitchens from six European countries. *Int. J. Food Microbiol.* **2021**, *347*, 109172:1–109172:13. [CrossRef]
30. Thomas, K.M.; de Glanville, M.A.; Barker, G.C.; Benschop, J.; Buza, J.J.; Cleaveland, S.; Davis, M.A.; French, N.P.; Mmbaga, B.T.; Prinsen, G.; et al. Prevalence of *Campylobacter* and *Salmonella* in African food animals and meat: A systematic review and meta-analysis. *Int. J. Food Microbiol.* **2020**, *315*, 108382:1–108382:22. [CrossRef]
31. OECD/FAO. *OECD-FAO Agricultural Outlook 2020–2029*; OECD Publishing: Paris, Italy; FAO: Rome, Italy, 2020; pp. 162–173, ISBN 978-92-64-58295-8.
32. Wang, Y. Broiler production and consumption in China. *Agric. Outlook* **2013**, *9*, 68–74. (In Chinese)
33. Prosser, D.J.; Wu, J.X.; Ellis, E.C.; Gale, F.; van Boeckel, T.P.; Wint, W.; Robinson, T.; Xiao, X.M.; Gilbert, M. Modelling the distribution of chickens, ducks, and geese in China. Agric. *Ecosyst. Environ.* **2011**, *141*, 381–389. [CrossRef]
34. Hwang, D.; Rothrock, M.J.; Pang, H.; Kumar, G.D.; Mishra, A. Farm management practices that affect the prevalence of *Salmonella* in pastured poultry farms. *LWT-Food Sci. Technol.* **2020**, *127*, 109423:1–109423:8. [CrossRef]
35. Ren, Q.Q. Research on cold chain distribution of fresh agricultural products under the environment of "Internet +"-taking Hainan as an example. *Agric. Econ.* **2017**, *6*, 124–126. (In Chinese)
36. Lianou, A.; Sofos, J.N. A review of the incidence and transmission of *Listeria monocytogenes* in ready-to-eat products in retail and food service environments. *J. Food Prot.* **2007**, *70*, 2172–2198. [CrossRef]
37. Wilcock, A.; Pun, M.; Khanona, J.; Aung, M. Consumer attitudes, knowledge and behaviour: A review of food safety issues. *Trends Food Sci. Technol.* **2004**, *15*, 56–66. [CrossRef]
38. Liu, P. Tracing and periodizing China's food safety regulation: A study on China's food safety regime change. *Regul. Gov.* **2010**, *4*, 244–260. [CrossRef]
39. Bucher, O.; Farrar, A.M.; Totton, S.C.; Wilkins, W.; Waddell, L.A.; Wilhelm, B.J.; McEwen, S.A.; Fazil, A.; Rajic, A. A systematic review-meta-analysis of chilling interventions and a meta-regression of various processing interventions for *Salmonella* contamination of chicken. *Prev. Vet. Med.* **2012**, *103*, 1–15. [CrossRef] [PubMed]
40. Regalado-Pineda, I.D.; Rodarte-Medina, R.; Resendiz-Nava, C.N.; Saenz-Garcia, C.E.; Castaneda-Serrano, P.; Nava, G.M. Three-Year Longitudinal Study: Prevalence of *Salmonella* Enterica in Chicken Meat is Higher in Supermarkets than Wet Markets from Mexico. *Foods* **2020**, *9*, 264. [CrossRef]
41. Marmion, M.; Ferone, M.T.; Whyte, P.; Scannell, A.G.M. The changing microbiome of poultry meat; from farm to fridge. *Food Microbiol.* **2021**, *99*, 103823:1–103823:16. [CrossRef]
42. Liu, R.; Xing, L.J.; Zhou, G.H.; Zhang, W.G. What is meat in China? *Anim. Front.* **2017**, *7*, 53–56. [CrossRef]
43. Capita, R.; Alvarez-Astorga, M.; Alonso-Calleja, C.; Moreno, B.; Garcia-Fernandez, M.D. Occurrence of salmonellae in retail chicken carcasses and their products in Spain. *Int. J. Food Microbiol.* **2003**, *81*, 169–173. [CrossRef]
44. Danyluk, M.D.; Schaffner, D.W. Quantitative assessment of the microbial risk of leafy greens from farm to consumption: Preliminary framework, data, and risk estimates. *J. Food Prot.* **2011**, *74*, 700–708. [CrossRef] [PubMed]
45. Pouillot, R.; Hoelzer, K.; Chen, Y.H.; Dennis, S. Estimating probability distributions of bacterial concentrations in food based on data generated using the most probable number (MPN) method for use in risk assessment. *Food Cont.* **2013**, *29*, 350–357. [CrossRef]

46. Yang, B.W.; Cui, Y.; Shi, C.; Wang, J.Q.; Xia, X.D.; Xi, M.L.; Wang, X.; Meng, J.H.; Alali, W.Q.; Walls, I.; et al. Counts, serotypes, and antimicrobial resistance of *Salmonella* isolates on retail raw poultry in the People's Republic of China. *J. Food Prot.* **2014**, *77*, 894–902. [CrossRef]
47. Huang, J.L.; Zong, Q.; Zhao, F.; Zhu, J.Q.; Jiao, X.A. Quantitative surveys of *Salmonella* and *Campylobacter* on retail raw chicken in Yangzhou, China. *Food Cont.* **2016**, *59*, 68–73. [CrossRef]
48. Wang, Y.; Yang, B.W.; Cui, Y.; Alali, W.Q.; Xia, X.D.; Xi, M.L.; Wang, X.; Shi, X.M.; Wang, D.P.; Meng, J.H. Subtyping of *Salmonella* isolates on retail raw chicken in China by pulsed-field gel electrophoresis and plasmid analysis. *Food Cont.* **2015**, *47*, 420–426. [CrossRef]
49. Ferrari, R.G.; Rosario, D.K.A.; Cunha-Neto, A.; Mano, S.B.; Figueiredo, E.E.S.; Conte, C.A. Worldwide epidemiology of *Salmonella* serovars in animal-based foods: A meta-analysis. *Appl. Environ. Microbiol.* **2019**, *85*, e00591-19:1–e00591-19:21. [CrossRef]
50. Gong, J.S.; Wang, C.M.; Shi, S.R.; Bao, H.D.; Zhu, C.H.; Kelly, P.; Zhuang, L.L.; Lu, G.W.; Dou, X.H.; Wang, R. Highly drug-resistant *Salmonella* enterica serovar Indiana clinical isolates recovered from broilers and poultry workers with diarrhea in China. *Antimicrob. Agents Chemother.* **2016**, *60*, 1943–1947. [CrossRef]
51. Foley, S.L.; Johnson, T.J.; Ricke, S.C.; Nayak, R.; Danzeisen, J. *Salmonella* pathogenicity and host adaptation in chicken-associated serovars. *Microbiol. Mol. Biol. Rev.* **2013**, *77*, 582–607. [CrossRef]
52. Xia, S.L.; Hendriksen, R.S.; Xie, Z.Q.; Huang, L.L.; Zhang, J.; Guo, W.S.; Xu, B.L.; Ran, L.; Aarestrup, F.W. Molecular Characterization and Antimicrobial Susceptibility of *Salmonella* Isolates from Infections in Humans in Henan Province, China. *J. Clin. Microbiol.* **2009**, *47*, 401–409. [CrossRef]
53. Hendriksen, R.S.; Vieira, A.R.; Karlsmose, S.; Wong, D.M.A.L.F.; Jensen, A.B.; Wegener, H.C.; Aarestrup, F.M. Global monitoring of *Salmonella* serovar distribution from the world health organization global foodborne infections network country data bank: Results of quality assured laboratories from 2001 to 2007. *Foodborne Pathog. Dis.* **2011**, *8*, 887–900. [CrossRef]
54. Foley, S.L.; Lynne, A.M.; Nayak, R. *Salmonella* challenges: Prevalence in swine and poultry and potential pathogenicity of such isolates. *J. Anim. Sci.* **2008**, *86*, E149–E162. [CrossRef] [PubMed]
55. Augulo, F.J.; Johnson, K.R.; Tauxe, R.V.; Cohen, M.L. Origins and Consequences of Antimicrobial-Resistant Nontyphoidal *Salmonella*: Implications for the Use of Fluoroquinolones in Food Animals. *Microb. Drug Resist.* **2000**, *6*, 77–83. [CrossRef]
56. Dallal, M.M.S.; Doyle, M.P.; Rezadehbashi, M.; Dabiri, H.; Sanaei, M.; Modarresi, S.; Bakhtiari, R.; Sharifiy, K.; Taremi, M.; Zali, M.R.; et al. Prevalence and antimicrobial resistance profiles of *Salmonella* serotypes, *Campylobacter* and *Yersinia* spp. isolated from retail chicken and beef, Tehran, Iran. *Food Cont.* **2010**, *21*, 388–392. [CrossRef]
57. Li, R.C.; Lai, J.; Wang, Y.; Liu, S.L.; Li, Y.; Liu, K.Y.; Shen, J.Z.; Wu, C.M. Prevalence and characterization of *Salmonella* species isolated from pigs, ducks and chickens in Sichuan province, China. *Int. J. Food Microbiol.* **2013**, *163*, 14–18. [CrossRef] [PubMed]
58. Bai, L.; Zhao, J.Y.; Gan, X.; Wang, J.; Zhang, X.L.; Cui, S.H.; Xia, S.L.; Hu, Y.J.; Yan, S.F.; Wang, J.H.; et al. Emergence and diversity of *Salmonella* enterica serovar Indiana isolates with concurrent resistance to ciprofloxacin and cefotaxime from patients and food-producing animals in China. *Antimicrob. Agents Chemother.* **2016**, *60*, 3365–3371. [CrossRef]

foods

MDPI

Article

Investigating Transcriptomic Induction of Resistance and/or Virulence in *Listeria monocytogenes* Cells Surviving Sublethal Antimicrobial Exposure

Eleni-Anna Kokkoni [1], Nikolaos Andritsos [1,2], Christina Sakarikou [1], Sofia Michailidou [1,3], Anagnostis Argiriou [1,3] and Efstathios Giaouris [1,*]

1 Department of Food Science and Nutrition, School of the Environment, University of the Aegean, Ierou Lochou 10 & Makrygianni, 81400 Myrina, Greece; fns15037@fns.aegean.gr (E.-A.K.); nickandritsos@aegean.gr (N.A.); chsakarikou@aegean.gr (C.S.); sofia_micha28@certh.gr (S.M.); argyriou@aegean.gr (A.A.)

2 Athens Analysis Laboratories S.A., Microbiology Laboratory, Nafpliou 29, 14452 Metamorfosi, Greece

3 Centre for Research and Technology Hellas (CERTH), Institute of Applied Biosciences, 57001 Thessaloniki, Greece

* Correspondence: stagiaouris@aegean.gr; Tel.: +30-22540-83115

Citation: Kokkoni, E.-A.; Andritsos, N.; Sakarikou, C.; Michailidou, S.; Argiriou, A.; Giaouris, E. Investigating Transcriptomic Induction of Resistance and/or Virulence in *Listeria monocytogenes* Cells Surviving Sublethal Antimicrobial Exposure. *Foods* **2021**, *10*, 2382. https://doi.org/10.3390/foods10102382

Academic Editors: Antonio Afonso Lourenco, Catherine Burgess and Timothy Ells

Received: 29 August 2021
Accepted: 5 October 2021
Published: 8 October 2021

Abstract: The potential transcriptomic induction of resistance and/or virulence in two *L. monocytogenes* strains belonging to the most frequent listeriosis-associated serovars (i.e., 1/2a and 4b), following their sublethal antimicrobial exposure, was studied through qPCR determination of the relative expression of 10 selected related genes (i.e., *groEL*, *hly*, *iap*, *inlA*, *inlB*, *lisK*, *mdrD*, *mdrL*, *prfA*, and *sigB*). To induce sublethal stress, three common antimicrobials (i.e., benzalkonium chloride, thymol, and ampicillin) were individually applied for 2 h at 37 °C against stationary phase cells of each strain, each at a sublethal concentration. In general, the expression of most of the studied genes remained either stable or was significantly downregulated following the antimicrobial exposure, with some strain-specific differences to be yet recorded. Thymol provoked downregulation of most of the studied genes, significantly limiting the expression of 6/10 and 4/10 genes in the strains of ser. 1/2a and ser. 4b, respectively, including those coding for the master regulators of stress response and virulence (SigB and PrfA, respectively), in both strains. At the same time, the two genes coding for the invasion internalin proteins (*InlA* and *InlB*), with crucial role in the onset of *L. monocytogenes* pathogenesis, were both importantly upregulated in ser. 4b strain. The results obtained increase our knowledge of the stress physiology of *L. monocytogenes* under certain sublethal antimicrobial conditions that could be encountered within the food chain and in clinical settings, and may assist in better and more effective mitigation strategies.

Keywords: *Listeria monocytogenes*; benzalkonium chloride; thymol; ampicillin; sublethal antimicrobial exposure; survival; gene expression; stress response; virulence

1. Introduction

Listeria monocytogenes is an important Gram-positive pathogenic bacterium provoking listeriosis, a rare but quite life-threatening foodborne disease mainly for those belonging to vulnerable groups, such as the elderly and immunocompromised [1]. Based on the latest available data for Europe, 2621 confirmed cases of human listeriosis were recorded in 2019, resulting in 1234 hospitalizations and eventually 300 deaths, presenting an enormous case fatality ratio of 17.6% [2]. In the United States, *L. monocytogenes* is estimated to cause approximately 1600 cases of foodborne illness annually, resulting in 1500 hospitalizations (i.e., 94% hospitalization rate) and more than 250 deaths, with a similar death rate to that recorded in Europe, which for the susceptible individuals is further increased to 25–30% [3]. *L. monocytogenes* is known as a highly versatile microorganism that can skillfully adjust its physiology to confront various stress conditions, including high acidity or alkalinity, high osmotic concentration, existence of reactive oxygen species (ROS), increased or low temperature, allowing this way its survival and persistence in a wide range of environmental, food-associated, and clinical conditions [4]. That remarkable adaptation to stress is accomplished through global changes in many cellular constituents, including modifications in gene expression and protein activities [5]. All those changes enable this soil-living bacterium to successfully switch from a harmless saphrophyte to a powerful intracellular pathogen [6].

Many of the survival mechanisms that are exploited by *L. monocytogenes* are known to be controlled by the stress-inducible alternative sigma factor B (σ^B), which is the master regulator of the general stress response (GSR) in that pathogen [7]. It is thus known that σ^B controls in *L. monocytogenes* the expression of more than 300 genes, while it seems that it plays the same important role in several other Gram-positive foodborne pathogens, such as *Bacillus cereus* and *Staphylococcus aureus* [8]. Following consumption of the contaminated food and the survival of *L. monocytogenes* under the hostile conditions of the gastrointestinal (GI) tract [9,10], the subsequent victorious transit of the bacterium through the intestinal epithelial barrier, its intracellular growth, further proliferation, and dissemination relies on multiple virulence factors, the expression of the majority of which is under the control of the master regulator of virulence PrfA [11,12]. Alarmingly, *L. monocytogenes* can not only survive long-term in a stationary phase outside the host without compromising its virulence [13], but at the same time a complex overlap and crosstalk between σ^B and PrfA regulons also exist at transcriptional, post-transcriptional, and protein activity levels. In this way bacterium succeeds achieving a peculiar balance and coordination between stress resistance and virulence skills, depending on the environment [14,15].

Up to now, many studies have selectively examined the expression of key stress response and/or virulence genes in *L. monocytogenes* cells that have either grown in foods such as fruits and vegetables [16,17], cheeses [18], raw and processed meats [19–22], and fish [23], or have been exposed to low temperatures, acid and/or salinity stresses [24–28], or even in a simulated gastrointestinal environment [29,30]. Undoubtedly, all these studies have provided valuable information on the physiology and pathogenesis of that bacterium under some critical food-associated circumstances, revealing in some cases a worrying increase in pathogenicity following such habituation [31]. It is also recognized that after repeated exposure to some antimicrobials, *L. monocytogenes* can adapt to them, and apart from surviving, these bacteria can also display cross-resistance to other antimicrobials and stresses other than those already adapted [32,33].

Indeed, sublethal antimicrobial concentrations could also be accidentally encountered following an ineffective sanitization program (e.g., due to the dilution of disinfectants in the environment, biodegradation, cellular entrapment in places that are not easily reached by the disinfectants, and biofilm formation) [34] or even applied on purpose. This last is the case for several chemical preservatives added to foods in low doses just to delay bacterial growth [35]. Riskily, sublethal concentrations of ampicillin have also been described to exist in the central nervous system (CNS), even following daily intravenous administration at high quantities (12 g), explaining the clinical failure of that antibiotic to treat this severe invasive case of listeriosis infection [36]. The stress-hardening that may appear in *L. monocytogenes* following such sublethal exposures should also contribute to the environmental persistence and spreading of that pathogen throughout the food chain [37]. However, only a few studies have investigated whether and in which way low concentrations of antimicrobial compounds can affect the physiology of that bacterium at the level of gene expression [38–40].

Considering all the above, the objective of the current study was to quantify the relative expression of some key stress response and/or virulence associated genes in two *L. monocytogenes* strains belonging to the most frequent listeriosis-associated serovars (i.e., 1/2a and 4b) [41], which survived after exposure to three common antimicrobials, belonging to different classes and which among others are used within the food industry and/or in clinical settings. These consisted of a general-purpose synthetic biocide (i.e., benzalkonium chloride; BAC), a natural terpenoid of plant origin (i.e., thymol; THY), and a broad-spectrum beta-lactam antibiotic (i.e., ampicillin; AMP). More specifically, BAC belongs to the family of quaternary ammonium compounds (QACs), which are membrane-active agents and among the most used disinfectants in industrial, healthcare, home, and cosmetics settings [42]. THY is found in rich quantities in the essential oils of thyme and oregano, as well as of several other related herbs, most native in the Mediterranean region, and this is well-known for its many biological and therapeutic properties, including broad-spectrum antimicrobial action [43]. Lastly, AMP is widely used to treat many bacterial infections, caused by either Gram-positive or -negative bacteria, inhibiting bacterial cell wall (peptidoglycan) biosynthesis [44]. In addition, this is currently included among the drugs of choice for the treatment of invasive listeriosis [45].

2. Materials and Methods

2.1. Bacterial Strains and Growth Conditions

The two tested *L. monocytogenes* strains were the foodborne AAL20066 (ser. 1/2a) and AAL20074 (ser. 4b) isolates deposited in the microbial culture collection of the Microbiology Laboratory in Athens Analysis Laboratories S.A. (AAL). Both strains were previously recovered from mixed fresh salads and were kept frozen long-term (at −80 °C) in Trypticase Soya Broth (TSB; Condalab, Torrejón de Ardoz, Madrid, Spain) containing 15% (*v/v*) glycerol. When needed for the experiments, each strain was streaked on to the surface of Tryptone Soya Agar (TSA; Oxoid, Thermo Fisher Specialty Diagnostics Ltd., Hampshire, UK) and incubated at 37 °C for 24 h (preculture). Working cultures were prepared by inoculating a colony from each preculture into 10 mL of fresh TSB and further incubating at 37 °C for 18 h. Bacteria from each of those final working cultures were collected by centrifugation (2000× *g* for 10 min at 4 °C), washed once with quarter-strength Ringer's solution (Lab M, Heywood, Lancashire, UK), and finally suspended in 5 mL of the same solution (ca. 10^9 CFU/mL). The purity of each cellular working suspension was verified through streaking on TSA plates.

2.2. Chemical Antimicrobials (BAC, THY and AMP)

BAC was bought from Acros Organics (Thermo Fisher Scientific, Geel, Belgium) (liquid, alkyl distribution from C_8H_{17} to $C_{16}H_{33}$), THY was purchased from Penta Chemicals (Radiová, Prague, Czech Republic) (powder min. 99.0%, molar mass: 150.22 g/mol), while AMP was acquired from Cayman Chemicals (Ann Arbor, MI, USA) (crystalline solid ≥ 95% purity, molar mass: 371.4 g/mol). The stock solution of BAC (1% *v/v*) was prepared in sterile distilled water (dH_2O), while those of THY and AMP (10% and 1% *w/v*, respectively) were prepared in absolute ethanol and were both subsequently filtrated by passing through disposable syringe filters (0.45 μm diameter; Whatman, Buckinghamshire, UK). All stock solutions were aliquoted and stored at −20 °C until needed for the experiments.

2.3. Determination of Minimum Inhibitory Concentration (MIC)

The MIC of AMP against the planktonic growth of each of the two bacterial strains was determined through the classical broth microdilution method, using sterile 96-well polystyrene flat-bottomed microtiter plates, as previously described [46]. In addition, the MICs of both BAC and THY had also been determined in that previous study. In sum, bacterial cultures of each strain (ca. 10^5 CFU/mL) in TSB, containing 10 different increasing concentrations of the antibiotic (ranging from 0.063 to 5 μg/mL), were statically incubated at 37 °C for 24 h and were then checked for turbidity (as a visible indication of bacterial growth). Wells containing inoculated medium with the bacteria without the antibiotic and wells containing only sterile medium were used as positive and negative growth controls, respectively. For each concentration, two replicate wells were used, while the experiment was thrice repeated starting from independent bacterial cultures.

2.4. Sublethal Antimicrobial Exposure and RNA Extraction

For each tested strain and antimicrobial, the freshly saline cellular suspension (prepared as described in Section 2.1) was aliquoted in two Eppendorf® tubes (2 mL in each one) and centrifuged (5000× *g* for 10 min at 4 °C). One of the two bacterial pellets was then suspended in 1 mL of the appropriate antimicrobial solution (i.e., 4.0 μg/mL BAC, 312.5 μg/mL THY, or 0.5 μg/mL AMP), while the second pellet was suspended in 1 mL of dH_2O to be used as the untreated control sample. In the case of THY and AMP testing, the dH_2O of the control sample also contained absolute ethanol at the concentration that existed in each working solution prepared for those two antimicrobials (i.e., 2812.5 and 50 μg/mL, for THY and AMP, respectively). Both samples (i.e., with the antimicrobial and its respective control) were incubated in a heating dry block for 2 h at 37 °C and were then immediately centrifuged (5000× *g* for 10 min at 4 °C). Supernatants were discarded and each pellet was washed with dH_2O through an additional centrifugation step (5000× *g* for 10 min at 4 °C) to remove any antimicrobial residues. It should be noted that this washing procedure was sufficient for the efficient neutralization of each disinfectant, as this had been confirmed in preliminary experiments (through agar plating). Washed pellets were then placed on ice and directly used for RNA extraction using the RiboPure™ -Bacteria Kit (Part Number: AM1925, Ambion, Life Technologies, Carlsbad, CA, USA). Eluted RNAs were treated with DNase I to remove any trace amounts of genomic DNA (gDNA), following the protocol guidelines, before measuring their absorbances at 260 and 280 nm to determine their concentrations and purities. One microgram of each extracted RNA sample was also run on electrophoresis (1.5% *w/v* TBE agarose gel; 100 V for 30 min) to verify its integrity, using the ssRNA Ladder (N0362S, 500–9000 bp, New England BioLabs Inc., Ipswich, MA, USA) as the molecular weight marker. The rest of each RNA sample was stored at −80 °C until its use as substrate for the subsequent reverse transcription (cDNA synthesis) reactions. Each antimicrobial exposure experiment was thrice repeated, starting each time from an independent bacterial culture and always using freshly prepared working antimicrobial solutions.

2.5. Reverse Transcription (cDNA Synthesis)

A cDNA synthesis was conducted starting from 500 ng of each RNA sample using the PrimeScript™ RT reagent Kit (Cat. #RR037A, Takara Bio Inc., Shiga, Japan). Both oligo dT and random hexamer primers were included in the reaction mixture (10 μL) at final concentrations of 25 and 50 pmol, respectively, according to the manufacturer's instructions. For each RNA sample, a no-reverse transcription control (NRTC), which did not contain the reverse transcriptase enzyme (PrimeScript RT Enzyme Mix I), was also prepared to evaluate (i.e., in the later qPCR reactions) the presence of any residual gDNA. All RT reactions were performed in a PeqStar 96 HPL Gradient Thermocycler (Peqlab, VWR International GmbH, Darmstadt, Germany) by initially incubating at 37 °C for 15 min (for the RT reaction) and subsequently at 85 °C for 5 s (to inactivate reverse transcriptase). All resulting cDNAs were stored at −20 °C until their use as substrates in the subsequent qPCR analyses.

2.6. qPCR for Quantitation of mRNA Transcripts

Each cDNA template was used to quantify the expression of each gene of interest (including the ten targets and two additional reference genes; Table 1), for each bacterial strain and antimicrobial treatment and in relation to the respective untreated control, in qPCR reactions prepared using the PowerUp™ SYBR™ Green Universal 2X Master Mix (Cat. No. A25780, Applied Biosystems, Thermo Fisher Scientific, Waltham, MA, USA). Each reaction mixture contained 10 μL of PowerUp™ SYBR™ Green 2X Master Mix, 400 nM of each primer, 10 ng of cDNA template and PCR-grade water to a total volume of 20 μL. A no-template control (NTC) was always included in each assay to exclude any external DNA contamination. Real-time PCR was conducted on a QuantStudio™ 5 Real-Time PCR Instrument (Applied Biosystems). The PCR program consisted of two initial 2-min incubations, first at 50 °C for the uracil-DNA glycosylase (UDG) activation and the second at 95 °C for the activation of the (hot-start) Dual-Lock™ DNA polymerase, followed by 40 cycles of denaturation at 95 °C for 1 s and primer annealing/extension at 60 °C for 30 s (fast cycling mode). At the end of the amplification protocol, a melting curve analysis was also performed to confirm the specificity of each qPCR reaction (excluding any nonspecific amplification). This consisted of an initial step at 95 °C for 15 s (1.6 °C /s), a second step at 60 °C for 1 min (1.6 °C /s), and a final step at 95 °C for 15 s (0.15 °C /s). The threshold cycle (C_T) for each reaction was calculated using the QuantStudio™ Design and Analysis Software v1.5.1 (Applied Biosystems). For each strain and antimicrobial treatment, the relative quantification of the expression of each target gene was finally performed using the classical comparative $\Delta\Delta C_T$ method [47] in relation to the untreated control samples (i.e., with no antimicrobial exposure). Two reference (internal control) genes (i.e., *tuf*, *gap*) were always included in each assay, and were both used in parallel for the normalization of the qPCR data for any differences in the amount of total cDNA added to each reaction [48]. Both had been found to present the most consistent expression at both strains (exposed at the different antimicrobial treatments) and had been selected in preliminary experiments from an initial pool of four potential candidates for such genes (also including *16S rRNA* and *rpoB*). The efficiency (%) of each qPCR reaction (i.e., of each primer pair) had been also initially determined [49] (Table 1). Each qPCR reaction was performed in triplicate, while the data derived from a total of 1296 qPCR reactions were analyzed. These were the result of 36 different RNA/cDNA samples (i.e., 2 bacterial strains × 3 antimicrobials × 2 treatments (with and without antimicrobial exposure) × 3 biological repetitions) × 12 genes/sample × 3 technical replicates/gene.

2.7. Statistical Analyses for Differential Gene Expression

For each tested bacterial strain and antimicrobial, unpaired two-tailed Student's *t*-tests were applied to the data to check for any significant difference in the expression of each target gene (expressed as $\log_2$(fold difference)) between the two treatments (i.e., with and without antimicrobial exposure). The same tests were also applied to check for any significant difference in the expression of each target gene between the two bacterial strains. All these tests were performed using the relevant function of Excel® module of the Microsoft® Office 365 suite (Redmond, WA, USA). Statistically significant expression differences were recorded at a P level of < 0.05. However, biologically significant ones were considered only those that in parallel presented a $|\log_2(\text{fold difference})| \geq 1$ between the two treatments [52].

Table 1. Sequences of the primers used for the in vitro quantitation of the mRNA transcripts of the ten target genes (*groEL, hly, iap, inlA, inlB, lisK, mdrD, mdrL, prfA, sigB*) and the two reference genes (*tuf, gap*). The amplicon size (bp) and amplification efficiency (%) of each primer pair, together with the regression coefficients (R^2) of the linear standard curves constructed for the determination of each PCR efficiency, are also shown.

s/n	Gene	Locus Tag †	Product Name	Gene Size (bp)	Primer Sequence ‡ (5′ → 3′)	Amplicon Size (bp)	Amplification Efficiency (%)	R^2
1	*groEL*	lmo2068	molecular chaperone GroEL	1629	F: AAGTCCAGCGTTATGTGCGA R: CGTAGCTGGTGGTGGTACTG	145	104.72	1.00
2	*hly*	lmo0202	listeriolysin O precursor	1590	F: TGCCAGGTAACGCGAGAAAT R: TGGTGCCCCAGATGGAGATA	135	93.96	1.00
3	*iap*	lmo0582	invasion associated secreted endopeptidase	1449	F: GCCAGAGCCGTGGATGTTAT R: TTCTGGCGCACAATACGCTA	178	113.63	0.99
4	*inlA*	lmo0433	internalin A	2403	F: AAATCCTGTGGCACCACCAA R: TTGTGCTGGCTGAATTCCCA	137	95.78	1.00
5	*inlB*	lmo0434	internalin B	1893	F: CGCGAAGCCAAAACACCAAT R: TTGGCGCTGACATAACGAGT	146	106.12	1.00
6	*lisK*	lmo1378	two-component sensor histidine kinase	1452	F: GATGTGCGTGATTACGGGGA R: CCGAGGCCATTACCACCTTT	113	105.96	1.00
7	*mdrD*	lmo0872	antibiotic resistance protein	1167	F: ATCGCCGATGTTTAGCGGAA R: CATTCGCAAAATGCCCACCA	113	108.26	0.98
8	*mdrL*	lmo2377	multidrug transporter	1212	F: CCGTTGCTTGCGCTTTATGT R: TCCCCATTTTCGCGTCATCA	117	94.03	0.97
9	*prfA*	lmo0200	listeriolysin positive regulatory protein	714	F: CTGAGCTATGTGCGATGCCA R: AGCTTGGCTCTATTTGCGGT	138	101.96	0.98
10	σ^B (*sigB*)	lmo0895	RNA polymerase sigma factor SigB	780	F: CTTCAAAGCTCGCCGCAAAT R: CCATCATCCGTACCACCAACA	182	105.40	1.00
11	*tuf*	lmo2653	elongation factor Tu	1188	F: CCAATGTTGTCGCCAGCTTC R: GCAACTGGACGTGTTGAACG	149	101.00	1.00
12	*gap*	lmo2459	glyceraldehyde-3-phosphate dehydrogenase	1011	F: AGCTGCTTCCATAGCTGCATT R: TTAGACGGAGCTGCTCAACG	114	95.88	0.96

† Corresponds to the NCBI Reference Sequence: NC_003210.1 of the complete genome of *L. monocytogenes* strain EGD-e [50]. ‡ Target-specific conserved primers were designed using Primer-BLAST software developed at National Center for Biotechnology Information (NCBI, Bethesda MD, 20894 USA; http://www.ncbi.nlm.nih.gov/tools/primer-blast; (Accessed on 29 August 2021) [51]).

3. Results and Discussion

All the antimicrobials applied here were previously verified for their strong killing efficiency against *L. monocytogenes* cells, as well as many other detrimental microorganisms [36,53,54]. Nevertheless, foodborne *L. monocytogenes* isolates displaying resistance to BAC [55,56] and enough times in parallel to other drugs, such as antibiotics and some other toxic compounds, have also been described [57,58]. Alarmingly, *L. monocytogenes* strains that are resistant to AMP have also been recovered from foods, mainly animal products probably due to the intensive use of antibiotics in animal farms [59–62]. Regarding THY and to the best of our knowledge, there are not any data available showing an increase in resistance or tolerance of *L. monocytogenes* cells following their sublethal habituation. Nevertheless, there are still some previous studies showing adaptive responses and increased survival of other bacteria following exposure to sublethal concentrations of even that natural monoterpenoid phenol [63,64]. The MIC of AMP against both bacterial strains was found equal to 0.125 μg/mL. This is a value similar to those described in the literature for that antibiotic and bacterial species [53,65]. Similarly, the MICs of BAC and THY previously determined equal to 2 and 78.1 μg/mL, respectively, against both strains [39], were similar to the ones previously reported for those compounds against that pathogenic species [56,66]. Surely, all those specific MIC values do not denote any resistance of the two strains employed here, thus confirming their initial sensitivity against all three antimicrobials. For the subsequent sublethal treatments, stationary phase cells of each serovar were exposed against a selected super-MIC (still sublethal) value of each antimicrobial. The specific concentrations tested had thus been previously shown to not cause any significant reduction in the numbers of viable and culturable cells of each strain (data not presented). Thus, all the subsequent RNA extractions were done starting from equal bacterial numbers (ca. 10^9 CFU), to minimize the variability between the different treatments. Antimicrobial exposure was done at 37 °C, which is in the range of optimum temperatures for the planktonic growth of *L. monocytogenes* cells (i.e., 30–37 °C) just for not causing any additional thermal stress to the bacteria, while those latter had been left to enter a non-growing stationary phase before the antimicrobial challenges to imitate the bacterial physiological state in which increased resistance against various stresses is normally established [67].

The $\log_2$(fold differences) in genes' expressions for both strains and all three antimicrobials are shown in Figure 1. In general, the expression of most of the studied genes remained either stable or was significantly downregulated following the antimicrobial exposure, with some strain-specific differences to be yet recorded. THY was the compound that provoked downregulation of most of the studied genes, significantly limiting the expression of 6/10 genes in one strain (ser. 1/2a), and 4/10 genes in the other strain (ser. 4b), including those coding for the master regulators of stress response and virulence (SigB and PrfA, respectively), in both strains (Figure 1 and Table S1). In agreement, subinhibitory THY concentration (0.50 mM) was previously described to reduce the expression of some key virulence genes in three *L. monocytogenes* strains and in parallel decrease their in vitro attachment to and invasion of human cells, motility, hemolysin production, and lecithinase activities [68]. Nevertheless, at the same time in the current study, the gene coding for the invasion surface protein internalin A (InlA), with crucial role in the onset of *L. monocytogenes* pathogenesis [69], was importantly (more than threefold) up regulated in ser. 4b strain (Figure 1B). Noteworthy, the same gene was also previously shown to be significantly overexpressed in the cells of another clinical isolate of *L. monocytogenes* belonging to the same serovar (Scott A strain) that survived exposure (for 1 h at 37 °C) to sublethal concentrations (40–100 μg/mL) of the essential oil of thyme [70].

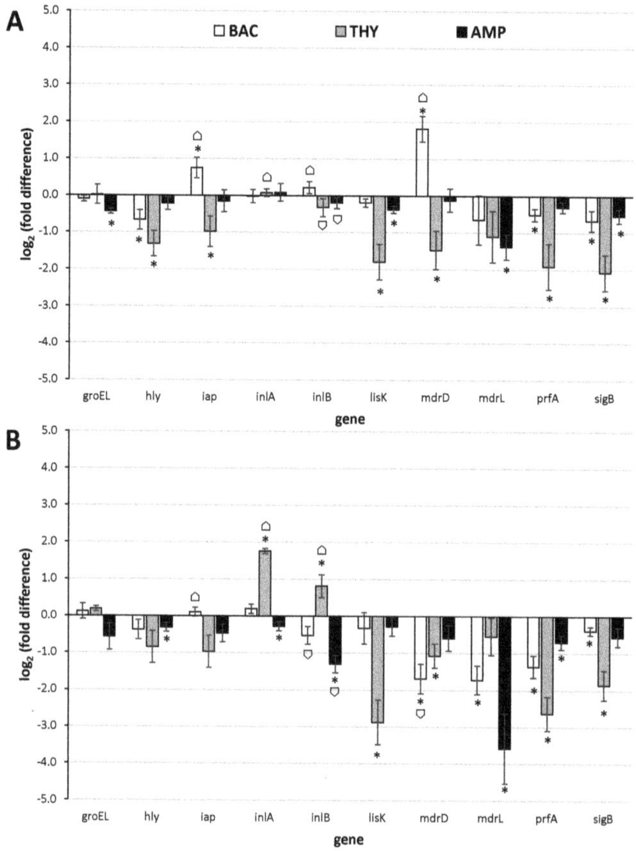

Figure 1. Relative quantification ($\log_2$(fold differences)) of the expressions of the 10 target genes (*groEL, hly, iap, inlA, inlB, lisK, mdrD, mdrL, prfA, sigB*) at the 2 *L. monocytogenes* strains (**A**) AAL20066 (ser. 1/2a) and (**B**) AAL20074 (ser. 4b), following their sublethal exposure (for 2 h at 37 °C) to BAC (4.0 μg/mL; □), THY (312.5 μg/mL; ▨) or AMP (0.5 μg/mL; ■), in comparison to the untreated controls (no antimicrobial exposure). Each bar represents the mean ± standard errors (n = 9). The statistically significant differences in genes' expressions relative to the controls appear as asterisks (*) above the bars, while ⌂ denote the statistically significant differences in genes' expressions between the two strains ($p < 0.05$).

Another gene with similar significant upregulation was that coding for the multidrug resistance transporter MdrD in ser. 1/2a strain following its exposure to BAC (Figure 1A). The expression of that gene was previously found to be significantly upregulated in *L. monocytogenes* cells during their intracellular growth in macrophages, over its level during growth in laboratory medium, thus suggesting an active role during infection [71]. In another study, the same gene was also found to be upregulated under acidic conditions (pH 5.0 vs. pH 7.3) [72]. Two other genes with statistically significant upregulation were *iap* in ser. 1/2a strain following exposure to BAC (Figure 1A), and *inlB* in ser. 4b stain following exposure to THY (Figure 1B). However, it should be noted that both recorded upregulations were slightly below the margin usually set for biologically significant differences (i.e., doubling or halving of mRNA transcripts in treated samples compared to the untreated ones; equal to a value of $|\log_2(\text{fold difference})| = 1$).

The *iap* gene of *L. monocytogenes* encodes the invasion-associated surface protein p60, a highly antigenic protein necessary for septum separation and known to affect adherence of *L. monocytogenes* cells to, and their uptake by, mammalian cells [73]. Interestingly, this gene has been found to be activated during growth of the pathogen in a dry-cured ham model system under osmotic stress and incubation at 15 °C [24], while in another study, it was worryingly confirmed that this gene was still expressed after 6 months of incubation of the pathogen in artisanal cheese at −20 °C [74]. Long-term adaptation of *L. monocytogenes* EGD-e strain (ser. 1/2a) to either acidic (pH 5.5) or NaCl (4.5% *w/v*) stress has also been found to induce transcription of *iap* [27]. The *inlB* is the second gene of the two-genes internalin operon (the other being *inlA*), which has been known for several years to play an important role for the entry of *L. monocytogenes* into epithelial cells [75]. The simultaneous upregulation of both *inlA* and *inlB* genes that was observed here following exposure of ser. 4b strain to THY is surely a case for concern. On the other hand, the expression of both those genes remained rather constant at ser. 1/2a strain, without being changed following the antimicrobial exposures (independently of the applied antimicrobial) (Figure 1A). The expression of both *iap* and internalin genes in a strain-dependent manner was previously shown, by microarray, during growth of three *L. monocytogenes* strains, belonging to different serovars (1/2a, 4b, and 3c), in meat juices [22].

The expression of *groEL*, *hly*, *lisK*, and *mdrL* genes was here significantly downregulated following the exposure of *L. monocytogenes* bacteria to at least one of the three antimicrobials (i.e., BAC, THY, and AMP) (Figure 1 and Table S1). The *groEL* encodes a molecular chaperone that is among the most highly conserved proteins in nature, and this is known to be involved in the cellular general stress response. In bacteria, GroEL has been found to be synthesized at high levels following their exposure to abusive environmental conditions [76]. However, in this work, the expression of this gene did not significantly change following the antimicrobial exposure, except in strain AAL20066 after its exposure to AMP (although still occurring in levels much lower those typically set for biologically significant differences). The *hly* is a key virulence determinant in *L. monocytogenes* encoding the hemolysin Listeriolysin O (LLO), which has been extensively characterized for its crucial role in pathogenesis of listeriosis by promoting cell-to-cell spread and thus efficient bacterial dissemination during infection [77]. The *lisK* encodes the histidine kinase of the two-component signal transduction system LisRK that is involved in the growth of *L. monocytogenes* at low temperatures, as well as in the response of this bacterium to a number of antimicrobial agents, such as ethanol, hydrogen peroxide, nisin, and cephalosporins [78,79]. Nevertheless, none of the three antimicrobials tested in the present study was able to induce expression of this gene. Lastly, *mdrL* encodes a major facilitator superfamily (MFS) efflux pump that is involved in tolerance of *L. monocytogenes* to BAC [80]. However, in this work, this gene was surprisingly found to be significantly downregulated following the exposure of AAL20074 strain to BAC, as well as following the exposure of both strains to AMP.

In addition to the upregulation of *iap* and *mdrD* (following exposure of ser. 1/2a strain to BAC) and *inlA* and *inlB* (following exposure of ser. 4b strain to THY), no other gene was found to be significantly induced here following the antimicrobial exposure (Figure 1 and Table S1). In addition, it is worth noting that the two genes *sigB* and *prfA* coding for the master regulators of stress response and virulence, respectively [14], were both significantly downregulated in almost all cases (except *prfA* in strain AAL20066 and *sigB* in strain AAL20074 whose expression, although decreased, did not significantly change following exposure to AMP). This is rather reassuring since it implies that, in general, *L. monocytogenes* are not likely to induce either resistance or virulence following the exposure to one of the three antimicrobials tested here. Nevertheless, there are some other previously published studies that showed an alarming increase in the expression of some key stress response and/or virulence-associated genes following sublethal exposure of cells of that pathogenic species to some common antimicrobials [38–40].

In one such study, Kastbjerg et al. (2010) developed an agar-based assay to examine the effect of 11 disinfectants used routinely in the food industry (left to act from 15 to 180 min), representing 4 different groups of active components, on the expression of promoters of

4 virulence genes (*prfA, plcA, inlA,* and *hly*) in *L. monocytogenes* strain EGD [38]. Northern blot analysis was also performed to validate transcript levels. Disinfectants with the same active ingredients were found to have a similar effect on gene expression. Thus, peroxides and chlorine compounds reduced the expression of virulence genes, whereas QACs (five products tested) induced the expression of these genes. In another similar study, Rodrigues et al. (2011) used qPCR methodology to study the expression of *prfA* and another stress-response gene (*clpC*) in surviving *L. monocytogenes* biofilm cells following their 15-min exposure to 4 disinfectants (sodium hypochlorite at 800 μg/mL, a commercial BAC-containing product again at 800 μg/mL, hydrogen peroxide at 9%, and triclosan at 0.4%) [39]. The results showed that the expression of both genes was significantly increased in the surviving cells compared to the controls. Using the same methodology, Tamburro et al. (2015) evaluated the relative expression of *mdrL, ladR, lde, sigB* and *bcrABC* genes in 20 *L. monocytogenes* strains of either food or clinical origin, following sublethal 5-min exposure to 10 μg/mL of BAC, finding a significant association between increased BAC resistance and both *mdrL* and *sigB* overexpression [40].

Surely, the way the genes are transcribed in each bacterium is a rather complex procedure, influenced by its genetic make-up, the (changing) environments (both past and present), and their mazy interactions [81]. It is also known that genes' expression may significantly vary between identically treated but different strains of the same bacterial species, or even stochastically among the cells within clonal populations [82]. Interestingly, that strain-dependent expression of stress response and virulence genes has been previously shown in *L. monocytogenes* [22,83] and was reconfirmed here for 4 out of the 10 tested genes (*iap, inlA, inlB,* and *mdrD*), also depending on the tested antimicrobial (Figure 1 and Table S1).

4. Conclusions

In general, the exposure of two foodborne *L. monocytogenes* strains, belonging to different listeriosis related serovars (i.e., 1/2a and 4b), to a selected sublethal concentration of each one of three common antimicrobials (i.e., BAC, THY or AMP) did not result in the transcriptomic induction of most of the key stress response and virulence-associated genes that were studied here. Nevertheless, the significant overexpression of the two genes of internalin operon (*inlA, inlB*) in one of the two strains (ser. 4b) following exposure to THY may be a cause for concern and should be further explored (e.g., in future in situ virulence studies employing cell cultures). In addition, the in-parallel implementation of high-throughput technologies able to globally explore and unravel the transcriptome of *L. monocytogenes* cells surviving biocidal actions of such and/or other common antimicrobials (e.g., through RNA sequencing; [84]) will increase our limited—for the time being—knowledge on the stress physiology of this important foodborne pathogenic bacterium, with hope to improve its control within the food chain and in clinical settings, ultimately protecting public health.

Supplementary Materials: The following are available online at https://www.mdpi.com/article/10.3390/foods10102382/s1, Table S1: Statistically significant changes (↑: up regulations; ↓: down regulations; $P < 0.05$) in the expressions of the 10 target genes (*groEL, hly, iap, inlA, inlB, lisK, mdrD, mdrL, prfA, sigB*) at the 2 *L. monocytogenes* strains AAL20066 (ser. 1/2a) and AAL20074 (ser. 4b), following their sublethal exposure (for 2 h at 37 °C) to BAC (4.0 μg/mL), THY (312.5 μg/mL) or AMP (0.5 μg/mL), in comparison to the untreated controls (no antimicrobial exposure). ∞: no significant change.

Author Contributions: Conceptualization, E.G.; methodology, E.G.; validation, E.-A.K.; formal analysis, E.G.; investigation, E.-A.K., N.A., C.S., and S.M.; resources, E.G.; data curation, E.-A.K. and E.G.; writing—original draft preparation, E.G.; writing—review and editing, N.A., S.M., A.A., and E.G.; visualization, E.G.; supervision, E.G.; project administration, E.G.; funding acquisition, E.G. All authors have read and agreed to the published version of the manuscript.

Funding: This research was funded in the context of the project "Transcriptional determination of combinatory induction of resistance (stress adaptation) and virulence in *Listeria monocytogenes* cells as a consequence of their exposure to antimicrobial compounds" (MIS 5049514) under the call for proposals "Supporting researchers with an emphasis on new researchers" (EDULLL 103). The project is co-financed by Greece and the European Union (European Social Fund- ESF) by the Operational Programme Human Resources Development, Education and Lifelong Learning 2014-2020.

Data Availability Statement: The data presented in this study are available on request from the corresponding author.

Conflicts of Interest: The authors declare no conflict of interest. The funders had no role in the design of the study; in the collection, analyses, or interpretation of data; in the writing of the manuscript, or in the decision to publish the results.

References

1. Jordan, K.; McAuliffe, O. *Listeria monocytogenes* in foods. *Adv. Food Nutr. Res.* **2018**, *86*, 181–213. [CrossRef]
2. EFSA; ECDC (European Food Safety Authority and European Centre for Disease Prevention and Control). The European Union One Health 2019 Zoonoses Report. *EFSA J.* **2021**, *19*, 6406. [CrossRef]
3. Scallan, E.; Hoekstra, R.M.; Angulo, F.J.; Tauxe, R.V.; Widdowson, M.A.; Roy, S.L.; Jones, J.L.; Griffin, P.M. Foodborne illness acquired in the United States—Major pathogens. *Emerg. Infect. Dis.* **2011**, *17*, 7–15. [CrossRef]
4. NicAogáin, K.; O'Byrne, C.P. The role of stress and stress adaptations in determining the fate of the bacterial pathogen *Listeria monocytogenes* in the food chain. *Front. Microbiol.* **2016**, *7*, 1865. [CrossRef]
5. Soni, K.A.; Nannapaneni, R.; Tasara, T. The contribution of transcriptomic and proteomic analysis in elucidating stress adaptation responses of *Listeria monocytogenes*. *Foodborne Pathog. Dis.* **2011**, *8*, 843–852. [CrossRef] [PubMed]
6. Tiensuu, T.; Guerreiro, D.N.; Oliveira, A.H.; O'Byrne, C.; Johansson, J. Flick of a switch: Regulatory mechanisms allowing *Listeria monocytogenes* to transition from a saprophyte to a killer. *Microbiology* **2019**, *165*, 819–833. [CrossRef] [PubMed]
7. Dorey, A.; Marinho, C.; Piveteau, P.; O'Byrne, C. Role and regulation of the stress activated sigma factor sigma B (σ^B) in the saprophytic and host-associated life stages of *Listeria monocytogenes*. *Adv. Appl. Microbiol.* **2019**, *106*, 1–48. [CrossRef] [PubMed]
8. Guerreiro, D.N.; Arcari, T.; O'Byrne, C.P. The σ^B-mediated general stress response of *Listeria monocytogenes*: Life and death decision making in a pathogen. *Front. Microbiol.* **2020**, *11*, 1505. [CrossRef] [PubMed]
9. Davis, M.L.; Ricke, S.C.; Donaldson, J.R. Establishment of *Listeria monocytogenes* in the gastrointestinal tract. *Microorganisms* **2019**, *7*, 75. [CrossRef]
10. Gahan, C.G.; Hill, C. *Listeria monocytogenes*: Survival and adaptation in the gastrointestinal tract. *Front. Cell. Infect. Microbiol.* **2014**, *4*, 9. [CrossRef] [PubMed]
11. Radoshevich, L.; Cossart, P. *Listeria monocytogenes*: Towards a complete picture of its physiology and pathogenesis. *Nat. Rev. Microbiol.* **2018**, *16*, 32–46. [CrossRef] [PubMed]
12. Rolhion, N.; Cossart, P. How the study of *Listeria monocytogenes* has led to new concepts in biology. *Future Microbiol.* **2017**, *12*, 621–638. [CrossRef]
13. Bruno, J.C., Jr.; Freitag, N.E. *Listeria monocytogenes* adapts to long-term stationary phase survival without compromising bacterial virulence. *FEMS Microbiol. Lett.* **2011**, *323*, 171–179. [CrossRef]
14. Gaballa, A.; Guariglia-Oropeza, V.; Wiedmann, M.; Boor, K.J. Cross talk between SigB and PrfA in *Listeria monocytogenes* facilitates transitions between extra- and intracellular environments. *Microbiol. Mol. Biol. Rev.* **2019**, *83*, e00034-19. [CrossRef]
15. Matereke, L.T.; Okoh, A.I. *Listeria monocytogenes* virulence, antimicrobial resistance and environmental persistence: A review. *Pathogens* **2020**, *9*, 528. [CrossRef]
16. Hadjilouka, A.; Molfeta, C.; Panagiotopoulou, O.; Paramithiotis, S.; Mataragas, M.; Drosinos, E.H. Expression of *Listeria monocytogenes* key virulence genes during growth in liquid medium, on rocket and melon at 4, 10 and 30 °C. *Food Microbiol.* **2016**, *55*, 7–15. [CrossRef]
17. Rieu, A.; Guzzo, J.; Piveteau, P. Sensitivity to acetic acid, ability to colonize abiotic surfaces and virulence potential of *Listeria monocytogenes* EGD-e after incubation on parsley leaves. *J. Appl. Microbiol.* **2010**, *108*, 560–570. [CrossRef] [PubMed]
18. Larsen, N.; Jespersen, L. Expression of virulence-related genes in *Listeria monocytogenes* grown on danish hard cheese as affected by NaCl Content. *Foodborne Pathog. Dis.* **2015**, *12*, 536–544. [CrossRef] [PubMed]
19. Mataragas, M.; Rovetto, F.; Bellio, A.; Alessandria, V.; Rantsiou, K.; Decastelli, L.; Cocolin, L. Differential gene expression profiling of *Listeria monocytogenes* in Cacciatore and Felino salami to reveal potential stress resistance biomarkers. *Food Microbiol.* **2015**, *46*, 408–417. [CrossRef]
20. Olesen, I.; Thorsen, L.; Jespersen, L. Relative transcription of *Listeria monocytogenes* virulence genes in liver pâtés with varying NaCl content. *Int. J. Food Microbiol.* **2010**, *141* (Suppl. 1), S60–S68. [CrossRef]
21. Pérez-Baltar, A.; Alía, A.; Rodríguez, A.; Córdoba, J.J.; Medina, M.; Montiel, R. Impact of water activity on the inactivation and gene expression of *Listeria monocytogenes* during refrigerated storage of pressurized dry-cured ham. *Foods* **2020**, *9*, 1092. [CrossRef] [PubMed]

22. Rantsiou, K.; Greppi, A.; Garosi, M.; Acquadro, A.; Mataragas, M.; Cocolin, L. Strain dependent expression of stress response and virulence genes of *Listeria monocytogenes* in meat juices as determined by microarray. *Int. J. Food Microbiol.* **2012**, *152*, 116–122. [CrossRef]
23. Zolfaghari, M.; Rezaei, M.; Mohabbati Mobarez, A.; Forozandeh Moghaddam, M.; Hosseini, H.; Khezri, M. Virulence genes expression in viable but non-culturable state of *Listeria monocytogenes* in fish meat. *Food Sci. Technol. Int.* **2020**, *26*, 205–212. [CrossRef]
24. Alía, A.; Rodríguez, A.; Andrade, M.J.; Gómez, F.M.; Córdoba, J.J. Combined effect of temperature, water activity and salt content on the growth and gene expression of *Listeria monocytogenes* in a dry-cured ham model system. *Meat Sci.* **2019**, *155*, 16–19. [CrossRef] [PubMed]
25. Makariti, I.P.; Printezi, A.; Kapetanakou, A.E.; Zeaki, N.; Skandamis, P.N. Investigating boundaries of survival, growth and expression of genes associated with stress and virulence of *Listeria monocytogenes* in response to acid and osmotic stress. *Food Microbiol.* **2015**, *45*, 231–244. [CrossRef] [PubMed]
26. Marras, L.; Carraro, V.; Sanna, C.; Sanna, A.; Ingianni, A.; Coroneo, V. Growth of *Listeria monocytogenes* in ready to eat salads at different storage temperatures and valuation of virulence genes expression. *Ann. Ig.* **2019**, *31*, 374–384. [CrossRef] [PubMed]
27. Olesen, I.; Vogensen, F.K.; Jespersen, L. Gene transcription and virulence potential of *Listeria monocytogenes* strains after exposure to acidic and NaCl stress. *Foodborne Pathog. Dis.* **2009**, *6*, 669–680. [CrossRef]
28. Werbrouck, H.; Vermeulen, A.; Van Coillie, E.; Messens, W.; Herman, L.; Devlieghere, F.; Uyttendaele, M. Influence of acid stress on survival, expression of virulence genes and invasion capacity into Caco-2 cells of *Listeria monocytogenes* strains of different origins. *Int. J. Food Microbiol.* **2009**, *134*, 140–146. [CrossRef]
29. Hadjilouka, A.; Gkolfakis, P.; Patlaka, A.; Grounta, A.; Vourli, G.; Paramithiotis, S.; Touloumi, G.; Triantafyllou, K.; Drosinos, E.H. In vitro gene transcription of *Listeria monocytogenes* after exposure to human gastric and duodenal aspirates. *J. Food Prot.* **2020**, *83*, 89–100. [CrossRef]
30. Jiang, L.; Olesen, I.; Andersen, T.; Fang, W.; Jespersen, L. Survival of *Listeria monocytogenes* in simulated gastrointestinal system and transcriptional profiling of stress- and adhesion-related genes. *Foodborne Pathog. Dis.* **2010**, *7*, 267–274. [CrossRef]
31. Bucur, F.I.; Grigore-Gurgu, L.; Crauwels, P.; Riedel, C.U.; Nicolau, A.I. Resistance of *Listeria monocytogenes* to stress conditions encountered in food and food processing environments. *Front. Microbiol.* **2018**, *9*, 2700. [CrossRef]
32. Gravesen, A.; Kallipolitis, B.; Holmstrøm, K.; Høiby, P.E.; Ramnath, M.; Knøchel, S. *pbp2229*-mediated nisin resistance mechanism in *Listeria monocytogenes* confers cross-protection to class IIa bacteriocins and affects virulence gene expression. *Appl. Environ. Microbiol.* **2004**, *70*, 1669–1679. [CrossRef]
33. Yu, T.; Jiang, X.; Zhang, Y.; Ji, S.; Gao, W.; Shi, L. Effect of benzalkonium chloride adaptation on sensitivity to antimicrobial agents and tolerance to environmental stresses in *Listeria monocytogenes*. *Front. Microbiol.* **2018**, *9*, 2906. [CrossRef]
34. Giaouris, E.; Simões, M. Pathogenic biofilm formation in the food industry and alternative control strategies. In *Handbook of Food Bioengineering, Foodborne Diseases*; Holban, A.M., Grumezescu, A.M., Eds.; Academic Press (Elsevier): Amsterdam, The Netherlands, 2018; Volume 15. [CrossRef]
35. Leistner, L. Basic aspects of food preservation by hurdle technology. *Int. J. Food Microbiol.* **2000**, *55*, 181–186. [CrossRef]
36. Pupo, I.; Lepe, J.A.; Smani, Y.; Aznar, J. Comparison of the in vitro activity of ampicillin and moxifloxacin against *Listeria monocytogenes* at achievable concentrations in the central nervous system. *J. Med. Microbiol.* **2017**, *66*, 713–720. [CrossRef] [PubMed]
37. Martínez-Suárez, J.V.; Ortiz, S.; López-Alonso, V. Potential impact of the resistance to quaternary ammonium disinfectants on the persistence of *Listeria monocytogenes* in food processing environments. *Front Microbiol.* **2016**, *7*, 638. [CrossRef]
38. Kastbjerg, V.G.; Larsen, M.H.; Gram, L.; Ingmer, H. Influence of sublethal concentrations of common disinfectants on expression of virulence genes in *Listeria monocytogenes*. *Appl. Environ. Microbiol.* **2010**, *76*, 303–309. [CrossRef] [PubMed]
39. Rodrigues, D.; Cerca, N.; Teixeira, P.; Oliveira, R.; Ceri, H.; Azeredo, J. *Listeria monocytogenes* and *Salmonella enterica* Enteritidis biofilms susceptibility to different disinfectants and stress-response and virulence gene expression of surviving cells. *Microb. Drug Resist.* **2011**, *17*, 181–189. [CrossRef] [PubMed]
40. Tamburro, M.; Ripabelli, G.; Vitullo, M.; Dallman, T.J.; Pontello, M.; Amar, C.F.; Sammarco, M.L. Gene expression in *Listeria monocytogenes* exposed to sublethal concentration of benzalkonium chloride. *Comp. Immunol. Microbiol. Infect. Dis.* **2015**, *40*, 31–39. [CrossRef] [PubMed]
41. Datta, A.R.; Burall, L.S. Serotype to genotype: The changing landscape of listeriosis outbreak investigations. *Food Microbiol.* **2018**, *75*, 18–27. [CrossRef]
42. Gerba, C.P. Quaternary ammonium biocides: Efficacy in application. *Appl. Environ. Microbiol.* **2015**, *81*, 464–469. [CrossRef]
43. Kowalczyk, A.; Przychodna, M.; Sopata, S.; Bodalska, A.; Fecka, I. Thymol and thyme essential oil-new insights into selected therapeutic applications. *Molecules* **2020**, *25*, 4125. [CrossRef]
44. Townsend, C.A. Convergent biosynthetic pathways to β-lactam antibiotics. *Curr. Opin. Chem. Biol.* **2016**, *35*, 97–108. [CrossRef] [PubMed]
45. Pagliano, P.; Arslan, F.; Ascione, T. Epidemiology and treatment of the commonest form of listeriosis: Meningitis and bacteraemia. *Infez. Med.* **2017**, *25*, 210–216.

46. Kostoglou, D.; Tsaklidou, P.; Iliadis, I.; Garoufallidou, N.; Skarmoutsou, G.; Koulouris, I.; Giaouris, E. Advanced killing potential of thymol against a time and temperature optimized attached *Listeria monocytogenes* population in lettuce broth. *Biomolecules* **2021**, *11*, 397. [CrossRef]
47. Schmittgen, T.D.; Livak, K.J. Analyzing real-time PCR data by the comparative C_T method. *Nat. Protoc.* **2008**, *3*, 1101–1108. [CrossRef] [PubMed]
48. Vandesompele, J.; De Preter, K.; Pattyn, F.; Poppe, B.; Van Roy, N.; De Paepe, A.; Speleman, F. Accurate normalization of real-time quantitative RT-PCR data by geometric averaging of multiple internal control genes. *Genome Biol.* **2002**, *3*, 1–12. [CrossRef]
49. Pfaffl, M.W. A new mathematical model for relative quantification in real-time RT-PCR. *Nucleic Acids Res.* **2001**, *29*, e45. [CrossRef]
50. Toledo-Arana, A.; Dussurget, O.; Nikitas, G.; Sesto, N.; Guet-Revillet, H.; Balestrino, D.; Loh, E.; Gripenland, J.; Tiensuu, T.; Vaitkevicius, K.; et al. The *Listeria* transcriptional landscape from saprophytism to virulence. *Nature* **2009**, *459*, 950–956. [CrossRef]
51. Ye, J.; Coulouris, G.; Zaretskaya, I.; Cutcutache, I.; Rozen, S.; Madden, T.L. Primer-BLAST: A tool to design target-specific primers for polymerase chain reaction. *BMC Bioinform.* **2012**, *13*, 134. [CrossRef] [PubMed]
52. Bickel, D.R. Degrees of differential gene expression: Detecting biologically significant expression differences and estimating their magnitudes. *Bioinformatics* **2004**, *20*, 682–688. [CrossRef]
53. Cusimano, M.G.; Di Stefano, V.; La Giglia, M.; Di Marco Lo Presti, V.; Schillaci, D.; Pomilio, F.; Vitale, M. Control of growth and persistence of *Listeria monocytogenes* and β-lactam-resistant *Escherichia coli* by thymol in food processing settings. *Molecules* **2020**, *25*, 383. [CrossRef]
54. Rodríguez-Melcón, C.; Riesco-Peláez, F.; García-Fernández, C.; Alonso-Calleja, C.; Capita, R. Susceptibility of *Listeria monocytogenes* planktonic cultures and biofilms to sodium hypochlorite and benzalkonium chloride. *Food Microbiol.* **2019**, *82*, 533–540. [CrossRef] [PubMed]
55. López-Alonso, V.; Ortiz, S.; Corujo, A.; Martínez-Suárez, J.V. Analysis of benzalkonium chloride resistance and potential virulence of *Listeria monocytogenes* isolates obtained from different stages of a poultry production chain in Spain. *J. Food Prot.* **2020**, *83*, 443–451. [CrossRef] [PubMed]
56. Minarovičová, J.; Véghová, A.; Mikulášová, M.; Chovanová, R.; Šoltýs, K.; Drahovská, H.; Kaclíková, E. Benzalkonium chloride tolerance of *Listeria monocytogenes* strains isolated from a meat processing facility is related to presence of plasmid-borne *bcrABC* cassette. *Antonie Van Leeuwenhoek* **2018**, *111*, 1913–1923. [CrossRef] [PubMed]
57. Mullapudi, S.; Siletzky, R.M.; Kathariou, S. Heavy-metal and benzalkonium chloride resistance of *Listeria monocytogenes* isolates from the environment of turkey-processing plants. *Appl. Environ. Microbiol.* **2008**, *74*, 1464–1468. [CrossRef]
58. Rakic-Martinez, M.; Drevets, D.A.; Dutta, V.; Katic, V.; Kathariou, S. *Listeria monocytogenes* strains selected on ciprofloxacin or the disinfectant benzalkonium chloride exhibit reduced susceptibility to ciprofloxacin, gentamicin, benzalkonium chloride, and other toxic compounds. *Appl. Environ. Microbiol.* **2011**, *77*, 8714–8721. [CrossRef]
59. Capita, R.; Felices-Mercado, A.; García-Fernández, C.; Alonso-Calleja, C. Characterization of *Listeria monocytogenes* originating from the Spanish meat-processing chain. *Foods* **2019**, *8*, 542. [CrossRef]
60. Maćkiw, E.; Stasiak, M.; Kowalska, J.; Kucharek, K.; Korsak, D.; Postupolski, J. Occurrence and characteristics of *Listeria monocytogenes* in ready-to-eat meat products in Poland. *J. Food Prot.* **2020**, *83*, 1002–1009. [CrossRef]
61. Maćkiw, E.; Korsak, D.; Kowalska, J.; Felix, B.; Stasiak, M.; Kucharek, K.; Antoszewska, A.; Postupolski, J. Genetic diversity of *Listeria monocytogenes* isolated from ready-to-eat food products in retail in Poland. *Int. J. Food Microbiol.* **2021**, *358*, 109397. [CrossRef]
62. Bouymajane, A.; Rhazi Filali, F.; Oulghazi, S.; Lafkih, N.; Ed-Dra, A.; Aboulkacem, A.; El Allaoui, A.; Ouhmidou, B.; Moumni, M. Occurrence, antimicrobial resistance, serotyping and virulence genes of *Listeria monocytogenes* isolated from foods. *Heliyon* **2021**, *7*, e06169. [CrossRef] [PubMed]
63. Di Pasqua, R.; Mamone, G.; Ferranti, P.; Ercolini, D.; Mauriello, G. Changes in the proteome of *Salmonella enterica* serovar Thompson as stress adaptation to sublethal concentrations of thymol. *Proteomics* **2010**, *10*, 1040–1049. [CrossRef] [PubMed]
64. Dubois-Brissonnet, F.; Naïtali, M.; Mafu, A.A.; Briandet, R. Induction of fatty acid composition modifications and tolerance to biocides in *Salmonella enterica* serovar Typhimurium by plant-derived terpenes. *Appl. Environ. Microbiol.* **2011**, *77*, 906–910. [CrossRef] [PubMed]
65. Lepe, J.A.; Rodríguez-Villodres, A.; Martín-Gutiérrez, G.; Luque, R.; Aznar, J. In vitro study of synergy of ampicillin with ceftriaxone against *Listeria monocytogenes*. *Rev. Esp. Quimioter.* **2019**, *32*, 465–468. [PubMed]
66. Cho, Y.; Kim, H.; Beuchat, L.R.; Ryu, J.H. Synergistic activities of gaseous oregano and thyme thymol essential oils against *Listeria monocytogenes* on surfaces of a laboratory medium and radish sprouts. *Food Microbiol.* **2020**, *86*, 103357. [CrossRef] [PubMed]
67. Finkel, S.E. Long-term survival during stationary phase: Evolution and the GASP phenotype. *Nat. Rev. Microbiol.* **2006**, *4*, 113–120. [CrossRef] [PubMed]
68. Upadhyay, A.; Johny, A.K.; Amalaradjou, M.A.; Ananda Baskaran, S.; Kim, K.S.; Venkitanarayanan, K. Plant-derived antimicrobials reduce *Listeria monocytogenes* virulence factors *in vitro*, and down-regulate expression of virulence genes. *Int. J. Food Microbiol.* **2012**, *157*, 88–94. [CrossRef] [PubMed]
69. Drolia, R.; Bhunia, A.K. Crossing the intestinal barrier via Listeria Adhesion Protein and Internalin A. *Trends Microbiol.* **2019**, *27*, 408–425. [CrossRef]

70. Braschi, G.; Serrazanetti, D.I.; Siroli, L.; Patrignani, F.; De Angelis, M.; Lanciotti, R. Gene expression responses of *Listeria monocytogenes* Scott A exposed to sub-lethal concentrations of natural antimicrobials. *Int. J. Food Microbiol.* **2018**, *286*, 170–178. [CrossRef]
71. Kaplan Zeevi, M.; Shafir, N.S.; Shaham, S.; Friedman, S.; Sigal, N.; Nir Paz, R.; Boneca, I.G.; Herskovits, A.A. *Listeria monocytogenes* multidrug resistance transporters and cyclic di-AMP, which contribute to type I interferon induction, play a role in cell wall stress. *J. Bacteriol.* **2013**, *195*, 5250–5261. [CrossRef]
72. Bowman, J.P.; Lee Chang, K.J.; Pinfold, T.; Ross, T. Transcriptomic and phenotypic responses of *Listeria monocytogenes* strains possessing different growth efficiencies under acidic conditions. *Appl. Environ. Microbiol.* **2010**, *76*, 4836–4850. [CrossRef]
73. Köhler, S.; Leimeister-Wächter, M.; Chakraborty, T.; Lottspeich, F.; Goebel, W. The gene coding for protein p60 of *Listeria monocytogenes* and its use as a specific probe for *Listeria monocytogenes*. *Infect. Immun.* **1990**, *58*, 1943–1950. [CrossRef]
74. Ben Slama, R.; Miladi, H.; Chaieb, K.; Bakhrouf, A. Survival of *Listeria monocytogenes* cells and the effect of extended frozen storage (−20 °C) on the expression of its virulence gene. *Appl. Biochem. Biotechnol.* **2013**, *170*, 1174–1183. [CrossRef] [PubMed]
75. Dramsi, S.; Kocks, C.; Forestier, C.; Cossart, P. Internalin-mediated invasion of epithelial cells by *Listeria monocytogenes* is regulated by the bacterial growth state, temperature and the pleiotropic activator *prfA*. *Mol. Microbiol.* **1993**, *9*, 931–941. [CrossRef]
76. Gahan, C.G.; O'Mahony, J.; Hill, C. Characterization of the *groESL* operon in *Listeria monocytogenes*: Utilization of two reporter systems (*gfp* and *hly*) for evaluating in vivo expression. *Infect. Immun.* **2001**, *69*, 3924–3932. [CrossRef] [PubMed]
77. Osborne, S.E.; Brumell, J.H. Listeriolysin O: From bazooka to Swiss army knife. *Philos. Trans. R. Soc. Lond. B Biol. Sci.* **2017**, *372*, 20160222. [CrossRef] [PubMed]
78. Pöntinen, A.; Markkula, A.; Lindström, M.; Korkeala, H. Two-component-system histidine kinases involved in growth of *Listeria monocytogenes* EGD-e at low temperatures. *Appl. Environ. Microbiol.* **2015**, *81*, 3994–4004. [CrossRef]
79. Cotter, P.D.; Guinane, C.M.; Hill, C. The LisRK signal transduction system determines the sensitivity of *Listeria monocytogenes* to nisin and cephalosporins. *Antimicrob. Agents Chemother.* **2002**, *46*, 2784–2790. [CrossRef] [PubMed]
80. Jiang, X.; Yu, T.; Xu, Y.; Wang, H.; Korkeala, H.; Shi, L. MdrL, a major facilitator superfamily efflux pump of *Listeria monocytogenes* involved in tolerance to benzalkonium chloride. *Appl. Microbiol. Biotechnol.* **2019**, *103*, 1339–1350. [CrossRef]
81. Browning, D.F.; Busby, S.J. Local and global regulation of transcription initiation in bacteria. *Nat. Rev. Microbiol.* **2016**, *14*, 638–650. [CrossRef]
82. Roberfroid, S.; Vanderleyden, J.; Steenackers, H. Gene expression variability in clonal populations: Causes and consequences. *Crit. Rev. Microbiol.* **2016**, *42*, 969–984. [CrossRef]
83. Arguedas-Villa, C.; Stephan, R.; Tasara, T. Evaluation of cold growth and related gene transcription responses associated with *Listeria monocytogenes* strains of different origins. *Food Microbiol.* **2010**, *27*, 653–660. [CrossRef] [PubMed]
84. Poulsen, L.D.; Vinther, J. RNA-Seq for bacterial gene expression. *Curr. Protoc. Nucleic Acid Chem.* **2018**, *73*, e55. [CrossRef] [PubMed]

foods

Article

Monitoring by a Sensitive Liquid-Based Sampling Strategy Reveals a Considerable Reduction of *Listeria monocytogenes* in Smeared Cheese Production over 10 Years of Testing in Austria

Peter Zangerl [1], Dagmar Schoder [2], Frieda Eliskases-Lechner [1], Abdoulla Zangana [2], Elisabeth Frohner [2], Beatrix Stessl [2] and Martin Wagner [2,3,*]

[1] Higher Federal Teaching and Research Institute in Tyrol for Agriculture and Nutrition as well as Food and Biotechnology, Rotholz 50, 6200 Strass im Zillertal, Austria; peter.zangerl@hblfa-tirol.at (P.Z.); frieda.eliskases-lechner@hblfa-tirol.at (F.E.-L.)

[2] Unit of Food Microbiology, Institute of Food Safety, Food Technology and Veterinary Public Health, Department for Farm Animals and Veterinary Public Health, University of Veterinary Medicine Vienna, Veterinärplatz 1, 1210 Vienna, Austria; dagmar.schoder@vetmeduni.ac.at (D.S.); Abdoulla.Zangana@vetmeduni.ac.at (A.Z.); Elisabeth.Frohner@vetmeduni.ac.at (E.F.); Beatrix.Stessl@vetmeduni.ac.at (B.S.)

[3] Austrian Competence Center for Feed and Food Quality, Safety and Innovation (FFOQSI), Technopark C, 3430 Tulln, Austria

* Correspondence: Martin.wagner@vetmeduni.ac.at; Tel.: +43-125-077-3500

Citation: Zangerl, P.; Schoder, D.; Eliskases-Lechner, F.; Zangana, A.; Frohner, E.; Stessl, B.; Wagner, M. Monitoring by a Sensitive Liquid-Based Sampling Strategy Reveals a Considerable Reduction of *Listeria monocytogenes* in Smeared Cheese Production over 10 Years of Testing in Austria. *Foods* **2021**, *10*, 1977. https://doi.org/10.3390/foods10091977

Academic Editors: Antonio Afonso Lourenco, Catherine Burgess and Timothy Ells

Received: 21 July 2021
Accepted: 19 August 2021
Published: 24 August 2021

Abstract: Most Austrian dairies and cheese manufacturers participated in a *Listeria* monitoring program, which was established after the first reports of dairy product-associated listeriosis outbreaks more than thirty years ago. Within the *Listeria* monitoring program, up to 800 mL of product-associated liquids such as cheese smear or brine are processed in a semi-quantitative approach to increase epidemiological sensitivity. A sampling strategy within cheese production, which detects environmental contamination before it results in problematic food contamination, has benefits for food safety management. The liquid-based sampling strategy was implemented by both industrial cheese makers and small-scale dairies located in the mountainous region of Western Austria. This report considers more than 12,000 *Listeria* spp. examinations of liquid-based samples in the 2009 to 2018 timeframe. Overall, the occurrence of L. monocytogenes in smear liquid samples was 1.29% and 1.55% (n = 5043 and n = 7194 tested samples) for small and industrial cheese enterprises, respectively. The liquid-based sampling strategy for Listeria monitoring at the plant level appears to be superior to solid surface monitoring. Cheese smear liquids seem to have good utility as an index of the contamination of cheese up to that point in production. A modelling or validation process should be performed for the new semi-quantitative approach to estimate the true impact of the method in terms of reducing Listeria contamination at the cheese plant level.

Keywords: *Listeria* spp.; *Listeria monocytogenes*; prevalence; detection; monitoring; smear

1. Introduction

Cheese products have been a possible source of outbreaks of listeriosis for many decades, especially smeared cheeses and those made from raw milk [1–3] (https://www.cdc.gov/Listeria/outbreaks/index.html; accessed on: 19 June 2021).

Cheeses made from goat or sheep milk are particularly likely to be *L. monocytogenes* positive (3.6–12.8%) [4]. This is also evident from a search of the portal for Food and Feed Safety Alerts (RASSF), where 39/90 *L. monocytogenes* notifications relate to cheeses made from goat or sheep milk (https://webgate.ec.europa.eu/rasff-window/screen/search; accessed on: 19 June 2021). Significant genetic diversity was identified among *L. monocytogenes* strains through the use of molecular epidemiology methods [5–10]. Other research groups noticed an increased occurrence of hypervirulent *L. monocytogenes* strains of genetic lineage I (serovar 1/2b, 4b, sequence type (ST)1, ST4, ST6) in the dairy niche [11,12]. In addition, *L. monocytogenes* genetic lineage II strains (e.g., ST7, ST14, ST204; ST451), including hypovirulent types (ST121, ST9) were reported to persist in the dairy processing environment, potentially due to the intra- and inter-species exchange of mobile genetic elements [6,13–18].

An important role in environmental adaptation is played by highly conserved plasmids circulating worldwide in a distinctive *L. monocytogenes* gene pool [9,19–21]. These more complex epidemiological considerations have a direct impact on surveillance used to verify the effectiveness of *L. monocytogenes* controls within food safety management systems.

Although milk is usually subjected to a heating process prior to processing, cheese can become contaminated during several process steps such as pressing, curing, ripening, and during cutting and packaging [22,23].

In food processing environments (FPEs), contamination is often related to *L. monocytogenes'* colonization of surfaces, including in the dairy sector [24].

Own-check systems are applied with a focus on testing end products and samples from the production environment according to EC regulation 2073/2005 [25]. In food processing environments (FPEs), contamination is often related to *L. monocytogenes'* colonization of surfaces, including in the dairy sector [25].

In particular, newly built manufacturing plants or plants undergoing reconstruction measures are at high risk of being colonized with *L. monocytogenes* [26,27].

In cases where *L. monocytogenes* is detected on the end product at unacceptable levels, withdrawals from the market or recalls are implemented to protect the safety of the consumer.

To minimize the risk of process contamination during cheese ripening via the cheese smear, this liquid-based sampling strategy was established, which is also applicable to brine or drain water samples [28] (Figure 1). Since the majority of soft, semi-hard and hard cheeses in Austria are surface-ripened, smear liquids are, in most cases, collected after the smearing process. Compared to product-contact surface-sampling using friction-swabs, these liquids constitute a matrix that provides a much broader representation of the contamination status by including both cheese components and contact with surfaces inside of the production equipment, e.g., smear robots [29]. Sampling of a non-homogenous solid product creates real challenges in terms of consistency and representativeness. *Listeria* contamination is more likely on the surface rind than inside the cheese matrix. Moreover, sampling of a batch of individual cheeses has potential for statistical biases unless true randomisation is rigorously adhered to [3,30]. Sampling biases are major concerns and the degree of harmonization among procedures is usually low (sampling frequency and sampling sites are usually less well standardized) [31]. The implementation of preventive food safety concepts by tailored food sector-specific sampling procedures provokes a deepened insight of the FBOs into the operation-specific status of contamination and facilitates a comparison of scenarios.

The monitoring of cheeses produced without smearing focuses on sampling liquids including brine, wash water (water used to clean production devices such as trolleys or trays) or drain water. Sampling events depend on ripening time and batch size and should be performed twice per month. For small-scale dairies, the sampling frequency should ensure that every cheese is included at least once during ripening. After detection of *L. monocytogenes* and *Listeria* spp. by ISO enrichment methods, PCR-based species differentiation should be performed on typical *Listeria* colonies isolated on selective agar [32,33]. Persistence of *L. innocua* was shown to occur more frequently than persistence of *L. monocytogenes* and is, therefore, seen as an indicator of inadequate hygiene [34,35].

If *L. monocytogenes* is detected, rigorous sanitation of the facility is essential. Additionally, the sample number is increased and testing entails end products and further environmental samples (e.g., tanks, racks, conveyor belts and ventilation). This step includes a microbiological investigation post sanitation to verify the efficiency of the measures taken. If desired, a facility inspection audits the internal traffic management and checks other elements of the prerequisite programs (PrPs) that are in place, such as the maintenance of buildings and rooms. The hygienic status of production is, therefore, checked stepwise at all production areas. At the heart of the monitoring and surveillance approach is the range of sample volume that is tested: 600 to 800 mL (two labs involved, method slightly deviates), 100 mL, 10 mL, and 1 mL of liquid (Figure 1). This semi-quantitative way of testing both low and high sample volumes substantially increases the epidemiological sensitivity of the method due to a higher quantity of sample matrix.

Indeed, directly after initial contamination of either the environment or the food, *L. monocytogenes* might be scarcely detectable in food business operations (FBOs) and testing of high volumes increases the likelihood of finding low contamination levels.

Therefore, the aim of this study was to present the alternative semi-quantitative liquid-based sampling strategy to increase the epidemiological sensitivity in the detection of *L. monocytogenes* and other *Listeria* species. For this purpose, the alternative method was implemented within the framework of *Listeria* monitoring, for both industrial cheese makers and small-scale dairies located in the mountainous region of Western Austria. By using this approach, more than 12,000 samples were tested during the period from 2009 to 2018.

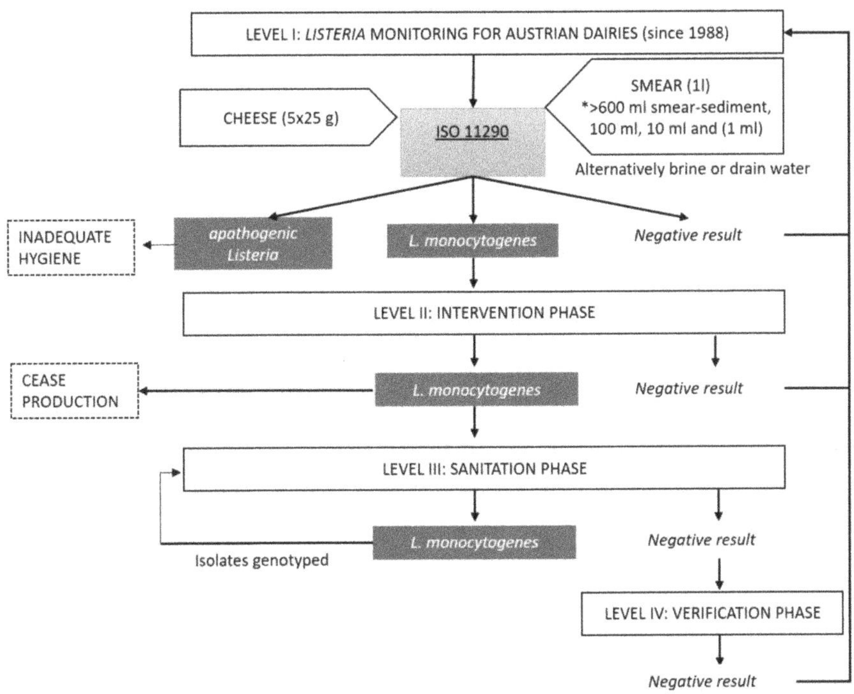

Figure 1. Flow chart displaying the structure of the Austrian *Listeria* monitoring and intervention program. Abbreviations: *, semi-quantitative liquid-based sample quantities.

2. Materials and Methods

2.1. Materials

Testing of cheeses for *L. monocytogenes* with a high level of confidence is limited by statistical biases. Investigation of smear liquid samples for monitoring purposes is a highly informative sampling strategy as all cheeses of a lot are usually treated with a smear liquid from the same tank. Therefore, analysis of the smear liquid allows for the contamination status of the entire cheese lot being stored for ripening. Sampling of smears is relatively simple and no cheese is damaged or spoiled by the sampling procedure [36] (Sampling scheme Figure 1).

2.2. Companies

According to the Austrian trade register for companies, around 80 professional cheese producers (this number does not include farm dairies directly marketing the product) exist in Austria (https://www.firmenbuchgrundbuch.at/ accessed on: 19 June 2021). Cheese making in Austria is conducted in operations that vary in size, ranging from small (products merchandised regionally) to industrial (products mostly merchandised across all of Austria and export markets such as the EU-27). Whereas some companies process a couple thousand liters of milk per year, industrial companies (spread over entire Austria) process tens of millions of liters. Small-scale cheese makers are mostly located in the Western parts of Austria. Many of them send their samples to the Higher Federal Teaching and Research Institute Tyrol (HBLFA) and, depending on the year, between 51 and 75 companies participate (see Table 1). The number of large industrial cheese producers that cooperate with the Institute of Food Safety, Food Technology and Veterinary Public Health (IFFV) ranges from 7 to 9, and these companies produce more than 80% of the industrially produced smeared soft and semi-hard cheeses in Austria.

Table 1. Numbers of small and industrial food establishments (FBOs) that tested positive for *L. monocytogenes* and other *Listeria* spp., which participated in the *Listeria* monitoring program (2009–2018).

	Small FBOs (HBLFA)		Industrial FBOs (IFFV)	
Year	*L. monocytogenes* Positive/Total *n* (%)	*Other listeria* spp. Positive/Total *n* (%)	*L. monocytogenes* Positive/Total *n* (%)	*Other listeria* spp. Positive/Total *n* (%)
2009	6/51 (11.8%)	7/51 (13.7%)	2/8 (25.0%)	2/8 (25.0%)
2010	8/64 (12.5%)	10/64 (15.6%)	1/9 (11.1%)	2/9 (22.2%)
2011	3/56 (5.4%)	8/56 (14.3%)	2/9 (22.2%)	3/9 (33.3%)
2012	2/63 (3.2%)	13/63 (20.6%)	0/9 (0%)	4/9 (44.4%)
2013	2/68 (0.3%)	11/68 (16.2%)	3/7 (42.9%)	3/9 (42.9%)
2014	2/73 (2.7%)	14/73 (19.2%)	2/8 (25.0%)	5/8 (62.5%)
2015	0/75 (0%)	13/75 (17.3%)	1/7 (14.3%)	2/7 (28.6%)
2016	2/74 (2.7%)	6/74 (8.1%)	2/7 (28.6%)	2/7 (28.6%)
2017	3/74 (4.1%)	12/74 (16.2%)	2/6 (33.3%)	2/6 (33.3%)
2018	2/75 (2.7%)	11/75 (14.7%)	1/7 (14.3%)	2/7 (28.6%)
Mean	3/67.3 (4.5%)	10.5/67.3 (15.6%)	1.6/7.7 (20.8%)	2/7.7 (28.6%)

Abbreviations: FBOs, food business operations supervised by Higher Federal Teaching and Research Institute Tyrol (HBLFA) and Institute of Food Safety, Food Technology and Veterinary Public Health (IFFV); *Listeria* spp., *Listeria* species other than *L. monocytogenes* differentiated by *iap* PCR [32].

2.3. Methods

A total of 12,237 smear liquid samples were examined in the years 2009–2018 (see Table 1) by both testing labs. Liquid smear samples were collected in two-month intervals from industrial cheesemakers. Small FBOs collected smear samples during cheese ripening, representing comparatively smaller batches. Sample volumes of 1 mL (IFFV only), 10 mL, 100 mL and 600 mL (IFFV) or 800 mL (HBLFA) are routinely investigated. The occurrence of *L. monocytogenes* in products, product-associated samples and in the processing environment is considered to be rather low and not equally distributed; therefore, the semi-quantitative enrichment protocol is assumed to increase the detection of *L. monocytogenes* in at least one of the enrichment steps [28]. One liter of liquid sample was divided into 4 preparations as follows: 600 or 800 mL were centrifuged at 4800 rpm for 30 min at 4 °C (Beckman Coulter, Brea, CA, USA). The sediment was completely transferred into 1 L Half-Fraser broth (Biokar Diagnostics-Solabia Group, Pantin Cedex, France). Subsequent preparation steps of the semi-quantitative approach included 100 mL, 10 mL, and 1 mL diluted 1:10 in Half-Fraser broth (Biokar Diagnostics-Solabia Group).

Sample enrichment in Half-Fraser broth and Fraser broth (both Biokar Diagnostics-Solabia Group) and strain isolation on Palcam Agar (Biokar Diagnostics-Solabia Group) and *Listeria* agar acc. Ottaviani and Agosti (ALOA; Merck KGaA, Darmstadt, Germany) was performed according to ISO 11290:1 [33]. In detail, for each semi-quantitative enrichment scenario (i.e., 600/800 mL, 100 mL, 10 mL, 1 mL), following 24 h incubation at 30 °C in Half-Fraser broth, aliquots of 100 µL were transferred to 10 mL Fraser broth and then incubated for 48 h at 37 °C.

In addition, at IFFV, polymerase chain reaction (PCR) assays targeting the *hly* gene (encoding the pore-forming cytolysin listeriolysin) and *iap* (invasion-associated protein p60) gene [31,37] were included for species confirmation (for technical details, see Asperger et al. [28]). This approach ensured that even a single *L. monocytogenes* colony that may have hidden in a plethora of other microorganisms, such as *Bacillus* spp. growing on PALCAM or chromogenic agar, would be detected [38].

The DNA extraction was performed directly from selective agar plates by rinsing the surface with 1 mL of 0.01 M Tris HCl buffer (Sigma Aldrich Corp., St. Louis, MO, USA). The suspension was centrifuged for 5 min at 8000 rpm and the pellet was suspended in 100 µL 0.01 M Tris HCl Buffer (Sigma Aldrich Corp.) and vortexed. In parallel, material from *L. monocytogenes* subcultures (1–2 colonies) was suspended in 100 µL Tris HCl Buffer. Subsequently, 400 µL Chelex® 100-Resin (BioRad, Hercules, CA, USA) was added to the bacterial suspension, heated for 10 min at 100 °C and centrifuged at 14,000 rpm for 5 s [39]. The DNA supernatant was transferred to Maxymum Recovery tubes (VWR International-Avantor, Radnor, PA, USA) and stored at −20 °C before downstream processing [31,37]. The PCR-amplicons were electrophoretically separated in a 1.5% agarose gel containing 0.5× Tris-Borate-EDTA (TBE) buffer and 3.5 µL peqGREEN DNA gel stain (VWR International-Avantor), at 120 V for 30 min. The DNA standard Thermo Scientific™ GeneRuler™ 100 bp (Thermo Fisher Scientific Inc., Waltham, MA, USA) was applied for fragment length comparison. The electrophoresis gels were photographed under UV light exposure (GelDoc 2000, BioRad, Hercules) and saved in tiff format for further comparison.

3. Results and Discussion

Listeria contamination is an adverse event for many food business operations (FBOs), and the entire dairy sector suffers whenever outbreaks occur. A survey of technical managers in food processing plants on *L. monocytogenes* risk outcomes by Evans et al. [40]. revealed interesting assessments. Participants perceived a medium risk (on a scale from 1 to 10; 5.5) of *Listeria* in their operations with a high level of control and a high level of responsibility. In this study, technical leaders expressed concern regarding *L. monocytogenes* and indicated that increased awareness of the pathogen would improve control actions. Installing *Listeria* environmental monitoring was considered essential in this regard [40].

A recent evaluation of monitoring approaches by Magdovitz et al. [41] showed that facilities prefer to test environmental monitoring zones 2 through 4 (non-food contact areas). Few facilities actively integrate raw material controls and intermediate products or product-associated samples into their sampling plan [41].

Many data are available for *Listeria* contamination scenarios in single FBOs, but little information is available for whole food production sectors such as smeared cheese manufacturing. EU baseline data on *L. monocytogenes* prevalence in cheese samples at the end of shelf-life showed a rate of 0.47%, with 0.06% of samples exceeding the level of 100 cfu/g [42].

The few studies that are focused on data across food producers and batches are somehow comparable to our data and are cited in the following paragraph. Data on liquid-based sampling concepts are not available from the literature. Barría et al. [43] studied 546 cheese and milk samples to establish a monitoring system in Chilean cheese factories. *L. monocytogenes* was identified in 19 cheeses (4.1%), with a prevalence similar to that reported in a Polish study (6.2% *L. monocytogenes*, 370 samples) [44]. In both studies, the monitoring system focused on cheese samples as no food contact surface (FCS) or non-food contact surface (NFCS) samples were included in the sampling plan. Another *Listeria* spp. pilot study in PDO Taleggio cheese processing revealed a mean prevalence of 23.1% *Listeria*-positive samples (n = 360 samples). The ripening and cutting equipment were identified as high-risk areas for *Listeria* contamination [45]. Other short-term monitoring datasets were published, with an overall *L. monocytogenes* prevalence of 4.6% in various food sectors [46]. A larger dataset based on pathogen monitoring in small cheese processing plants (4430 samples; 6.03% *Listeria* spp.) suggested running routine sampling plans for at least 6 months and then evaluating appropriate sampling sites inclusively for *Listeria* occurrence [34].

In general, cheese surfaces are more likely to be contaminated by *L. monocytogenes* than the internal areas of the cheese. This was also the outcome of a baseline study, conducted at a national level, where Gorgonzola and Taleggio were the most frequently contaminated cheeses. Transmission of *L. monocytogenes* from contaminated cheese rind to the cheese interior during cutting or packaging is possible [47]. Therefore, product-associated samples, such as smear liquids and surface scrapings, should be considered in a *Listeria* monitoring program.

Our data from the cheese smear liquid-based monitoring showed, in small cheese producers (mainly soft and semi-soft cheeses), an average *Listeria* spp. (other than *L. monocytogenes*) and *L. monocytogenes* contamination of 15.6% and 4.5%, respectively, During the sampling period, an average of 67 out of 75 FBOs were *Listeria* spp. positive. Numbers for industrial cheesemakers show that an average of eight FBOs participated in the program, where means of 20.8% *L. monocytogenes* and 28.6% *Listeria* spp. (other than *L. monocytogenes*) were detected.

The *L. monocytogenes* contamination ranged from 0 to 12.5% and from 0 to 33.3% in small and industrial FBOs during 2009 to 2018, respectively. *Listeria* spp. other than *L. monocytogenes*, which were differentiated by the PCR approach [32], ranged from 8.1 to 20.6% in small FBOs and from 22.2 to 44.4% in industrial FBOs (Table 1), indicating that the latter was more highly contaminated with the potential pathogen. The industrial FBOs were higher contaminated with *L. monocytogenes* in comparison to small FBOs. Similar observations were made by Muhterem et al. [25], where the FPE of industrial cheesemakers indicated a higher *L. monocytogenes* contamination of up to 26% compared to farm cheesemakers (up to 6.4%). In total, *Listeria* spp. was detected in 4.19% (513 out of 12.237) of all smear liquid samples examined, whereas the percentage of *L. monocytogenes*-positive samples was 1.45% (178 out of 12.237 samples). The higher frequency of *Listeria* spp. (other than *L. monocytogenes*) contamination is an important indicator of necessary hygiene improvement measures to prevent *L. monocytogenes* from successfully establishing itself as a zoonotic pathogen in a FPE [48]. This value for *Listeria* spp.-associated contamination was substantially lower in comparison to samples that were tested at the IFFV between 1990 and 1999 (industrial cheese makers only: 14.09%) [28]. If calculated based on years, the prevalence of *L. monocytogenes* in smears was 0–4.4% (average: 1.29%) and 0–6% (average: 1.55%) for the small and the industrial cheese establishments, respectively (see Table 2).

Table 2. The number of smear liquid samples tested and the rate of *L. monocytogenes* and other *Listeria* spp.-positive results found.

Year	Small Dairys (Western Austria; HBLFA)					Industrial Cheesemakers (IFFV)				
	n	*L. monocytogenes*	(%)	Other *Listeria* spp.	(%)	*n*	*L. monocytogenes*	(%)	Other *Listeria* spp.	(%)
2009	475	19	4	13	2.7	189	5	2.1	13	6.9
2010	620	27	4.4	12	1.9	503	3	0.6	68	13.5
2011	394	3	0.8	10	2.5	881	12	1.4	27	3.1
2012	441	2	0.5	23	5.2	774	0	0.0	70	9.0
2013	441	3	0.7	21	4.8	711	3	0.4	22	3.1
2014	516	2	0.4	22	4.3	702	2	0.3	19	2.7
2015	523	0	0.0	21	4.0	1535	24	1.6	14	0.9
2016	512	3	0.6	9	1.8	634	8	1.3	14	2.2
2017	544	3	0.6	24	4.4	752	45	6.0	46	6.1
2018	577	5	0.9	14	2.4	513	9	1.8	51	9.9
Total	5043	67	1.29	169	3.4	7194	111	1.55	344	5.74

Abbreviations: Small dairys and industrial cheesemakers supervised by Higher Federal Teaching and Research Institute Tyrol (HBLFA) and Institute of Food Safety, Food Technology and Veterinary Public Health (IFFV); *Listeria* spp., *Listeria* species other than *L. monocytogenes* differentiated by *iap* PCR [32].

This is of interest as the industrial cheese producers included in this study mainly used pasteurized milk, while the small producers tended to use raw milk for the production of traditional specialty cheeses.

Since the occurrence of *L. monocytogenes* contamination was similar for both categories (Table 2), we confirmed that heat treatment of milk had little impact on the presence of *L. monocytogenes* in the smears and that, in the majority of our observations, cheese is more likely to become contaminated after coagulation [18,35,49].

Inclusion of high sample volumes was found to increase the detection sensitivity of the method as applied at both institutes. At HBLFA, 11.98% of samples tested positive in 800 mL and 100 mL but not in 10 mL, and 19.4% (n = 13) of all positive findings were found in the highest sample volume only (data not shown). Only 26.9% (n = 18) of all samples tested positive in 800 mL, 100 mL and 10 mL. From the fact that more than 30% of the positive events were observed in volumes of $\geq$100 mL only, we conclude that *L. monocytogenes* contamination levels are often very low at the beginning of a contamination event. Data also suggest that testing only 25 mL of cheese-associated fluids (which is commonly the case in other countries) does not provide enough epidemiologic sensitivity to detect low-level contamination.

This assumption would be interesting to compare in the performance testing of the ISO method versus alternative liquid-based sampling strategies with higher sample volumes. Some samples revealed *L. monocytogenes* detection in either 10 mL or 100 mL but not in 800 mL. This effect could have been caused by a not-yet-understood antiListerial potential of the smear microbiota in some samples, testing too soon following the use of protective cultures against *L. monocytogenes* (e.g., phages), and extremely high numbers of accompanying flora after centrifugation of 600 and 800 mL, respectively [50–52]. Unpublished results on the inhibitory effects of smear samples on *Listeria* showed a highly variable pattern, ranging from a decrease in numbers of *L. monocytogenes* by 3 log units in some samples to a proliferation capacity of up to 4 log CFU/mL in other samples (Part, pers. communication). We conclude that testing of high volumes only is not sufficient to detect a contamination event; therefore, the more extensive approach of testing more than one sampling volume should be incorporated. Findings from small cheese producers were consistent with results that were found with samples originating from industrial cheese plants. Twenty-four percent (n = 22) of positive results were found in the high sample volume (600 mL) only. As many as 16.5% of smear liquid samples were found to be positive in sample volumes of 600 mL and 100 mL. Another 15.4% of the samples were positive in 600 mL, 100 mL and 10 mL. The smear monitoring conducted at IFFV also incorporated

1 mL samples. Being positive in 1 mL was thought to be a cause for concern as a higher number of *L. monocytogenes* might be present in the smear liquid and, subsequently, on the cheese. In 12% (n = 11) of all positive smear liquid samples, *L. monocytogenes* was found in all sample volumes (600 mL, 100 mL, 10 mL and 1 mL). As with the data provided by HBFLA, the findings at IFFV are inconclusive in some cases. In 24% of the positive results, *L. monocytogenes* was detected in 100 mL of sample volume only.

Although the first food-associated outbreaks were reported from USA and Canada in the early 1980s, a game changer for the national dairy industry was the Swiss Vacherin Mont d'Or outbreak in 1983–1987 [53]. Austrian companies began testing cheese brine and smears in 1988 to improve *L. monocytogenes* detection during production. From 1992 to 1994, a 30 to 40% positive test rate for *L. monocytogenes* was observed. Within a decade of increased measures, prevalence decreased to a detection rate of <5% [28]. The liquid-based sampling strategy also shows successful detection of *L. monocytogenes* in our approach, and possibly a positive impact in terms of avoiding false negatives and product withdrawals or recalls. This positive development of improved awareness of possible *L. monocytogenes* contamination occurred in spite of an ongoing restructuring of the dairy sector in Austria, which reduced the number of industrial cheese dairies from >50 in 1990 to less than 10 in 2019.

In line with the economic growth of some major players, the amount of produced cheese (soft, semisoft and hard cheese) quintupled from 1990 to 2019 (< 30,000 tons in 1995 to 131,000 tons per year in 2018).

The monitoring of results such as those achieved by the *Listeria* monitoring program is a prerequisite for the timely detection of potential safety hazards, including the contamination of cheese environments with *L. monocytogenes*. Frequent monitoring aids early *L. monocytogenes* detection, and prevents contamination and the placing of contaminated food on the market [31]. That there is a considerable likelihood for introduction is evidenced by the fact that, at least once, positive *Listeria* spp. results were revealed over all the years from a majority of the participants in the program. If contamination remains unaffected by routine hygiene measures, *Listeria* is spread within the production area through daily in-plant manipulations.

In the long run, *Listeria* spp. colonizes niches within the FBO, where the hygienic pressure is not high enough to prevent them from surviving, thereby allowing *Listeria* spp. to survive.

Experience in recent years has repeatedly confirmed that testing higher sample volumes effectively complements other hygiene inspection techniques, such as swabbing or contact sliding.

In accordance with the testing of product-associated liquids, environment-derived liquid samples such as drain water samples encompass the contamination status of large plant areas. The use of large volumes of liquid in our semi-quantitative sampling approach potentially reduces the false negative test results that can occur when using smaller volumes or simple contact sliding.

Investigation of smear liquid is beneficial as this substrate is used on entire cheese batches for extended production periods. Therefore, with respect to cheese processing, the microbiological investigation of smear liquid is an appropriate parameter in any safety program dealing with smear-ripened cheeses.

Preventing foodborne hazards along the food processing chain is supported by an intelligent sampling strategy that may differ among food sectors and professionals. For *L. monocytogenes* environmental testing, mostly swab and sponge-based friction sampling methods are used [54]. The decrease in the *L. monocytogenes* detection rate, as seen in Austrian cheese factories in recent years, coincides with an increased understanding and acceptance of food safety parameters by the cheese producers, which was in part contributed to by a high-profile cheese-borne outbreak of listeriosis [55].

The consideration of a preventive QS certification system is important within the context of the explicit obligations placed on food business operators through EU food law to undertake such monitoring both against microbiological criteria in food and, in the case of *L. monocytogenes*, within the food production environment, to validate the effectiveness of their food safety management systems. Official control analyses serve a different purpose, and are required to be risk based, as opposed to representing food production, which generally occurs at much lower frequency. For example, in Austria, 35,000 food samples are annually taken by public authorities by a factor of >5.

4. Conclusions

The increasing trend of listeriosis incidence in Austria, from a mean value of 0.17 per 100,000 inhabitants from 2000 to 2005 to a mean value of 0.4 from 2009 to 2018 (Austrian Agency for Health and Food Safety—AGES, 2018; https://www.ages.at/download/0/0/c38f0d95e095fe7e74162ddae9052a4c532450db/fileadmin/AGES2015/Themen/Krankheitserreger_Dateien/Zoonosen/Zoonosenbroschuere_2018_1o_Din-A4_BF.pdf; accessed on 9 June 2021), emphasizes the requirements for effective strategies that meet the control needs of the national public health system and food manufacturers. The liquid-based sampling strategy within a *Listeria* monitoring program at the plant level appears to be superior to solid surface monitoring. Cheese smear liquids seem to have good utility as an index of the contamination of cheese up to that point in production. Multiple volumes of liquid phase, as implemented with our semi-quantitative approach, seem to improve the likelihood of detection, which is consistent with improved epidemiological sensitivity. Monitoring results show a downward trend in *Listeria* prevalence within this matrix, at least for industrial cheese production, which is thereby consistent with improved hygiene in cheese processing environments and cheese products. Modeling or performance testing of this new semi-quantitative approach against the ISO method would be important to more concretely assess the potential for *Listeria* minimization in cheese production.

Author Contributions: P.Z. and M.W. conceived and designed the experiments. F.E.-L., A.Z. and E.F. performed the experiments. P.Z., M.W., D.S. and B.S. analyzed the data. P.Z. and M.W. drafted the manuscript. B.S. and D.S. reviewed the manuscript. All authors have read and agreed to the published version of the manuscript.

Funding: This work was partly funded by the Austrian Competence Centre for Feed and Food Quality, Safety and Innovation (FFoQSI). The COMET-K1 competence centre FFoQSI is funded by the Austrian ministries BMVIT and BMDW, and the Austrian provinces Niederoesterreich, Upper Austria and Vienna, within the scope of COMET—Competence 301 Centers for Excellent Technologies. The COMET program is handled by the Austrian Research Promotion Agency FFG.

Institutional Review Board Statement: Not applicable.

Informed Consent Statement: Not applicable.

Data Availability Statement: Not applicable.

Acknowledgments: We thank the FBOs for their constructive cooperation. Furthermore, we thank Clair Firth for proofreading the manuscript.

Conflicts of Interest: The authors declare no conflict of interest.

References

1. Desai, A.N.; Anyoha, A.; Madoff, L.C.; Lassmann, B. Changing epidemiology of *Listeria monocytogenes* outbreaks, sporadic cases, and recalls globally: A review of ProMED reports from 1996 to 2018. *Int. J. Infect. Dis.* **2019**, *84*, 48–53. [CrossRef] [PubMed]
2. Melo, J.; Andrew, P.; Faleiro, M. *Listeria monocytogenes* in cheese and the dairy environment remains a food safety challenge: The role of stress responses. *Food Res. Int.* **2015**, *67*, 75–90. [CrossRef]
3. Rudolf, M.; Scherer, S. High incidence of *Listeria monocytogenes* in European red smear cheese. *Int. J. Food Microbiol.* **2001**, *63*, 91–98. [CrossRef]
4. Gonzales-Barron, U.; Gonçalves-Tenório, A.; Rodrigues, V.; Cadavez, V. Foodborne pathogens in raw milk and cheese of sheep and goat origin: A meta-analysis approach. *Curr. Opin. Food Sci.* **2017**, *18*, 7–13. [CrossRef]

5. Amato, E.; Filipello, V.; Gori, M.; Lomonaco, S.; Losio, M.N.; Parisi, A.; Huedo, P.; Knabel, S.J.; Pontello, M. Identification of a major *Listeria monocytogenes* outbreak clone linked to soft cheese in Northern Italy–2009–2011. *BMC Infect. Dis.* **2017**, *17*, 1–7. [CrossRef]
6. Cabal, A.; Pietzka, A.; Huhulescu, S.; Allerberger, F.; Ruppitsch, W.; Schmid, D. Isolate-Based Surveillance of *Listeria monocytogenes* by Whole Genome Sequencing in Austria. *Front. Microbiol.* **2019**, *10*, 2282. [CrossRef]
7. Finazzi, G.; Filipello, V.; Gori, M.; Scaltriti, E.; Bracchi, C.; Menozzi, I.; Tanzi, E.; Bolzoni, L. A *Listeria monocytogenes* ST325 clone is widespread in the Lombardy Region dairy processing plants. *Eur. J. Public Health* **2020**, *30* (Suppl. 5), 166–229. [CrossRef]
8. Nüesch-Inderbinen, M.; Bloemberg, G.V.; Müller, A.; Stevens, M.J.; Cernela, N.; Kollöffel, B.; Stephan, R. Listeriosis Caused by Persistence of *Listeria monocytogenes* Serotype 4b Sequence Type 6 in Cheese Production Environment. *Emerg. Infect. Dis.* **2021**, *27*, 284–288. [CrossRef]
9. Muhterem-Uyar, M.; Ciolacu, L.; Wagner, K.-H.; Wagner, M.; Schmitz-Esser, S.; Stessl, B. New Aspects on *Listeria monocytogenes* ST5-ECVI Predominance in a Heavily Contaminated Cheese Processing Environment. *Front. Microbiol.* **2018**, *9*, 64. [CrossRef]
10. Schmitz-Esser, S.; Müller, A.; Stessl, B.; Wagner, M. Genomes of sequence type 121 *Listeria monocytogenes* strains harbor highly conserved plasmids and prophages. *Front. Microbiol.* **2015**, *6*, 380. [CrossRef]
11. Bechtel, T.D.; Gibbons, J.G. Population Genomic Analysis of *Listeria monocytogenes* From Food Reveals Substrate-Specific Genome Variation. *Front. Microbiol.* **2021**, *12*. [CrossRef]
12. Maury, M.M.; Bracq-Dieye, H.; Huang, L.; Vales, G.; Lavina, M.; Thouvenot, P.; Disson, O.; Leclercq, A.; Brisse, S.; Lecuit, M. Hypervirulent *Listeria monocytogenes* clones' adaption to mammalian gut accounts for their association with dairy products. *Nat. Commun.* **2019**, *10*, 1–13. [CrossRef]
13. Castro, H.; Douillard, F.; Korkeala, H.; Lindström, M. Diverse mobile genetic elements support the persistence of *Listeria monocytogenes* on dairy farms. *bioRxiv* **2021**. [CrossRef]
14. Guidi, F.; Orsini, M.; Chiaverini, A.; Torresi, M.; Centorame, P.; Acciari, V.; Salini, R.; Palombo, B.; Brandi, G.; Amagliani, G.; et al. Hypo- and Hyper-Virulent *Listeria monocytogenes* Clones Persisting in Two Different Food Processing Plants of Central Italy. *Microorganisms* **2021**, *9*, 376. [CrossRef]
15. Matereke, L.T.; Okoh, A.I. *Listeria monocytogenes* Virulence, Antimicrobial Resistance and Environmental Persistence: A Review. *Pathogens* **2020**, *9*, 528. [CrossRef] [PubMed]
16. Palaiodimou, L.; Fanning, S.; Fox, E.M. Genomic insights into persistence of Listeria species in the food processing environment. *J. Appl. Microbiol.* **2021**. [CrossRef] [PubMed]
17. Rychli, K.; Wagner, E.M.; Ciolacu, L.; Zaiser, A.; Tasara, T.; Wagner, M.; Schmitz-Esser, S. Comparative genomics of human and non-human *Listeria monocytogenes* sequence type 121 strains. *PLoS ONE* **2017**, *12*, e0176857. [CrossRef]
18. Stessl, B.; Fricker, M.; Fox, E.; Karpiskova, R.; Demnerová, K.; Jordan, K.; Ehling-Schulz, M.; Wagner, M. Collaborative Survey on the Colonization of Different Types of Cheese-Processing Facilities with *Listeria monocytogenes*. *Foodborne Pathog. Dis.* **2014**, *11*, 8–14. [CrossRef] [PubMed]
19. Jennison, A.V.; Masson, J.J.; Fang, N.-X.; Graham, R.M.; Bradbury, M.; Fegan, N.; Gobius, K.; Graham, T.M.; Guglielmino, C.; Brown, J.L.; et al. Analysis of the *Listeria monocytogenes* Population Structure among Isolates from 1931 to 2015 in Australia. *Front. Microbiol.* **2017**, *8*, 603. [CrossRef]
20. Naditz, A.L.; Dzieciol, M.; Wagner, M.; Schmitz-Esser, S. Plasmids contribute to food processing environment–associated stress survival in three *Listeria monocytogenes* ST121, ST8, and ST5 strains. *Int. J. Food Microbiol.* **2019**, *299*, 39–46. [CrossRef] [PubMed]
21. Schmitz-Esser, S.; Anast, J.M.; Cortes, B.W. A Large-Scale Sequencing-Based Survey of Plasmids in *Listeria monocytogenes* Reveals Global Dissemination of Plasmids. *Front. Microbiol.* **2021**, *12*, 510. [CrossRef] [PubMed]
22. Falardeau, J.; Trmčić, A.; Wang, S. The occurrence, growth, and biocontrol of *Listeria monocytogenes* in fresh and surface-ripened soft and semisoft cheeses. *Compr. Rev. Food Sci. Food Saf.* **2021**. [CrossRef] [PubMed]
23. Sauders, B.D.; D'Amico, D.J. *Listeria monocytogenes* cross-contamination of cheese: Risk throughout the food supply chain. *Epidemiology Infect.* **2016**, *144*, 2693–2697. [CrossRef] [PubMed]
24. Muhterem-Uyar, M.; Dalmasso, M.; Bolocan, A.S.; Hernandez, M.; Kapetanakou, A.; Kuchta, T.; Manios, S.G.; Melero, B.; Minarovičová, J.; Nicolau, A.I.; et al. Environmental sampling for *Listeria monocytogenes* control in food processing facilities reveals three contamination scenarios. *Food Control* **2015**, *51*, 94–107. [CrossRef]
25. Commission Regulation (EC) No 2073/2005 on microbiological criteria for foodstuffs. *Off. J. Eur. Union L.* **2005**, *338*, 1–26.
26. Alvarez-Molina, A.; Cobo-Díaz, J.F.; López, M.; Prieto, M.; de Toro, M.; Alvarez-Ordóñez, A. Unraveling the emergence and population diversity of *Listeria monocytogenes* in a newly built meat facility through whole genome sequencing. *Int. J. Food Microbiol.* **2021**, *340*, 109043. [CrossRef]
27. Melero, B.; Stessl, B.; Manso, B.; Wagner, M.; Esteban-Carbonero, Ó.J.; Hernandez, M.; Rovira, J.; Rodriguez-Lázaro, D. *Listeria monocytogenes* colonization in a newly established dairy processing facility. *Int. J. Food Microbiol.* **2019**, *289*, 64–71. [CrossRef]
28. Asperger, H.; Wagner, M.; Brandl, E. An approach towards public health and foodborne human listeriosis—The Austrian Listeria monitoring. *Berl. Munchener Tierarztliche Wochenschr.* **2001**, *114*, 446–452. [PubMed]
29. Jordan, K.; Leong, D.; Ordóñez, A.Á. National *Listeria* Monitoring Programmes. In *Listeria monocytogenes in the Food Processing Environment*; Springer: Cham, Germany, 2015; pp. 73–75.
30. Schvartzman, M.S.; Gonzalez-Barron, U.; Butler, F.; Jordan, K. Modeling the growth of *Listeria monocytogenes* on the surface of smear- or mold-ripened cheese. *Front. Cell. Infect. Microbiol.* **2014**, *4*, 90. [CrossRef]

31. Spanu, C.; Jordan, K. *Listeria monocytogenes* environmental sampling program in ready-to-eat processing facilities: A practical approach. *Compr. Rev. Food Sci. Food Saf.* **2020**, *19*, 2843–2861. [CrossRef]
32. Bubert, A.; Hein, I.; Rauch, M.; Lehner, A.; Yoon, B.; Goebel, W.; Wagner, M. Detection and Differentiation of *Listeria* spp. by a Single Reaction Based on Multiplex PCR. *Appl. Environ. Microbiol.* **1999**, *65*, 4688–4692. [CrossRef]
33. *ISO 11290-1:2017 Microbiology of the Food Chain—Horizontal Method for the Detection and Enumeration of* Listeria Monocytogenes *and of Listeria Spp.—Part 1: Detection Method;* International Organization for Standardization: Geneva, Switzerland, 2017.
34. Beno, S.M.; Stasiewicz, M.; Andrus, A.D.; Ralyea, R.D.; Kent, D.J.; Martin, N.H.; Wiedmann, M.; Boor, K. Development and Validation of Pathogen Environmental Monitoring Programs for Small Cheese Processing Facilities. *J. Food Prot.* **2016**, *79*, 2095–2106. [CrossRef]
35. Kaszoni-Rückerl, I.; Mustedanagic, A.; Muri-Klinger, S.; Brugger, K.; Wagner, K.-H.; Wagner, M.; Stessl, B. Predominance of Distinct Listeria Innocua and *Listeria monocytogenes* in Recurrent Contamination Events at Dairy Processing Facilities. *Microorganisms* **2020**, *8*, 234. [CrossRef]
36. Wagner, M.; Stessl, B. Sampling the Food-Processing Environment: Taking Up the Cudgel for Preventive Quality Management in Food Processing (FP). In *Listeria monocytogenes;* Humana: New York, NY, USA, 2021; pp. 233–242.
37. Border, P.M.; Howard, J.J.; Plastow, G.; Siggens, K.W. Detection of Listeria species and *Listeria monocytogenes* using polymerase chain reaction. *Lett. Appl. Microbiol.* **1990**, *11*, 158–162. [CrossRef]
38. Stessl, B.; Luf, W.; Wagner, M.; Schoder, D. Performance testing of six chromogenic ALOA-type media for the detection of *Listeria monocytogenes. J. Appl. Microbiol.* **2009**, *106*, 651–659. [CrossRef]
39. Walsh, P.S.; Metzger, D.A.; Higuchi, R. Chelex 100 as a Medium for Simple Extraction of DNA for PCR-Based Typing from Forensic Material. *BioTechniques* **2013**, *54*, 134–139. [CrossRef] [PubMed]
40. Evans, E.; Samuel, E.; Redmond, E.; Taylor, H. Exploring *Listeria monocytogenes* perceptions in small and medium sized food manufacturers: Technical leaders' perceptions of risk, control and responsibility. *Food Control* **2021**, *126*, 108078. [CrossRef]
41. Magdovitz, B.F.; Gummalla, S.; Thippareddi, H.; Harrison, M.A. Evaluating Environmental Monitoring Protocols for *Listeria* spp. and *Listeria monocytogenes* in Frozen Food Manufacturing Facilities. *J. Food Prot.* **2019**, *83*, 172–187. [CrossRef]
42. European Food Safety Authority-Efsa. Analysis of the baseline survey on the prevalence of *Listeria monocytogenes* in certain ready-to-eat foods in the EU, 2010-2011 Part A: *Listeria monocytogenes* prevalence estimates. *EFSA J.* **2013**, *11*. [CrossRef]
43. Barría, C.; Singer, R.S.; Bueno, I.; Estrada, E.; Rivera, D.; Ulloa, S.; Fernández, J.; Mardones, F.O.; Moreno-Switt, A.I. Tracing *Listeria monocytogenes* contamination in artisanal cheese to the processing environments in cheese producers in southern Chile. *Food Microbiol.* **2020**, *90*, 103499. [CrossRef]
44. Pyz-Łukasik, R.; Gondek, M.; Winiarczyk, D.; Michalak, K.; Paszkiewicz, W.; Piróg-Komorowska, A.; Policht, A.; Ziomek, M. Occurrence of *Listeria monocytogenes* in Artisanal Cheeses from Poland and Its Identification by MALDI-TOF MS. *Pathogens* **2021**, *10*, 632. [CrossRef]
45. Tirloni, E.; Bernardi, C.; Pomilio, F.; Torresi, M.; De Santis, E.P.L.; Scarano, C.; Stella, S. Occurrence of *Listeria* spp. and *Listeria monocytogenes* Isolated from PDO Taleggio Production Plants. *Foods* **2020**, *9*, 1636. [CrossRef]
46. Leong, D.; Alvarez-Ordóñez, A.; Ejordan, K. Monitoring occurrence and persistence of *Listeria monocytogenes* in foods and food processing environments in the Republic of Ireland. *Front. Microbiol.* **2014**, *5*, 436. [CrossRef]
47. Iannetti, L.; Acciari, V.A.; Antoci, S.V.; Addante, N.; Bardasi, L.; Bilei, S.; Calistri, P.; Cito, F.; Cogoni, P.; D'Aurelio, R.; et al. *Listeria monocytogenes* in ready-to-eat foods in Italy: Prevalence of contamination at retail and characterisation of strains from meat products and cheese. *Food Control.* **2016**, *68*, 55–61. [CrossRef]
48. Rückerl, I.; Muhterem-Uyar, M.; Muri-Klinger, S.; Wagner, K.-H.; Wagner, M.; Stessl, B.L. monocytogenes in a cheese processing facility: Learning from contamination scenarios over three years of sampling. *Int. J. Food Microbiol.* **2014**, *189*, 98–105. [CrossRef]
49. Schaffner, E.; Mühlemann, M.; Spahr, U.; Schällibaum, M. Quantification of the probability of milk contamination by *Listeria monocytogenes* during manufacture of hard cheese. *Rev. D'epidemiologie et de Sante Publique* **2003**, *51*, 493–503.
50. Fister, S.; Fuchs, S.; Stessl, B.; Schoder, D.; Wagner, M.; Rossmanith, P. Screening and characterisation of bacteriophage P100 insensitive *Listeria monocytogenes* isolates in Austrian dairy plants. *Food Control* **2016**, *59*, 108–117. [CrossRef]
51. Bockelmann, W.; Koslowsky, M.; Goerges, S.; Scherer, S.; Franz, C.M.; Heller, K.J. Growth inhibition of *Listeria monocytogenes* by bacteriocin-producing Staphylococcus equorum SE3 in cheese models. *Food Control* **2017**, *71*, 50–56. [CrossRef]
52. Zilelidou, E.A.; Skandamis, P.N. Growth, detection and virulence of *Listeria monocytogenes* in the presence of other microorganisms: Microbial interactions from species to strain level. *Int. J. Food Microbiol.* **2018**, *277*, 10–25. [CrossRef] [PubMed]
53. Büla, C.; Bille, J.; Glauser, M.P. An Epidemic of Food-Borne Listeriosis in Western Switzerland: Description of 57 Cases Involving Adults. *Clin. Infect. Dis.* **1995**, *20*, 66–72. [CrossRef]
54. Brauge, T.; Barre, L.; Leleu, G.; André, S.; Denis, C.; Hanin, A.; Frémaux, B.; Guilbaud, M.; Herry, J.-M.; Oulahal, N.; et al. European survey and evaluation of sampling methods recommended by the standard EN ISO 18593 for the detection of *Listeria monocytogenes* and Pseudomonas fluorescens on industrial surfaces. *FEMS Microbiol. Lett.* **2020**, *367*, 057. [CrossRef] [PubMed]
55. Fretz, R.; Sagel, U.; Ruppitsch, W.; Pietzka, A.T.; Stöger, A.; Huhulescu, S.; Heuberger, S.; Pichler, J.; Much, P.; Pfaff, G.; et al. Listeriosis outbreak caused by acid curd cheese 'Quargel', Austria and Germany 2009. *Eurosurveillance* **2010**, *15*, 19477. [CrossRef] [PubMed]

foods

MDPI

Article

Fate of *Salmonella* Typhimurium and *Listeria monocytogenes* on Whole Papaya during Storage and Antimicrobial Efficiency of Aqueous Chlorine Dioxide Generated with HCl, Malic Acid or Lactic Acid on Whole Papaya

Lianger Dong and Yong Li *

Department of Human Nutrition, Food and Animal Sciences, University of Hawaii, 1955 East-West Road, Honolulu, HI 96822, USA; ldong@hawaii.edu
* Correspondence: liyong@hawaii.edu; Tel.: +1-808-956-6408; Fax: +1-808-956-4024

Citation: Dong, L.; Li, Y. Fate of *Salmonella* Typhimurium and *Listeria monocytogenes* on Whole Papaya during Storage and Antimicrobial Efficiency of Aqueous Chlorine Dioxide Generated with HCl, Malic Acid or Lactic Acid on Whole Papaya. *Foods* **2021**, *10*, 1871. https://doi.org/10.3390/foods10081871

Academic Editors: Antonio Afonso Lourenco, Catherine Burgess, Timothy Ells and Susana Casal

Received: 24 May 2021
Accepted: 6 August 2021
Published: 12 August 2021

Abstract: Papaya-associated foodborne illness outbreaks have been frequently reported worldwide. The goal of this study was to evaluate the behavior of *Salmonella* Typhimurium and *Listeria monocytogenes* on whole papaya during storage and sanitizing process. Fresh green papayas were inoculated with approximately 7 log CFU of *S.* Typhimurium and *L. monocytogenes* and stored at 21 or 7 °C for 14 days. Bacteria counts were determined on day 0, 1, 7, 10 and 14. Fresh green papayas inoculated with approximately 8 log CFU of the bacteria were treated for 5 min with 2.5, 5 and 10 ppm aqueous chlorine dioxide (ClO_2). The ClO_2 solutions were generated by mixing sodium chlorite with an acid, which was HCl, lactic acid or malic acid. The detection limit of the enumeration method was 2.40 log CFU per papaya. At the end of storage period, *S.* Typhimurium and *L. monocytogenes* grew by 1.88 and 1.24 log CFU on papayas at 21 °C, respectively. Both bacteria maintained their initial population at inoculation on papayas stored at 7 °C. Higher concentrations of ClO_2 reduced more bacteria on papaya. 10 ppm ClO_2, regardless the acid used to generate the solutions, inactivated *S.* Typhimurium to undetectable level on papaya. 10 ppm ClO_2 generated with HCl, lactic acid and malic acid reduced *L. monocytogenes* by 4.40, 6.54 and 8.04 log CFU on papaya, respectively. Overall, ClO_2 generated with malic acid showed significantly higher bacterial reduction than ClO_2 generated with HCl or lactic acid. These results indicate there is a risk of survival and growth for *S.* Typhimurium and *L. monocytogenes* on papaya at commercial storage conditions. Aqueous ClO_2 generated with malic acid shows effectiveness in inactivating the pathogenic bacteria on papaya.

Keywords: whole papaya; *Salmonella* Typhimurium; *Listeria monocytogenes*; survival; aqueous chlorine dioxide; malic acid; shelf-life

1. Introduction

Papaya (*Carica papaya*) is one of the major tropical agricultural commodities amongst banana, mango, avocado and pineapple [1]. Annual global papaya production has increased by approximately 90% since 2000 and reached 13.7 million metric tons in 2019 [2]. The top three papaya-producing countries are India, Brazil and Mexico, among which 99% of Mexican papayas are exported to the United States [2]. However, along with the increased papaya demand and production worldwide, foodborne illness outbreaks linked to papaya have also been emerging in recent years [3,4]. In particular, outbreaks associated with whole fresh papaya have been frequently reported in the U.S. from 2011 to 2019, which affected the papaya industry in both US and Mexico [4,5]. Papaya grows best in tropic environments at 21–33 °C where the survival and growth of pathogenic bacteria are favored [6]. Microbial contamination of papaya might happen at any step of the production chain where the fruits are in contact with water, soil, harvest equipment and human handling [7]. *Salmonella* Litchfield was detected on whole papayas associated with an outbreak in Australia between 2006 and 2007, and other *Salmonella* serotypes of Chester, Eastborne and Poona were detected in farm water samples [3]. In multiple cases reported in the U.S., whole papayas were contaminated by *Salmonella* serotypes of Agona, Uganda, Newport, etc. [5]. Therefore, papaya seems to be susceptible to *Salmonella* contamination. In addition, *Listeria monocytogenes* is one of the concerned foodborne pathogenic bacteria associated with fresh produce due to its nature of being present in the environment and its ability to grow at refrigeration temperature [8]. *L. monocytogenes*-caused multistate outbreaks in the U.S. were linked to whole cantaloupe and caramel apple [9,10]. *L. monocytogenes* was also found to be able to survive or grow on the surfaces of apple, mango, kiwifruit and cherry tomato under various storage conditions [11–14].

Studies have reported the survival and growth of foodborne pathogenic bacteria in fresh-cut papaya and papaya pulp [15–19]. However, little is known regarding whole fresh papaya. There are differences between fresh-cut and whole fruits in terms of pH, nutrient availability and native microflora composition. For example, *S.* Typhimurium and *L. monocytogenes* decreased by approximately 2–2.5 log CFU over 20 days on whole mango at 25 °C; however, these bacteria grew on cut mango [12]. The growth of *L. monocytogenes* was inhibited on intact jalapeño pepper stored at 7 °C for 14 days, but it grew in the internal cavity of jalapeño pepper at the same storage condition [20]. It is important to note that even when the skin part of fruit is inedible or usually not eaten, pathogenic bacteria surviving on the surface may further cross-contaminate wash water and other fruits that are rinsed in the same batch, internalize into the flesh or transfer to fruit flesh during cutting [21,22]. Information of pathogenic bacteria behavior on whole papaya would assist regulatory and industrial agencies in the assessment and prevention of papaya microbiological safety issues.

Once contaminated, fresh fruits cannot be thermally disinfected and would likely be distributed to the market. Therefore, washing and sanitizing is a critical step in the post-harvest process to prevent cross-contamination and reduce pathogens. Chlorine-based bleach at a concentration of 50–200 ppm is the most widely used sanitizer in fresh produce handling and processing [23]. However, the effectiveness of chlorine varies at different pH and is reduced significantly in the presence of organics, and there are concerns regarding the carcinogenetic by-products such as trihalomethanes formed in the reactions between chlorine and organics [24]. Chlorine dioxide (ClO_2) is approved by FDA for fresh produce washing with a maximum residue of 3 ppm in the wash water [24]. The antimicrobial efficacy of ClO_2 is less prone to low pH and the presence of organics than chlorine [25]. ClO_2 also forms fewer carcinogenetic by-products than chlorine when chlorinated [24]. Despite the advantages, ClO_2 is reduced to chlorite (ClO^{2-}), chlorate (ClO^{3-}) and chloride (Cl^-) to some extend [26]. The United State Environmental Protection Agency (EPA) sets the Maximum Residual Disinfectant Level (MRDL) of ClO_2 in public drinking water to be 0.8 mg/L and the Maximum Contaminant Level (MCL) of ClO^{2-} to be 1.0 mg/L [27]. ClO_2 has been studied in sanitizing a wide variety of fresh produce, such as lettuce, cantaloupe, alfalfa sprouts and blueberries [23,28–30]. No ClO_2, ClO^{2-} or ClO^{3-} residues were detected in Mulberry fruit treated by 60 ppm aqueous ClO_2 for 15 min [31]. Cantaloupes, oranges, tomatoes and apples treated with 5 ppm gaseous ClO_2 for 10 min showed very minimal ClO^{2-} residue on the fruits with a maximum of 0.36 mg/kg; however, lettuce and alfalfa sprouts had high ClO^{2-} residue of 16.5–1259.6 mg/kg [32]. Acidified sodium chlorite was used to reduce microbial contamination in shredded green papaya [33]. Ozone was used to reduce the microbial load and improve the nutritional values of fresh-cut papaya [34]. Gu et al. investigated the efficiency of chlorine or peracetic acid in the inactivation and cross-contamination prevention of *Salmonella* spp. on Maradol papayas [35]. Inactivation of pathogenic bacteria by ClO_2 has not been investigated on whole papayas.

Aqueous ClO_2 can be made by mixing an acid with sodium chlorite ($NaClO_2$) [36]. Hydrochloric acid (HCl) is a commonly used acid in ClO_2 generation [30–32,36]. Kim et al. [37] reported ClO_2 solutions formed from organic acids, including acetic acid, citric acid and lactic acid, were more stable and more lethal to *Bacillus cereus* spores than ClO_2 formed using HCl. Our previous study has also shown that aqueous ClO_2 generated by mixing $NaClO_2$ with organic acids, including citric acid, lactic acid and malic acid, had higher antimicrobial efficacy against common foodborne pathogenic bacteria on Romaine lettuce than ClO_2 generated with inorganic acids [38]. For example, 5 min treatments with 5 ppm ClO_2 generated with lactic acid, citric acid and malic acid reduced S. Typhimurium on Romaine lettuce by 0.92, 1.39 and 1.37 log CFU/g, respectively, whereas lettuce treated with ClO_2 generated with HCl and sodium bisulfate reduced *S.* Typhimurium by 0.71 and 1.14 log CFU/g, respectively [38].

In numerous studies investigating the survival of foodborne pathogenic bacteria on fresh produce or decontamination of fresh produce using sanitizers, procedures used to recover and quantify bacteria cells from fresh produce vary. The ununiformed procedures make it difficult to compare and accurately interpret results of different studies [39]. For example, pummeling using a stomacher resulted in higher bacteria recovery than pulsifying, sonication and shaking by hand from iceberg lettuce, perilla leaves, cucumber and green pepper, while a lower level of bacteria was recovered from cherry tomato due to its acidity [40]. Sample preparation method, bacteria type and produce type may affect the efficiency of bacteria recovery and hence further affect the accuracy of a microbiological method. So far, there has been no recommendation of sample preparation methods specifically for whole papaya.

This study aimed to optimize homogenization parameters and enumeration methods for recovering *S.* Typhimurium and *L. monocytogenes* from papaya surface. It also sought to evaluate the behaviors of these pathogenic bacteria on whole papaya during storage and sanitizing process. Obtaining information in this regard would assist the papaya industry in selecting optimal sanitizer type, usage concentration and treatment time for papaya washing and sanitizing.

2. Materials and Methods

2.1. Bacterial Strains and Cell Cultures

Salmonella Typhimurium (ATCC 14028) and *Listeria monocytogenes* (F2365) were obtained from Food Microbiology Lab at the University of Hawaii at Manoa and stored in trypticase soy broth (TSB; Becton Dickinson, Franklin Lakes, NJ, USA) containing 50% glycerol at −80 °C. Working cultures were prepared by transferring 50 μL of stock culture into 5 mL of sterile TSB and incubating at 37 °C for 24 h. Working cultures were transferred twice in TSB before each experiment.

2.2. Preparation of Papayas and Inocula

Fresh papayas (*Carica papaya* L.cv. Rainbow Solo) were purchased on the day of experimentations on separate occasions from local grocery stores in Honolulu, USA. Non-injured whole papayas at mature green/color break stage were selected according to the maturity chart [41]. Papayas were rinsed with tap water and dried on a lab bench at room temperature for 1 h. Then an area of 2.5×2.5 cm^2 on the middle part of the fruit surface was marked with a thin-line non-toxic marker (Sharpie, Oak Brook, IL, USA). The marked whole papayas were placed on sterile Petri dishes in a biosafety hood before experimenting. *S.* Typhimurium and *L. monocytogenes* cultures were diluted with 0.1% peptone water (Becton Dickinson, Franklin Lakes, NJ, USA) to desired concentrations. 100 μL of the inoculum was spot inoculated on the marked area and the papayas were dried under a biosafety hood. For Sections 2.3 and 2.4, approximately 10^7 log CFU of *S.* Typhimurium or *L. monocytogenes* inocula were used, and the papayas were dried for 1 h to initiate the attachment before every experiment [42]. For Section 2.5, approximately 10^8 log CFU of the inocula were used, and the papayas were dried for two hours to ensure attachment and initiate colonization before being washed with sanitizer solutions [42].

2.3. Optimization of Recovery Method for Counting Bacteria Cells on Papaya Surface

2.3.1. Recovery Method

Optimization of homogenization parameters is essential for accurate assessment of bacterial behavior on fruit surfaces. The goal of this experiment was to maximize the number of bacteria cells recovered from the papaya surface. After inoculation and drying as described above, the skin of the inoculated area was excised with a sterile knife and placed in a sterile stomacher bag. Bacterial cells were collected by homogenizing the skin under different conditions described as follows. Tested homogenization buffers included phosphate buffered saline (PBS, pH 7.4), 0.1% peptone water (PEPT), PBS + 0.2% Tween 80 (PBS + T) and 0.1% peptone water + 0.2% Tween 80 (PEPT + T). 25 mL of each buffer was separately added into the stomacher bag containing the excised skin and homogenized at 150 or 250 rpm for 1 or 5 min using a stomacher (Seward Stomacher®, Model 400 Circulator, West Sussex, UK). After homogenization, the homogenate was serially diluted with 0.1% peptone water and plated on selective agar or using the agar overlay method. The agar overlay method was to plate the serially diluted homogenate on Plate Count Agar (PCA, Becton Dickinson, Franklin Lakes, NJ, USA) and incubating the plate at 37 °C for 1 h to ensure the recovery of injured cells, followed by pouring warm selective agar at 55 °C over the PCA [43]. The agar plates were incubated at 37 °C for 24 h and then analyzed for bacterial counts. The selective agar for *S.* Typhimurium and *L. monocytogenes* were xylose lysine deoxycholate agar (XLD, Becton Dickinson, Franklin Lakes, NJ, USA) and modified oxford agar (MOX, Becton Dickinson, Franklin Lakes, NJ, USA), respectively. Bacterial colonies were counted and populations were expressed as log CFU/papaya. The detection limit was 2.40 log CFU/papaya.

2.3.2. PH of Papaya Skin Homogenate as Affected by Homogenization Parameters

Papayas were prepared as described in Section 2.2 except that they were not inoculated with pathogenic bacteria. The skin of the marked area was cut and homogenized with buffer in a stomacher bag under the conditions described above. Papaya skin was also homogenized with water as a control. pH of the homogenate was measured using a pH meter (Model pH 6+, Oakton Instruments, Vernon Hills, IL, USA).

2.4. Behavior of Pathogenic Bacteria on Whole Papayas Stored at Different Temperatures

After harvesting and packing, papayas are usually stored at 7–13 °C before being distributed to grocery stores [44]. At grocery stores and customers' homes, papayas are usually placed at room temperature (21–25 °C). Hence, we selected 21 and 7 °C to simulate the two papaya storage scenarios. Inoculated whole papayas were individually placed in large sterile beakers and stored at 21 and 7 °C for 14 days. One papaya was randomly sampled, with the skin of the inoculated area being sterilely excised and collected for bacteria count on storage days 0, 7, 10 and 14. The papaya that was inoculated and dried for 1 h on the day of inoculation was considered as the sample on day 0. To determine bacterial population on papaya, the excised skin was homogenized using the optimized method from Section 2.3, which was homogenizing in PBS + T buffer at 250 rpm for 1 min for both *S.* Typhimurium and *L. monocytogenes*. Subsequently, the homogenates were serially diluted with 0.1% peptone water and plated using the agar overlay method described above. After incubation, bacterial colonies were counted and populations were expressed as log CFU/papaya.

2.5. ClO_2 Treatment on Whole Papayas

2.5.1. Preparation of Aqueous ClO_2

Aqueous ClO_2 solutions were made on-site using a previous method [38]. Briefly, ClO_2 stock solutions were prepared by mixing 10 mL of 4.0% $NaClO_2$ (Fisher Scientific, Waltham, MA, USA) with 10 mL of 1 M HCl (Fisher Scientific, Waltham, MA, USA), lactic acid (VWR Chemicals, Radnor, PA, USA) or malic acid (Fisher Scientific) in aluminum foil-covered bottles. After reacting for 1 min, 100 mL of distilled water was added into the bottles.

The final mixture was set at 21 °C for 20 min before being placed in a refrigerator at 4 °C. We previously investigated the generation kinetics and the stability of ClO_2 [38]. As organic acids release hydrogen ions slowly, it took one week to achieve equilibrium. During the 14-day-experimentation, the ClO_2 concentration increased till up to day seven and then remained stable for those generated with organic acids. For ClO_2 generated with HCl, the reaction was quick and the concentration remained stable for up to eight days and eventually decreased. Therefore, the stock solutions were all stored for seven days to allow the completion of the reaction in malic acid- and lactic acid-produced ClO_2 solutions and ensure no loss of the effectiveness of HCl-produced ClO_2 solutions. On the day of experimentation, the concentration of ClO_2 in each stock solution was measured using Chlordioxid-Test kit (EMD Millipore Corp., Burlington, MA, USA). The stock solutions were diluted with distilled water to 2.5, 5 and 10 ppm to treat papayas. The pH of each diluted solution was determined.

2.5.2. Washing Papayas with Aqueous ClO_2 and Individual Acid Solutions

To wash artificially contaminated papayas, each papaya was inoculated with S. Typhimurium or *L. monocytogenes* as described in Section 2.2 and then submerged into a sterile container containing 1 L of ClO_2 made with HCl, lactic acid or malic acid at concentrations of 2.5, 5 and 10 ppm. The submerged papayas were mildly stirred at a rate of 150 rpm for 5 min [45]. Subsequently, the washed fruits were dried under a biosafety hood for 15 min. After drying, the marked surface was sterilely cut and homogenized in 25 mL of PBS + T buffer at 250 rpm for 1 min. The homogenate was serially diluted and plated by the agar overlay method with XLD and MOX agar for the selection of S. Typhimurium and *L. monocytogenes*, respectively. Bacterial populations were expressed as log CFU/papaya, and the detection limit was 2.40 log CFU/papaya. Washing with distilled water and 200 ppm bleach (sodium hypochlorite (NaClO), pH 6.5) diluted from Clorox® (6.0% NaClO, The Clorox Company, Oakland, CA, USA) served as the control treatments.

Acid solutions were prepared by adjusting 1 L of distilled water individually with 1 M HCl, 1 M lactic acid or 1 M malic acid to the pH of 10 ppm ClO_2 made with the corresponding acid. Papayas inoculated with S. Typhimurium or *L. monocytogenes* were washed with the acid solutions, and the remaining bacteria were collected and enumerated following the procedures described above.

2.5.3. ClO_2 Residue on Papaya Surface after Washing

Papayas were washed with tap water and dried on a lab bench for 1 h. Subsequently, the papayas were washed with 1 L of ClO_2 made with HCl, lactic acid or malic acid at concentrations of 5, 10 and 20 ppm. After drying for 15 min, the papayas were placed in 1-gallon Ziploc bags containing 100 mL distilled water. The papayas surfaces were hand massaged and rinsed thoroughly for 2 min, followed by filtering the rinse water into a flask [46]. 10 mL of the filtrate was collected and measured for ClO_2 concentration using Chlordioxid-Test kit. The detection limit was 0.02 mg/L in the undiluted filtrate. The ClO_2 concentration was converted into mg/kg papaya.

2.6. Statistical Analysis

All experiments were conducted in three independent replicates. Bacterial cultures were separately grown following the same procedure for each replicate. ClO_2 solutions were prepared freshly for each replicate. Data were reported as mean ± standard deviation (SD). Analysis of variance and Tukey's multiple comparison test were performed using SSPS software (IBM® SPSS® Statistics 24.0 for Windows, IBM Corp., Armonk, NY, USA). A significance level of 0.05 was used to determine the differences between the means of treatment groups.

3. Results and Discussion

3.1. Recovery of S. Typhimurium and L. monocytogenes Cells from Whole Papaya Surface as Affected by Homogenization Parameters and Enumeration Methods

Statistical analysis revealed no interactions among homogenization parameters, and only buffer significantly affected the bacterial count ($p < 0.05$). For *S.* Typhimurium (Table 1), papayas homogenized in buffers with the non-ionic surfactant Tween 80 resulted in significantly higher bacteria counts than those homogenized in peptone water alone. Tween 80 interrupts the hydrophobic interactions between bacteria cells and papaya surface and promotes the detachment of cells [47]. Papayas homogenized in the combination of PBS and Tween 80 (PBS + T) had the highest *S.* Typhimurium counts; an average of 5.36 log CFU was recovered from the initial inoculum of approximately 7 log CFU. Among all treatments, homogenization at 150 rpm for 5 min using XLD plating resulted in the highest recovery of 5.64 log CFU from papaya surface. For *L. monocytogenes* (Table 2), homogenization in PBS + T collected significantly more cells than in PBS alone ($p < 0.05$). Homogenization time, speed or plating method did not play a significant role in the collection. Homogenization at 150 rpm for 5 min by the agar overlay method resulted in the highest count of 5.09 log CFU. However, homogenization at 250 rpm for 1 min also resulted in relatively high *L. monocytogenes* counts. Homogenization at 250 rpm for 1 min was chosen for collecting *S.* Typhimurium and *L. monocytogenes* from papaya surface to maintain the time efficiency and consistency of the experiment. Even though the agar overlay method did not result in significantly higher bacteria counts than using selective agar alone, incubating on non-selective media before adding selective media would help recover bacteria cells injured by sanitizers [43]. It is an essential step to avoid over-estimation of the antimicrobial efficiency of sanitizers. Therefore, homogenizing the inoculated papaya piece in PBS + T at 250 rpm for 1 min was chosen, and the homogenate was decided to be plated by overlaying selective agar on PCA.

Table 1. *S.* Typhimurium population (log CFU) recovered from papaya surface as affected by homogenization buffer, time (min), speed (rpm) and enumeration methods *.

Buffer	1 Min				5 Min				Average
	150 Rpm		250 Rpm		150 Rpm		250 Rpm		
	XLD	PCA + XLD	XLD	PCA + XLD	XLD	PCA + XLD	XLD	PCA + XLD	
PBS	5.31 ± 0.65	5.34 ± 0.84	4.91 ± 0.38	4.95 ± 0.44	5.14 ± 0.29	5.38 ± 0.18	4.86 ± 0.82	4.81 ± 0.99	5.11 ± 0.57 ^{a,b}
PEPT	4.77 ± 0.50	4.80 ± 0.43	4.99 ± 0.52	5.05 ± 0.46	4.57 ± 0.47	4.85 ± 0.67	4.74 ± 0.32	4.54 ± 0.22	4.77 ± 0.40 ^{b}
PBS + T	5.08 ± 0.26	5.18 ± 0.30	5.43 ± 0.38	5.55 ± 0.34	5.64 ± 0.46	5.39 ± 0.49	5.29 ± 0.10	5.31 ± 0.07	5.36 ± 0.33 ^{a}
PEPT + T	5.08 ± 0.87	5.25 ± 0.74	5.26 ± 0.47	5.45 ± 0.41	5.07 ± 0.34	5.49 ± 0.28	5.39 ± 0.50	5.41 ± 0.24	5.28 ± 0.45 ^{a}

* "PBS", "PEPT", "PBS + T" and "PEPT + T" stand for phosphate buffered saline, 0.1% peptone water, PBS with 0.2% Tween 80 and 0.1% peptone water with 0.2% Tween 80, respectively. Enumeration methods "XLD" and "PCA + XLD" stand for xylose lysine deoxycholate agar and plate count agar overlaid with XLD, respectively. Numbers are mean ± standard deviation (n = 3). No significant interactions were found between the factors. Means in the same column with different superscripts are significantly different ($p < 0.05$).

Table 2. *L. monocytogenes* population (log CFU) recovered from papaya surface as affected by homogenization buffer, time (min), speed (rpm) and enumeration methods *.

Buffer	1 Min				5 Min				Average
	150 Rpm		250 Rpm		150 Rpm		250 Rpm		
	MOX	PCA + MOX	MOX	PCA + MOX	MOX	PCA + MOX	MOX	PCA + MOX	
PBS	4.23 ± 0.49	4.49 ± 0.75	4.79 ± 0.17	4.76 ± 0.20	4.60 ± 0.81	4.50 ± 0.74	4.36 ± 0.58	4.35 ± 0.59	4.51 ± 0.52 b
PEPT	5.02 ± 0.21	4.55 ± 0.62	4.84 ± 0.55	4.67 ± 0.45	4.89 ± 0.76	4.30 ± 0.37	4.57 ± 0.60	4.61 ± 0.45	4.68 ± 0.49 a,b
PBS + T	4.97 ± 0.29	4.61 ± 0.33	4.97 ± 0.13	4.93 ± 0.15	4.92 ± 0.48	5.09 ± 0.19	5.08 ± 0.3	4.94 ± 0.01	4.94 ± 0.25 a
PEPT + T	4.96 ± 0.63	4.54 ± 1.03	4.85 ± 0.58	4.88 ± 0.55	4.62 ± 0.83	4.38 ± 0.89	4.55 ± 1.04	4.49 ± 0.97	4.66 ± 0.73 a,b

* "PBS", "PEPT", "PBS + T" and "PEPT + T" stand for phosphate buffered saline, 0.1% peptone water, PBS with 0.2% Tween 80 and 0.1% peptone water with 0.2% Tween 80, respectively. Enumeration methods "MOX" and "PCA + MOX" stand for modified Oxford agar and plate count agar overlaid with MOX, respectively. Numbers are mean ± standard deviation (n = 3). No significant interactions were found between the factors. Means in the same column with different superscripts are significantly different ($p < 0.05$).

pH values of the above-mentioned homogenates were measured with uninoculated samples to compare buffering capacity between homogenization buffers. Even with careful excision, papaya flesh attached to the skin could acidify the homogenate. Papaya flesh has a pH of 4.87–5.7 [16,18]. This pH range does not inhibit the growth of *S.* Typhimurium or *L. monocytogenes*; however, it could influence the recovery of cells injured by desiccation [43]. Tian et al. incubated sublethally injured *E. coli* O157:H7 cells in nutrient broth at pH 4.0, 5.0, 6.0, 7.2 and 8.0. They found that the cells showed no significant recovery at pH 4.0 and 8.0 whereas the cells recovered by 0.48, 0.49 and 0.72 log CFU/mL in pH 5.0, 6.0 and 7.2, respectively, indicating that pH even at relatively high levels (5.0 and 6.0) did affect the recovery of sublethally injured cells [48]. Shown in Table 3, homogenizing papaya skin in different buffers resulted in significant differences in homogenate acidity in a descent order of PBS, PBS + T, PEPT, water and PEPT + T ($p < 0.05$). The initial pH value of each buffer was measured with PBS, PBS + T and water being neutral whereas PEPT and PEPT + T being slightly acidic (pH = 6.5–6.7). PBS is known for its high buffering capacity, whereas water and peptone water have little buffering capacity. When mixed with the papaya juice, the pH of water and peptone water decreased to 5.89–6.26. The pH of the homogenate may affect the state of cells, and this is consistent with the higher cell counts observed in PBS + T. Peptone water is often used in studies involving fresh produce [20,23,49]. Researchers should carefully select homogenization buffers since peptone water alone may lead to experimental errors in studies with acidic produce.

Table 3. pH of papaya skin homogenate as affected by homogenization buffer type, time (min) and speed (rpm) *.

Buffer	1 Min		5 Min		Average
	150 Rpm	250 Rpm	150 Rpm	250 Rpm	
PBS	7.19 ± 0.08	7.20 ± 0.09	7.21 ± 0.06	7.22 ± 0.09	7.21 ± 0.07 a
PEPT	6.32 ± 0.07	6.19 ± 0.28	6.37 ± 0.25	6.18 ± 0.05	6.26 ± 0.18 b
PBS + T	7.11 ± 0.06	7.08 ± 0.06	7.44 ± 0.56	7.12 ± 0.06	7.19 ± 0.29 a
PEPT + T	5.88 ± 0.27	5.87 ± 0.21	5.79 ± 0.06	6.03 ± 0.23	5.89 ± 0.20 c
Water	6.05 ± 0.22	6.03 ± 0.17	6.10 ± 0.14	5.82 ± 0.08	6.00 ± 0.17 c

* "PBS", "PEPT", "PBS + T" and "PEPT + T" stand for phosphate buffered saline, 0.1% peptone water, PBS with 0.2% Tween 80 and 0.1% peptone water with 0.2% Tween 80, respectively. Numbers are mean ± standard deviation (n = 3). No significant interactions were found between the factors. Means in the same column with different superscripts are significantly different ($p < 0.05$).

3.2. Behavior of Pathogenic Bacteria on Whole Papayas Stored at Different Temperatures

With about 7 log CFU of initial inocula, 5.46 and 4.67 log CFU *S.* Typhimurium and *L. monocytogenes* were detected on papaya surfaces on day 0, respectively (Figure 1). Bacteria response to environmental stress differently. *Salmonella* showed higher desiccation tolerance than *L. monocytogenes* in powdered infant formula and desiccated shredded coconut [50,51]. *S.* Typhimurium had an interesting survival and growth pattern. At 21 °C, the population increased gradually to 7.34 log CFU on day 14. At 7 °C, *S.* Typhimurium level decreased to 4.10 log CFU on day 7 and then increased to 6.18 log CFU at the end of the storage period (Figure 1A). Intrinsic factors of fruit, including surface roughness, surface hydrophobicity, nutrient and moisture availability and background flora, may affect the behavior of foodborne pathogenic bacteria on the fruit [8]. At ambient temperature, *S. enterica* level remained stable on whole mangos stored at 20–22 °C for nine days [52]. *Salmonella* was reduced by about 5 and 2 log CFU at high (~7 log) and low (~4 log) inoculation levels, respectively, on whole kiwifruits stored at room temperatures for 10 days [14]. On whole cucumbers stored at 23 °C, *Salmonella* level significantly increased by 1.7 log CFU within the first day of inoculation and remained stable for four days [53]. Looking at the fruit type alone, at commercial cold storage conditions (7–12 °C), *S.* Typhimurium level did not significantly change on whole papaya or mango at the end of the storage period [54]. However, *Salmonella* tended to decrease over time on other fruits, such as passionfruit, strawberry, cucumber and peppers [53–56]. Different from other tropical fruits, sugar accumulates on papaya surfaces as ripening progresses, which provides more nutrients for the attached microorganisms. Naturally present yeast may also aid the growth of *S.* Typhimurium by their saccharolytic interactions with the compounds permeated through papaya skin [57].

L. monocytogenes showed a major increase from 4.67 to 5.60 log CFU during the 1st day of storage at 21 °C, and then gradually grew to 5.91 log CFU in the following 13 days. At 7 °C, *L. monocytogenes* level remained stable on papayas for 14 days (Figure 1B). The behavior of *L. monocytogenes* on fruits varies. *L. monocytogenes* grew on whole cucumbers stored at 4 °C and grew on fresh Gala apples stored at 5 °C and 25 °C [53,57]. However, on Granny Smith apples, 1.5 log CFU and 0.5–1.2 log CFU reductions were observed at 25 and 22 °C, respectively, in two studies [13,57]. The reductions of *L. monocytogenes* on whole cantaloupe and mango were also reported [12,49]. Aside from the intrinsic differences of the fruits, initial inoculation levels and the carrying capacity of the fruit may contribute to the varied behavior of *L. monocytogenes* [8,18]. Approximately three-fold more *L. monocytogenes* died on whole kiwi fruits inoculated with 7 log CFU than those inoculated with 4 log CFU at room temperature over 10 days [14]. In the case of organic Granny Smith apples, *L. monocytogenes* decreased by 1.8 and 0.7 log CFU at inoculation levels of 6.3 and 3.0 log CFU, respectively, at 22 °C over two weeks [13]. Papayas could have a higher carrying capacity than the above-mentioned fruits, leading to the growth of *L. monocytogenes* on papayas even at a relatively high inoculation level. Regardless, *L. monocytogenes* is known for its ability to adapt to cold temperatures through mechanisms of alternating membrane fatty acid composition, synthesizing cold shock proteins and cold acclimation proteins and activating energy providing pathways such as glycolysis [58].

S. Typhimurium and *L. monocytogenes* showed abilities to survive and grow on papaya, and hence effective sanitation methods are essential for papaya production.

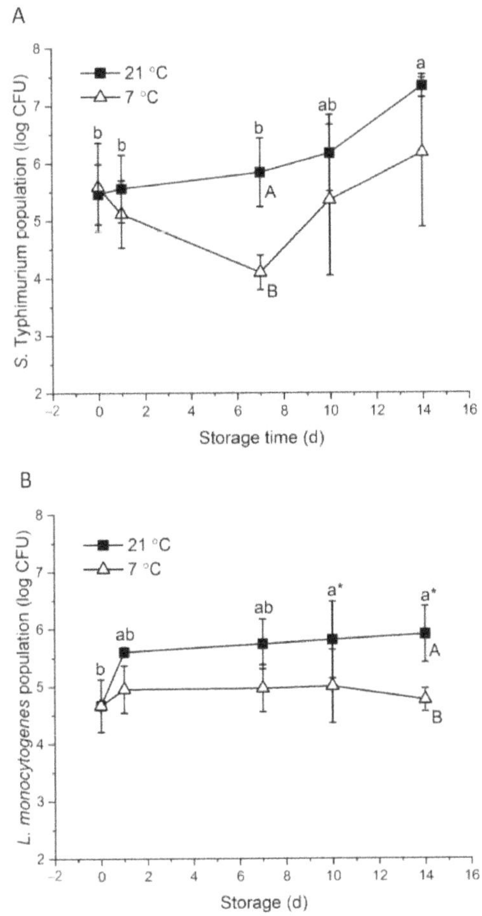

Figure 1. Behavior of *S.* Typhimurium (**A**) and *L. monocytogenes* (**B**) on whole papayas at 21 and 7 °C. Error bars are standard deviations ($n = 3$). Different lower-case letters horizontally indicate significant differences between the means of different time points at each temperature ($p < 0.05$). Different upper-case letters vertically indicate significant differences ($p < 0.05$) between the means of different temperatures at the same time point. "a*" means p values were marginal, which were 0.058 and 0.059 on day 10 and day 14, respectively, compared with day 0.

3.3. Inactivation of *S.* Typhimurium and *L. monocytogenes* on Whole Papayas Using Aqueous ClO_2

Figure 2A shows *S.* Typhimurium reduction by water, aqueous ClO_2, and bleach on whole papayas. 10 ppm of ClO_2 was significantly more effective than 2.5 and 5 ppm ($p < 0.05$). 10 ppm of ClO_2 reduced *S.* Typhimurium from the initial inoculation level of 7.5 log CFU to an undetectable level. 200 ppm of bleach achieved the same result. Malic acid-produced ClO_2 reduced *S.* Typhimurium by 6.23 and 6.90 log CFU at 2.5 and 5 ppm, respectively. HCl- and lactic acid-produced ClO_2 reduced *S.* Typhimurium by 4.20 and 5.05 log CFU, and 3.89 and 4.67 log CFU at 2.5 and 5 ppm, respectively. Overall, ClO_2 solutions generated with malic acid inactivated significantly higher numbers of *S.* Typhimurium than the solutions generated with HCl or lactic acid ($p < 0.05$). 1.74–2.01 and 0.86–1.97 log CFU/cm^2 *Salmonella* was inactivated in 100 ppm free chlorine and 80 ppm peracetic acid with scrubbing by sponges/microfiber, respectively [35]. Comparing with these results, the microbial reduction on papayas achieved by immersing in ClO_2 for 5 min seems more effective.

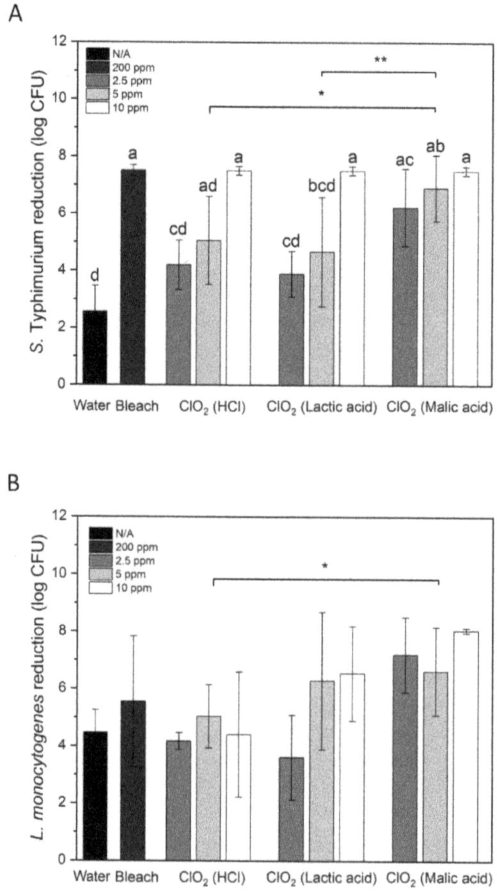

Figure 2. *S.* Typhimurium (**A**) and *L. monocytogenes* (**B**) reduction by water, 200 ppm bleach, and aqueous ClO_2 generated by mixing $NaClO_2$ with HCl, lactic acid or malic acid on whole papayas. Error bars are standard deviations ($n = 3$). Bars labeled with different letters indicate significant differences between the means of treatments ($p < 0.05$). Lines labeled with "*" indicate significant differences between ClO_2 groups made with different acids ("*", $p < 0.05$; "**", $p < 0.01$).

Water treatment only removed 2.56 log CFU of *S.* Typhimurium from papaya surface, whereas 4.47 log CFU of *L. monocytogenes* was removed by water (Figure 2). This may be partially due to that *S.* Typhimurium attached stronger to papaya surfaces than *L. monocytogenes*. In a study conducted by Collignon and Korsten [42], *S.* Typhimurium adhered to peach immediately after contact, whereas *L. monocytogenes* required 60 s for the adhesion. Higher numbers of *S.* Typhimurium cells were observed in one hour than *L. monocytogenes* on peach.

ClO_2 produced with HCl did not show higher effectiveness in reducing *L. monocytogenes* than water (Figure 2B). ClO_2 produced using lactic acid had increased bacterial reductions than HCl-produced ClO_2 at 5 and 10 ppm but with large variations. Malic acid-produced ClO_2 showed the highest *L. monocytogenes* reduction among all ClO_2 treatments. However, there was no significant difference between the three tested concentrations. The group treated with ClO_2 made with malic acid showed statistically higher bacterial reduction than the group treated with ClO_2 made with HCl ($p < 0.05$). 2.5, 5 and 10 ppm of malic acid-generated ClO_2 reduced *L. monocytogenes* by 7.20, 6.63 and 8.04 log CFU, respectively. These reductions were higher than the *L. monocytogenes* reductions on apples, lettuce, strawberries and cantaloupe treated with 5 ppm ClO_2 made with phosphoric acid (~5.6 log CFU) [59]. *L. monocytogenes*-contaminated papayas treated with 200 ppm bleach also showed a relatively large variation with an average reduction of 5.5 log CFU, which was lower than all samples treated with malic acid-generated ClO_2. However, the concentration of bleach was much higher than that of ClO_2, indicating the high antimicrobial efficiency of ClO_2. This result agrees with the higher reduction of *L. monocytogenes* on blueberries treated with 10 ppm ClO_2 (1.7 log CFU/g) than those treated with 200 ppm chlorine (1.0 log CFU/g) for 5 min [23].

ClO_2 generated with malic acid inactivated significantly more *S.* Typhimurium and *L. monocytogenes* than ClO_2 generated with HCl. This result is consistent with our previous observation of the high antimicrobial efficiency of ClO_2 generated with organic acids. In particular, malic acid-generated ClO_2 had higher efficacy in killing *S.* Typhimurium and *L. monocytogenes* than HCl-, sodium bisulfate-, citric acid- and lactic acid- generated ClO_2 [38]. This conclusion was drawn from experiments conducted on bacteria cell suspensions and Romaine lettuce. We hypothesized that synergistic effects between ClO_2 and the excessive organic acids in the ClO_2 solutions may contribute to the high antimicrobial efficiency of organic acid-generated ClO_2. We treated contaminated papayas with individual acid solutions to confirm this hypothesis. Since the pH of ClO_2 decreased with the increase of its concentration (data not shown), pH values corresponding to 10 ppm ClO_2 were selected for the decontamination experiments with acids alone. This means the pH of HCl, lactic acid and malic acid solutions were adjusted to 3.15, 3.42 and 3.36, respectively. *S.* Typhimurium on papayas treated with the acids was reduced by 2.45–3.01 log CFU, which was not significantly different from the samples treated with water (Table 4, $p > 0.05$). Similarly, *L. monocytogenes* on papayas treated with the acids was reduced by 3.58–3.91 log CFU and was not significantly different from the samples treated with water ($p > 0.05$). Hence these results confirmed the high antimicrobial effect of ClO_2 solutions made with malic acid and lactic acid was contributed little by the excessive organic acids, but rather a synergistic effect between ClO_2 and organic acids. The combination treatment of 2.0% lactic acid and 10 ppm ClO_2 resulted in higher reductions of *S.* Typhimurium and *L. monocytogenes* on blueberries than the treatments by each sanitizer alone [60]. On papaya, ClO_2 produced with lactic acid interestingly had similar killing effects to ClO_2 produced with HCl, yet ClO_2 produced with malic acid still performed better than that with HCl. In many studies, lactic acid was either better or as good as malic acid in the inactivation of pathogens when used alone as the sanitizers [61,62]. The synergistic effect somehow altered the antimicrobial efficiency of lactic acid and malic acid. Another factor may contribute to the altered antimicrobial efficacy of the organic-acid-generated ClO_2 compared with HCl-generated ClO_2 is the intermediate compounds produced in the ClO_2 solutions. ClO_2 solution is a mixture of pure ClO_2 and oxidative chlorine compounds such as ClO^{2-}, ClO^{3-}, free chlorine (Cl_2), hypochlorous acid (HOCl) and hypochlorite ion (OCl^-) [32]. These oxy-species varies in oxidation capacity and stability. Since the *pKa* values of lactic acid and malic acid are different, ClO_2 solutions generated with the two organic acids reach equilibrium differently and have different intermediate compound compositions. Measures of the intermediate compound compositions and their chemical oxygen demand would help further understand the mechanisms underlining the different antimicrobial efficacies between various ClO_2 solutions.

Table 4. *S.* Typhimurium and *L. monocytogenes* reduction (log CFU) by water, HCl, lactic acid and malic acid on whole papayas *.

Acid	S. Typhimurium	L. monocytogenes
Water	2.41 ± 0.24	3.86 ± 0.09
HCl	3.01 ± 0.42	3.58 ± 0.19
Lactic acid	2.77 ± 0.18	3.64 ± 0.43
Malic acid	2.45 ± 0.15	3.91 ± 0.43

* Numbers are mean ± standard deviation (n = 3). No statistical significance was found between treatments within each column.

Additionally, CFR Sec. 173.300 specifies that ClO_2 can be used in fresh produce wash with a rinse procedure, and ClO_2 residue in the wash water of the applied fresh produce should not exceed 3 ppm [25]. EPA also specifies that ClO_2 is allowed to rinse fruits and vegetables at a concentration of 5 ppm [63]. Some literature also suggests that the residue on the washed produce should not exceed 3 ppm [64,65]. In this study, the ClO_2 residue on papayas after being treated with 5, 10 and 20 ppm ClO_2 solutions ranged from 8.0×10^{-5} to 6.2×10^{-3} mg/kg, which were far below 3 ppm (Table 5). These numbers were also far below the EPA regulation of 0.8 mg/L ClO_2 residue in public drinking water [27]. Tomatoes and strawberries treated with 0.5 ppm gaseous ClO_2 for 10 min had 0.09 and 0.37 mg/kg ClO_2 residue [29]. ClO_2 residue on produce treated with gaseous ClO_2 was much higher than ClO_2 residue on papayas treated with aqueous ClO_2, providing insights into safety concerns in the application of ClO_2. However, future studies of ClO^{2-} reside on food matrix treated with ClO_2 should be carried out as ClO^{2-} and ClO^{3-} are harmful disinfection by-products (DPBs) that can cause anemia and thyroid dysfunction in animals [26].

Table 5. ClO_2 residue (mg/kg) on papaya surface after being washed with ClO_2 *.

Acid Used to Generate ClO_2	Concentration of ClO_2 Wash Solution		
	5 ppm	10 ppm	20 ppm
HCl	$7.8\times10^{-4} \pm 1.4\times10^{-3}$	$<3.7\times10^{-7}$	$<4.0\times10^{-7}$
Lactic acid	$<3.6\times10^{-7}$	$8.0\times10^{-5} \pm 1.4\times10^{-4}$	$6.2\times10^{-3} \pm 3.9\times10^{-3}$
Malic acid	$<3.6\times10^{-7}$	$<3.3\times10^{-7}$	$<3.6\times10^{-7}$

* Numbers are mean ± standard deviation (n = 3).

4. Conclusions

To provide potential solutions to the emerging issue of foodborne illness outbreaks associated with whole papayas, this study investigated the survival of *S.* Typhimurium and *L. monocytogenes* on whole papaya during storage at 21 and 7 °C and determined the efficiency of aqueous ClO_2 in inactivating the two pathogenic bacteria on whole papaya. Temperature played a significant role in the survival and growth of the bacteria on the fruit. *S.* Typhimurium grew by 1.88 log CFU on whole papaya in 14 days at 21 °C and remained at the initial inoculation level at 7 °C. *L. monocytogenes* grew by 0.93 log CFU on papaya during the 1st day of storage at 21 °C; the level remained stable thereafter at 21 °C and at 7 °C. The acid used to produce aqueous ClO_2 influenced the killing efficiency of ClO_2 against these pathogenic bacteria. ClO_2 solutions generated with malic acid reduced significantly higher levels of *S.* Typhimurium and *L. monocytogenes* than the solution generated with HCl. Methodology wise, we optimized the methods for recovering pathogenic bacteria cells from papaya surface, which was a crucial step evaluating bacterial behavior on fresh produce. This study also provided information on ClO_2 residue on the washed papayas. These results give insights on risk assessment and management of microbiological safety issues associated with whole papaya. Further studies including the intermediate compound compositions in various ClO_2 solutions and the residue of DPBs on ClO_2 treated food matrix are suggested to better understand the antimicrobial mechanisms and safety concerns regarding using aqueous ClO_2.

Author Contributions: Conceptualization, L.D. and Y.L.; methodology, L.D. and Y.L.; investiga-tion, L.D.; formal analysis, L.D.; resources, Y.L.; writing—original draft preparation, L.D.; writ-ing—review and editing, L.D. and Y.L.; supervision, Y.L. Both authors have read and agreed to the published version of the manuscript.

Funding: This work was supported by the State of Hawaii Department of Agriculture and the Hawaii Farm Bureau Grant No. 64921 and the United States Department of Agriculture-Agricultural Research Service Agreement No. 58-2040-8-010.

Institutional Review Board Statement: Not applicable.

Informed Consent Statement: Not applicable.

Conflicts of Interest: The authors declare no conflict of interest.

References

1. Evans, E.A.; Ballen, F.H.; Crane, J.H. *An Overview of US Papaya Production, Trade, and Consumption*; Electronic Data Information Source (EDIS) FE914; University of Florida: Gainesville, FL, USA, 2012; Volume 9, pp. 1–8.
2. FAOSTAT. 2019. Available online: http://www.fao.org/faostat/en/#data/QC/visualize (accessed on 23 May 2021).
3. Gibbs, R.; Pingault, N.; Mazzucchelli, T.; O'Reilly, L.; MacKenzie, B.; Green, J.; Mogyorosy, R.; Stafford, R.; Bell, R.; Hiley, L.; et al. An Outbreak of *Salmonella* enterica Serotype Litchfield Infection in Australia Linked to Consumption of Contaminated Papaya. *J. Food Prot.* **2009**, *72*, 1094–1098. [CrossRef]
4. Hassan, R.; Whitney, B.; Williams, D.L.; Holloman, K.; Grady, D.; Thomas, D.; Omoregie, E.; Lamba, K.; Leeper, M.; Gieraltowski, L.; et al. Multistate outbreaks of *Salmonella* infections linked to imported Maradol papayas—United States, December 2016–September 2017. *Epidemiol. Infect.* **2019**, *147*, e265. [CrossRef]
5. FDA. Letter to Papaya Grower, Harvesters, Packers, Distributors, Exporters, Importers, and Retailers Concerning Food-Borne Illness Outbreaks Tied to Papayas. 2019. Available online: https://www.fda.gov/media/130271/download (accessed on 23 May 2021).
6. de Oliveira, J.G.; Vitória, A.P. Papaya: Nutritional and pharmacological characterization, and quality loss due to physiological disorders. An overview. *Food Res. Int.* **2011**, *44*, 1306–1313. [CrossRef]
7. Strawn, L.K.; Schneider, K.R.; Danyluk, M.D. Microbial Safety of Tropical Fruits. *Crit. Rev. Food Sci. Nutr.* **2011**, *51*, 132–145. [CrossRef] [PubMed]
8. Marik, C.M.; Zuchel, J.; Schaffner, D.W.; Strawn, L.K. Growth and Survival of *Listeria monocytogenes* on Intact Fruit and Vegetable Surfaces During Postharvest Handling: A Systematic Literature Review. *J. Food Prot.* **2019**, *83*, 108–128. [CrossRef] [PubMed]
9. CDC. Multistate Outbreak of Listeriosis Linked to Whole Cantaloupes from Jensen Farms. Colorado | Listeria | CDC. 2012. Available online: https://www.cdc.gov/listeria/outbreaks/cantaloupes-jensen-farms/index.html (accessed on 23 May 2021).
10. CDC. Multistate Outbreak of Listeriosis Linked to Commercially Produced, Prepackaged Caramel Apples Made from Bidart Bros. Apples | Listeria | CDC. 2015. Available online: https://www.cdc.gov/listeria/outbreaks/caramel-apples-12-14/index.html (accessed on 23 May 2021).
11. Poimenidou, S.V.; Chatzithoma, D.-N.; Nychas, G.-J.; Skandamis, P.N. Adaptive Response of *Listeria monocytogenes* to Heat, Salinity and Low pH, after Habituation on Cherry Tomatoes and Lettuce Leaves. *PLoS ONE* **2016**, *11*, e0165746. [CrossRef] [PubMed]
12. Rangel-Vargas, E.; Luna-Rojo, A.M.; Cadena-Ramírez, A.; Torres-Vitela, R.; Gomez-Aldapa, C.A.; Villarruel-López, A.; Téllez-Jurado, A.; Villagómez-Ibarra, J.R.; Reynoso-Camacho, R.; Castro-Rosas, J. Behavior of 11 Foodborne Bacteria on Whole and Cut Mangoes var. Ataulfo and Kent and Antibacterial Activities of Hibiscus sabdariffa Extracts and Chemical Sanitizers Directly onto Mangoes Contaminated with Foodborne Bacteria. *J. Food Prot.* **2018**, *81*, 743–753. [CrossRef]
13. Sheng, L.; Edwards, K.; Tsai, H.-C.; Hanrahan, I.; Zhu, M.-J. Fate of *Listeria monocytogenes* on Fresh Apples under Different Storage Temperatures. *Front. Microbiol.* **2017**, *8*, 1396. [CrossRef]
14. Yuan, J.; Wang, L. Survival of *Escherichia coli* O157:H7, *Salmonella* spp., and *Listeria monocytogenes* on Fresh and Sliced Green and Golden Kiwifruits. *Foodborne Pathog. Dis.* **2018**, *15*, 560–567. [CrossRef] [PubMed]
15. Feng, K.; Hu, W.; Jiang, A.; Xu, Y.; Sarengaowa; Li, X.; Bai, X. Growth Potential of *Listeria Monocytogenes* and *Staphylococcus Aureus* on Fresh-Cut Tropical Fruits. *J. Food Sci.* **2015**, *80*, M2548–M2554. [CrossRef] [PubMed]
16. Penteado, A.L.; Leitão, M.F. Growth of *Listeria monocytogenes* in melon, watermelon and papaya pulps. *Int. J. Food Microbiol.* **2004**, *92*, 89–94. [CrossRef] [PubMed]
17. Penteado, A.L.; Leitão, M.F. Growth of *Salmonella* Enteritidis in melon, watermelon and papaya pulp stored at different times and temperatures. *Food Control* **2004**, *15*, 369–373. [CrossRef]
18. Strawn, L.K.; Danyluk, M.D. Fate of *Escherichia coli* O157:H7 and *Salmonella* spp. on fresh and frozen cut mangoes and papayas. *Int. J. Food Microbiol.* **2010**, *138*, 78–84. [CrossRef] [PubMed]
19. Singh, A.; Yemmireddy, V. Fate of *Salmonella* spp. in fresh-cut papaya (*Carica papaya* L.) at different storage temperature and relative humidity. *LWT* **2021**, *148*, 111810. [CrossRef]

20. Huff, K.; Boyer, R.; Denbow, C.; O'Keefe, S.; Williams, R. Effect of Storage Temperature on Survival and Growth of Foodborne Pathogens on Whole, Damaged, and Internally Inoculated Jalapeños (*Capsicum annuum* var. *annuum*). *J. Food Prot.* **2012**, *75*, 382–388. [CrossRef]
21. Penteado, A.L.; Eblen, B.S.; Miller, A.J. Evidence of *Salmonella* Internalization into Fresh Mangos during Simulated Postharvest Insect Disinfestation Procedures. *J. Food Prot.* **2004**, *67*, 181–184. [CrossRef]
22. Perez-Rodriguez, F.; Begum, M.; Johannessen, G. Study of the cross-contamination and survival of *Salmonella* in fresh apples. *Int. J. Food Microbiol.* **2014**, *184*, 92–97. [CrossRef]
23. Tadepalli, S.; Bridges, D.F.; Driver, R.; Wu, V.C. Effectiveness of different antimicrobial washes combined with freezing against *Escherichia coli* O157:H7, *Salmonella* Typhimurium, and *Listeria monocytogenes* inoculated on blueberries. *Food Microbiol.* **2018**, *74*, 34–39. [CrossRef]
24. U.S. Food and Drug Administration. CFR173.300. CFR—Code of Federal Regulations Title 21. 2020. Available online: https://www.accessdata.fda.gov/scripts/cdrh/cfdocs/cfcfr/cfrsearch.cfm?fr=173.300 (accessed on 23 May 2021).
25. Gómez-López, V.M.; Rajkovic, A.; Ragaert, P.; Smigic, N.; Devlieghere, F. Chlorine dioxide for minimally processed produce preservation: A review. *Trends Food Sci. Technol.* **2009**, *20*, 17–26. [CrossRef]
26. Van Haute, S.; Tryland, I.; Escudero, C.; Vanneste, M.; Sampers, I. Chlorine dioxide as water disinfectant during fresh-cut iceberg lettuce washing: Disinfectant demand, disinfection efficiency, and chlorite formation. *LWT* **2017**, *75*, 301–304. [CrossRef]
27. U.S. Environmental Protection Agency. National Primary Drinking Water Regulations. Available online: https://www.epa.gov/ground-water-and-drinking-water/national-primary-drinking-water-regulations#one (accessed on 11 July 2021).
28. Adhikari, A.; Chhetri, V.; Bhattacharya, D.; Cason, C.; Luu, P.; Suazo, A. Effectiveness of daily rinsing of alfalfa sprouts with aqueous chlorine dioxide and ozonated water on the growth of *Listeria monocytogenes* during sprouting. *Lett. Appl. Microbiol.* **2019**, *69*, 252–257. [CrossRef] [PubMed]
29. Banach, J.; van Overbeek, L.; Groot, M.N.; van der Zouwen, P.; van der Fels-Klerx, H. Efficacy of chlorine dioxide on *Escherichia coli* inactivation during pilot-scale fresh-cut lettuce processing. *Int. J. Food Microbiol.* **2018**, *269*, 128–136. [CrossRef] [PubMed]
30. Ni Tan, J.; Hwang, C.-A.; Huang, L.; Wu, V.C.H.; Hsiao, H.-I. In Situ Generation of Chlorine Dioxide for Decontamination of *Salmonella, Listeria monocytogenes*, and Pathogenic *Escherichia coli* on Cantaloupes, Mung Beans, and Alfalfa Seeds. *J. Food Prot.* **2020**, *83*, 287–294. [CrossRef]
31. Chen, Z.; Zhu, C.; Han, Z. Effects of aqueous chlorine dioxide treatment on nutritional components and shelf-life of mulberry fruit (*Morus alba* L.). *J. Biosci. Bioeng.* **2011**, *111*, 675–681. [CrossRef] [PubMed]
32. Trinetta, V.; Vaidya, N.; Linton, R.; Morgan, M. Evaluation of Chlorine Dioxide Gas Residues on Selected Food Produce. *J. Food Sci.* **2010**, *76*, T11–T15. [CrossRef] [PubMed]
33. Boonyaritthongchai, P.; Techavuthiporn, C.; Cumsingnok, T. *Effect of Acidified Sodium Chlorite and Packaging on Microbial Reduction and Quality Maintenance of Shredded Green Papaya*; Acta Hortic: Istanbul, Turkey, 2018; Volume 1292, pp. 287–292. [CrossRef]
34. Yeoh, W.K.; Ali, A.; Forney, C. Effects of ozone on major antioxidants and microbial populations of fresh-cut papaya. *Postharvest Biol. Technol.* **2014**, *89*, 56–58. [CrossRef]
35. Gu, G.; Bolten, S.; Mendes-Oliveira, G.; Zhou, B.; Teng, Z.; Pearlstein, D.; Luo, Y.; Millner, P.; Nou, X. *Salmonella* inactivation and sponge/microfiber mediated cross-contamination during papaya wash with chlorine or peracetic acid as sanitizer. *Food Microbiol.* **2020**, *95*, 103677. [CrossRef]
36. Gordon, G.; Rosenblatt, A.A. Chlorine Dioxide: The Current State of the Art. *Ozone Sci. Eng.* **2005**, *27*, 203–207. [CrossRef]
37. Kim, H.; Kang, Y.; Beuchat, L.R.; Ryu, J.-H. Production and stability of chlorine dioxide in organic acid solutions as affected by pH, type of acid, and concentration of sodium chlorite, and its effectiveness in inactivating Bacillus cereus spores. *Food Microbiol.* **2008**, *25*, 964–969. [CrossRef]
38. Dong, L.; Li, Y. Comparison of aqueous chlorine dioxide generated with different acids on reducing foodborne pathogenic bacteria. In Proceedings of the International Association for Food Protection 2020 Annual Meeting, Des Moines, IA, USA, 26–28 October 2020.
39. Lang, M.M.; Harris, L.J.; Beuchat, L.R. Evaluation of Inoculation Method and Inoculum Drying Time for Their Effects on Survival and Efficiency of Recovery of *Escherichia coli* O157:H7, *Salmonella*, and *Listeria monocytogenes* Inoculated on the Surface of Tomatoes. *J. Food Prot.* **2004**, *67*, 732–741. [CrossRef]
40. Kim, S.-R.; Yoon, Y.; Kim, W.-I.; Park, K.-H.; Yun, H.-J.; Chung, D.H.; Yun, J.C.; Ryu, K.Y. Comparison of Sample Preparation Methods for the Recovery of Foodborne Pathogens from Fresh Produce. *J. Food Prot.* **2012**, *75*, 1213–1218. [CrossRef] [PubMed]
41. Tripathi, S.; Suzuki, J.Y.; Carr, J.B.; McQuate, G.T.; Ferreira, S.A.; Manshardt, R.M.; Pitz, K.Y.; Wall, M.M.; Gonsalves, D. Nutritional composition of Rainbow papaya, the first commercialized transgenic fruit crop. *J. Food Compos. Anal.* **2011**, *24*, 140–147. [CrossRef]
42. Collignon, S.; Korsten, L. Attachment and Colonization by *Escherichia coli* O157:H7, *Listeria monocytogenes*, *Salmonella* enterica subsp. *enterica* serovar Typhimurium, and *Staphylococcus aureus* on Stone Fruit Surfaces and Survival through a Simulated Commercial Export Chain. *J. Food Prot.* **2010**, *73*, 1247–1256. [CrossRef] [PubMed]
43. Wu, V.C.H. A review of microbial injury and recovery methods in food. *Food Microbiol.* **2008**, *25*, 735–744. [CrossRef] [PubMed]
44. Zhou, L.; Paull, R.E.; Chen, N.J. *Papaya: Postharvest Quality-Maintenance Guidelines. Fruit, Nut, and Beverage Crops*; UH–CTAHR Extension Publication: Honolulu, HI, USA, 2014; Volume 34, pp. 2–4. Available online: https://www.ctahr.hawaii.edu/oc/freepubs/pdf/F_N-34.pdf (accessed on 23 May 2021).

45. Visvalingam, J.; Holley, R.A. Evaluation of chlorine dioxide, acidified sodium chlorite and peroxyacetic acid for control of *Escherichia coli* O157:H7 in beef patties from treated beef trim. *Food Res. Int.* **2018**, *103*, 295–300. [CrossRef]
46. Wu, V.C.; Rioux, A. A simple instrument-free gaseous chlorine dioxide method for microbial decontamination of potatoes during storage. *Food Microbiol.* **2010**, *27*, 179–184. [CrossRef]
47. Brandl, M.T.; Huynh, S. Effect of the Surfactant Tween 80 on the Detachment and Dispersal of *Salmonella* enterica Serovar Thompson Single Cells and Aggregates from Cilantro Leaves as Revealed by Image Analysis. *Appl. Environ. Microbiol.* **2014**, *80*, 5037–5042. [CrossRef]
48. Tian, X.; Yu, Q.; Shao, L.; Li, X.; Dai, R. Sublethal injury and recovery of *Escherichia coli* O157:H7 after ohmic heating. *Food Control* **2018**, *94*, 85–92. [CrossRef]
49. Ma, Q.; Zhang, Y.; Critzer, F.; Davidson, P.M.; Zhong, Q. Quality attributes and microbial survival on whole cantaloupes with antimicrobial coatings containing chitosan, lauric arginate, cinnamon oil and ethylenediaminetetraacetic acid. *Int. J. Food Microbiol.* **2016**, *235*, 103–108. [CrossRef] [PubMed]
50. Koseki, S.; Nakamura, N.; Shiina, T. Comparison of Desiccation Tolerance among *Listeria monocytogenes*, *Escherichia coli* O157:H7, *Salmonella* enterica, and *Cronobacter sakazakii* in Powdered Infant Formula. *J. Food Prot.* **2015**, *78*, 104–110. [CrossRef]
51. Dhowlaghar, N.; Tang, J.; Zhu, M.-J. Thermal inactivation of *Salmonella*, *Listeria monocytogenes* and *Enterococcus faecium* NRRL B-2354 in desiccated shredded coconut. *LWT* **2021**, *149*, 111851. [CrossRef]
52. Mathew, E.N.; Muyyarikkandy, M.S.; Kuttappan, D.; Amalaradjou, M.A. Attachment of *Salmonella* enterica on Mangoes and Survival Under Conditions Simulating Commercial Mango Packing House and Importer Facility. *Front. Microbiol.* **2018**, *9*, 1519. [CrossRef]
53. Bardsley, C.A.; Truitt, L.N.; Pfuntner, R.C.; Danyluk, M.D.; Rideout, S.L.; Strawn, L.K. Growth and Survival of *Listeria monocytogenes* and *Salmonella* on Whole and Sliced Cucumbers. *J. Food Prot.* **2019**, *82*, 301–309. [CrossRef]
54. Behrsing, J.; Jaeger, J.; Horlock, F.; Kita, N.; Franz, P.; Premier, R. Survival of *Listeria innocua*, *Salmonella* salford and *Escherichia coli* on the surface of fruit with inedible skins. *Postharvest Biol. Technol.* **2003**, *29*, 249–256. [CrossRef]
55. Castro-Rosas, J.; Gomez-Aldapa, C.; Acevedo-Sandoval, O.A.; Ramírez, C.A.G.; Villagomez-Ibarra, J.R.; Hernández, N.C.; Villarruel-Lopez, A.; Torres-Vitela, M.D.R. Frequency and Behavior of *Salmonella* and *Escherichia coli* on Whole and Sliced Jalapeño and Serrano Peppers. *J. Food Prot.* **2011**, *74*, 874–881. [CrossRef]
56. Knudsen, D.M.; Yamamoto, S.A.; Harris, L.J. Survival of *Salmonella* spp. and *Escherichia coli* O157:H7 on Fresh and Frozen Strawberries. *J. Food Prot.* **2001**, *64*, 1483–1488. [CrossRef]
57. Salazar, J.K.; Carstens, C.K.; Bathija, V.M.; Narula, S.S.; Parish, M.; Tortorello, M.L. Fate of *Listeria monocytogenes* in Fresh Apples and Caramel Apples. *J. Food Prot.* **2016**, *79*, 696–702. [CrossRef] [PubMed]
58. Chan, Y.C.; Wiedmann, M. Physiology and Genetics of *Listeria Monocytogenes* Survival and Growth at Cold Temperatures. *Crit. Rev. Food Sci. Nutr.* **2008**, *49*, 237–253. [CrossRef] [PubMed]
59. Rodgers, S.L.; Cash, J.N.; Siddiq, M.; Ryser, E.T. A Comparison of Different Chemical Sanitizers for Inactivating *Escherichia coli* O157:H7 and *Listeria monocytogenes* in Solution and on Apples, Lettuce, Strawberries, and Cantaloupe. *J. Food Prot.* **2004**, *67*, 721–731. [CrossRef] [PubMed]
60. Tadepalli, S.; Bridges, D.F.; Anderson, R.; Zhang, R.; Wu, V.C. Synergistic effect of sequential wash treatment with two different low-dosage antimicrobial washes in combination with frozen storage increases *Salmonella* Typhimurium and *Listeria monocytogenes* reduction on wild blueberries. *Food Control* **2019**, *102*, 87–93. [CrossRef]
61. Almasoud, A.; Hettiarachchy, N.; Rayaprolu, S.; Babu, D.; Kwon, Y.M.; Mauromoustakos, A. Inhibitory effects of lactic and malic organic acids on autoinducer type 2 (AI-2) quorum sensing of *Escherichia coli* O157:H7 and *Salmonella* Typhimurium. *LWT* **2016**, *66*, 560–564. [CrossRef]
62. Mohan, A.; Pohlman, F. Role of organic acids and peroxyacetic acid as antimicrobial intervention for controlling *Escherichia coli* O157:H7 on beef trimmings. *LWT* **2016**, *65*, 868–873. [CrossRef]
63. EPA. Reregistration Eligibility Decision (RED) for Chlorine Dioxide and Sodium Chlorite (Case 4023). 2006. Available online: https://www3.epa.gov/pesticides/chem_search/reg_actions/reregistration/red_PC-020503_3-Aug-06.pdf (accessed on 23 May 2021).
64. Mathew, E.N.; Muyyarikkandy, M.S.; Bedell, C.; Amalaradjou, M.A. Efficacy of Chlorine, Chlorine Dioxide, and Peroxyacetic Acid in Reducing *Salmonella* Contamination in Wash Water and on Mangoes Under Simulated Mango Packinghouse Washing Operations. *Front. Sustain. Food Syst.* **2018**, *2*, 18. [CrossRef]
65. Pao, S.; Kelsey, D.F.; Long, W. Spray washing of tomatoes with chlorine dioxide to minimize *Salmonella* on inocu-lated fruit surfaces and cross-contamination from revolving brushes†. *J. Food Prot.* **2009**, *72*, 2448–2452. [CrossRef]

foods

MDPI

Review

Salmonella, Food Safety and Food Handling Practices

Olugbenga Ehuwa [1], Amit K. Jaiswal [1,2,*] and Swarna Jaiswal [1,2]

[1] School of Food Science and Environmental Health, College of Sciences and Health, Technological University Dublin—City Campus, Central Quad, Grangegorman, D07 EWV4 Dublin, Ireland; D19124038@mytudublin.ie (O.E.); swarna.jaiswal@TUDublin.ie (S.J.)
[2] Environmental Sustainability and Health Institute (ESHI), Technological University Dublin—City Campus, Grangegorman, D07 H6K8 Dublin, Ireland
* Correspondence: amit.jaiswal@TUDublin.ie

Abstract: Salmonellosis is the second most reported gastrointestinal disorder in the EU resulting from the consumption of *Salmonella*-contaminated foods. Symptoms include gastroenteritis, abdominal cramps, bloody diarrhoea, fever, myalgia, headache, nausea and vomiting. In 2018, *Salmonella* accounted for more than half of the numbers of foodborne outbreak illnesses reported in the EU. *Salmonella* contamination is mostly associated with produce such as poultry, cattle and their feeds but other products such as dried foods, infant formula, fruit and vegetable products and pets have become important. Efforts aimed at controlling *Salmonella* are being made. For example, legislation and measures put in place reduced the number of hospitalizations between 2014 and 2015. However, the number of hospitalizations started to increase in 2016. This calls for more stringent controls at the level of government and the private sector. Food handlers of "meat processing" and "Ready to Eat" foods play a crucial role in the spread of *Salmonella*. This review presents an updated overview of the global epidemiology, the relevance of official control, the disease associated with food handlers and the importance of food safety concerning salmonellosis.

Keywords: food safety; food handling; food hygiene; *Salmonella*; Salmonellosis; foodborne illness

Citation: Ehuwa, O.; Jaiswal, A.K.; Jaiswal, S. *Salmonella*, Food Safety and Food Handling Practices. *Foods* **2021**, *10*, 907. https://doi.org/10.3390/foods10050907

Academic Editors: Antonio Afonso Lourenco, Catherine Burgess and Timothy Ells

Received: 8 February 2021
Accepted: 19 April 2021
Published: 21 April 2021

1. Introduction

Food poisoning due to pathogens is a major issue of public health concern worldwide with countries expending many resources to overcome it. Bacterial food infections are a source of worry for developed and developing countries. In Europe, *Salmonella* and *Campylobacter* are the most important causes of foodborne illness [1,2]. The European Centre for Disease Prevention and Control, ECDC, [3] asserts that aside from campylobacteriosis which had 246,571 reported cases, *Salmonella* is responsible for the highest number of human infections causing illnesses in 91,857 people in the EU in 2018. A foodborne outbreak is defined as an "incident during which at least two people contract the same illness from the same contaminated food or drink" [3]. There were 5146 reported foodborne outbreaks in 2018 from the EU Member States resulting in illnesses to 48,365 people. *Salmonella* alone accounted for 33% of these outbreaks.

Salmonellosis is linked to the consumption of *Salmonella*-contaminated food products mostly from poultry, pork and egg products. Poor hand washing and contact with infected pets are some of the contamination routes [4]. When infective doses are ingested, the pathogen causes sickness by colonizing the intestinal tract. The *Salmonella* outbreak in Slovakia, Spain and Poland that resulted in 1581 cases was directly linked to infected eggs [4]. It is increasingly becoming a major concern with the global push towards ready-to-eat food products [5]. This group of products is of greater concern because of the minimal heating they are subjected to. The fact they can be consumed without high heat treatment further increases the risk.

This review presents an updated overview of the global epidemiology, the relevance of official control, the disease association with food handlers and the importance of food safety to salmonellosis. Furthermore, numerous control measures for salmonellosis have been discussed.

2. Salmonella

Salmonella is a Gram-negative bacterium that uses flagella for movement. Salmonellosis is regarded as a foodborne infection of the gastrointestinal tract and has been reported to have high incidence rates. The causative organism can pass from the faeces of an infected person or animal to healthy ones [6]. There are more than 2500 recognized serotypes [7].

Salmonella is known to survive for extended periods in low moisture food products [8]. Table 1 shows how long different serotypes survive in dry products. Its ability to survive in low moisture environments is a problem with spices and herbs that are used globally because if contaminated, these organisms survive for extended periods. Worldwide trade of spices and herbs means these organisms could travel and break geographical barriers [9].

Table 1. *Salmonella* survival times in low water activity environments.

Food	*Salmonella* Serotypes	Survival Times	Reference
Dried milk products	S. Infantis, S. Typhimurium, S. Eastbourne	≤10 months	[10]
Desiccated plastic surface Pasta	S. Typhimurium *SL 1344*, S. Infantis, S. Typhimurium, S. Eastbourne	<100 weeks ≤12 months	[11] [12]
Milk chocolate	S. Infantis, S. Typhimurium, S. Eastbourne	>9 months at 20 °C	[13]
Bitter chocolate	S. Eastbourne	≤9 months at 20 °C	[13]
Halva	S. Enteritidis	>8 months at refrigeration temp	[14]
Peanut butter	S. Agona, S. Enteritidis, S. Michigan, S. Montevideo, S. Typhimurium	≤24 weeks at 5 °C ≤6 weeks at 21 °C	[15,16]
Paprika powder	multiple serotypes	>8 months	[17]

2.1. Occurrence of Salmonella

Salmonellae live in the gastrointestinal tracts of domestic and wild animals [18]. A study by Munck et al. [4] identified nine potential sources of *Salmonella*: avian, bio solids-soil-compost, companion animals, equine, poultry, porcine, reptile, ruminant, and wildlife. Wild birds have been known to be a reservoir of these bacteria. The organism resides in the intestines of infected birds and may not cause obvious clinical symptoms except intermittent fevers. Migratory birds are a particular concern. For example, there are several points in the Ukraine where these migratory birds' nest on their journeys between Europe to Africa and Asia [19]. These areas are considered hot spots for *Salmonella* from where the pathogen is distributed to different parts of the world.

Domestic animals are also *Salmonella* reservoirs. In 2019, it was estimated that about 12 million people, that is 40% of the households, in the UK owned pets. Dogs and cats are top on the list but exotic pets such as reptiles, birds, etc. are also kept more frequently [20]. As early as the 1940s, it was proven that humans can get *Salmonella* from reptiles [21]. Bjelland et al. [22] found that 43% of Norwegian reptiles shed *Salmonella*. The Centre for Food Security and Public Health [23] indicated that 93,000 human cases resulted from human association with reptiles. Table 2 gives an overview of salmonellosis cases associated with pets and domesticated animals. Salmonellosis is chiefly a foodborne infection but 7% of human salmonellosis is related to reptiles [23]. These reptiles carry the bacteria in their intestinal tract and shed them through their faeces. This is especially a problem when children are involved with these pets as children belong to a high-risk group. Finlay et al. [21] indicated that *Salmonella* cannot be eliminated from reptiles with the use of antibiotics, as a treatment only increase their antibiotic resistance. Humans, especially infected food handlers, and contaminated environments are also major reservoirs of *Salmonella* [24].

Table 2. *Salmonella* outbreaks involving pets/pet foods.

Salmonella Strains	Pet/Pet Food Product	Cases	Locations Affected	References
S. Typhimurium	Small Pet Turtles	34 reported cases and 11 Hospitalizations	9	[25]
S. Oranienburg	Small Pet Turtles	26 reported cases and 8 Hospitalizations	14	[26]
S. Cerro *S.* Derby *S.* London *S.* Infantis *S.* Newport *S.* Rissen	Pig Ear Pet Treats	154 reported cases and 35 hospitalizations	35	[27]
Salmonella spp.	Backyard Poultry	1134 reported cases, 219 hospitalizations and 2 deaths	49	[28]
Salmonella spp.	Poultry in Backyard Flocks	1120 reported cases, 249 hospitalizations and 1 death	48	[29]
S. Reading	Paws Ground Turkey Food for Pets	90 reported cases	26	[30]
Salmonella spp.	Reptiles	449 hospitalizations	Ireland	[31]

2.2. Epidemiology and Pathogenicity

The severity of *Salmonella* infections is dependent on the specific strain responsible for the infection and on the health status of the host. Children below the age of 5, the elderly and immunocompromised adults represent a specific group that is more susceptible to salmonellosis [32].

Salmonellosis is often characterized by stomach flu (gastroenteritis). This illness is accompanied by nausea, vomiting, abdominal cramps and bloody diarrhoea. It is also associated with headache, feverish conditions and myalgia. The continuous loss of body fluids may result in dehydration especially for infants and the elderly [23]. Salmonellosis is a self-limiting illness that ceases in a week, but deaths have been recorded especially in vulnerable population groups such as very young, elderly and immunocompromised persons [32]. Kurtz, Goggins and McLachlan [33] assert that in cases where salmonellosis becomes systemic, enteric fevers can arise after gastroenteritis and enterocolitis have waned. Enteric fever is a common symptom when *S.* Typhi is the causative organism. These cases are characterized by fever, anorexia, headache, lethargy, myalgia, constipation, and other non-specific symptoms. When resulting in septicemia or meningitis, the disease can be fatal.

Reactive arthritis (ReA) or Reiter's syndrome is a reactive inflammation of the joints that occurs after a gastrointestinal or genitourinary infection. However, its pathogenesis is currently not fully understood [34]. It affects adults between the ages of 20–40 and symptoms may include: painful joint inflammations, eye inflammation, discomfort in urination, swollen toes and fingers, lower back pain, rash on soles and palms, etc. ReA occurs due to *Salmonella* infection in 12 cases per 1000 globally [35]. In both the USA and Europe, ReA has followed salmonellosis in about 15–17% of self-reported patients [36].

There is no agreement on the role of genetics and the risk of having this disease. However, some studies have shown a correlation between the possession of the HLA-B27 surface antigens and the severity of the disease [32].

2.3. Food Products Associated with Salmonella

Salmonella Agona is a less known *Salmonella* serovar. Between the years 2007–2016, it was responsible for 13 outbreaks resulting in 636 illnesses that required hospitalization in the EU. Nine of these outbreaks were due to the consumption of contaminated foods (Table 3). Chicken was responsible for two outbreaks in 2013, red meat for one outbreak in 2014, pork for one outbreak in 2012, unspecified poultry meat for an outbreak in 2007, mixed foods and bakery products were both vehicles for different outbreaks in 2017 [37].

Table 3. Food products involved in *Salmonella* outbreaks in Europe and United States.

Salmonella Strain	Food Product	Cases	Locations Affected	References
S. Javiana	Pre-cut fruits	165 reported cases and 73 hospitalizations	14	[25]
S. Newport	Red Onions	640 reported cases and 85 hospitalizations	43	[38]
S. Javiana	Fruit Mix	165 reported cases and 73 hospitalizations	14	[39]
S. Uganda	Cavi Brand Whole, Fresh Papayas	81 reported cases and 27 hospitalizations	9	[40]
S. Newport	Frozen Raw Tuna	15 reported cases and 2 hospitalizations	8	[41]
S. Carrau	Pre-Cut Melons	137 reported cases and 38 hospitalizations	10	[42]
S. Uganda	Fresh Papayas	81 reported cases and 27 hospitalizations	9	[43]
S. Dublin	Reblochon (bovine raw-milk cheese)	83 reported cases and 41 hospitalizations and 10[b] deaths	France	[44]
S. Agona	infant milk products	37 case and 18 were hospitalized	France	[45]
S. Infantis	Raw chicken products	129 reported cases and 25 hospitalizations	32	[46]
S. Bovismorbificans	uncooked ham products	57 cases and 15 hospitalizations	Netherlands	[47]
S. Mbandaka	Kellogg's Honey Smacks Cereal	135 reported cases and 34 hospitalizations	36	[48]
S. Enteritidis PT14b*	Egg and chicken products	287 reported cases and 78 hospitalizations	North West and South of England	[49]

*b: Information provided by the National Reference Centre for *Salmonella* (NRC), without confirmation that cause of death was attributable to *Salmonella* infection.

In accordance with EU Zoonosis Directive 2003/99/EC, Member States are required to report sources and trends of zoonosis, zoonotic agents and foodborne outbreaks [50]. In 2016, *S.* Agona were isolated from 25 units of foods in 4 Member States and a non-Member State. Approximately 68% of these samples were from meat from poultry. Other isolates were from beef (3), pork (1), cheese from unpasteurized milk (1) and dried seeds (1) [50]. In the same year, 242 units of animals tested positive for *S.* Agona from chicken (209) and turkey (25). These were reported by 11 Member States and two non-Member States. Between the years 2004 and 2015, 608 units tested positive for *S.* Agona in different animal feeds. A majority of them were related to oil seeds or fruit origin (243), then those feeds sourced from land animals (64), another 64 came from unspecified feed sources, feeds from marine animals (43), pet foods (30) while feed for poultry accounted for 28 [37]. However, *S.* Agona occurs less in eggs and its products, fish and its products and fruits and vegetables. There was no report of it being present in "foodstuffs intended for special nutritional uses" and "infant formula" [37]. In the United States, the two most common strains remain *Salmonella* Typhimurium and *Salmonella* Enteritidis [51] but according to outbreaks reported by the CDC in 2019, other strains have been responsible for several foodborne illnesses, leading to hospitalizations and death as reported on (Table 3).

2.4. Salmonella and Vegetable Produce

Traditionally, plants are not recognized as hosts for human pathogens such as *Salmonella* but in the last few decades, the niches for these organisms have changed [52]. *Salmonella* produces periplasmic enzymes with the ability to break plant surface barriers. However, the penetration of these enzymes into plant systems is dependent on pectin and polygalacturonate processing (level of ripening) and physiological wounds [21,53].

Members of the Enterobacteriaceae family are capable of penetrating the stomata of plant leaves [54], hydratodes [55] and roots [56]. Plants contaminated pre- or post-harvest do not exhibit signs of spoilage [57] while the organisms contaminate the produce whether pre-harvest or post-harvest [58].

On the farm, produce is exposed to *Salmonella* by contact with wildlife, contaminated irrigation water, untreated manure [55,59–63]. Poor hygiene by fieldworkers, use of mobile toilets and hand-washing stations increase the risk of pathogen dissemination at pre-harvest [64] and during harvest [65]. After harvest, contamination of produce is mainly duc to poor hygienic practices [63,66].

In the United States, food poisoning outbreaks from raw eggs and seafood is on a decline while outbreaks due to fruits and vegetables keep increasing [15,67], even though field surveys carried out in the United States indicated that *Salmonella* contamination is low during pre-harvest production. Fruits and vegetables have been associated with 130 outbreaks since 1996 [15,42,67,68]. Bennett et al. [69] noted that tomatoes specifically were implicated in 15 multi-state outbreaks of salmonellosis between 1990 and 2010. Traceback analysis suggested that contamination happened during the production or processing stages.

Devleesschauwer et al. [70] noted that although salmonellosis outbreaks due to fruits and vegetables have been well documented, their occurrence, however, remains sporadic. Moreover, Devleesschauwer et al. [70] also stated that for outbreaks involving fruits and vegetables to occur, a multitude of factors must come together. These factors include the presence of vectors, level of crop maturity, physiological defects, presence of native biota that may inhibit or promote human pathogens, type of irrigation practised, etc. The role of environmental conditions and farm practices is also essential in determining the factors that make plants susceptible to *Salmonella* proliferation both pre and post-harvest. The study carried out by Devleesschauwer et al. [70] confirmed that harvesting tomatoes when still green significantly reduces *Salmonella* infestation, as does harvesting after a period of high humidity. Pre-harvest application of copper, iron, potassium, nitrogen or foliar sprays did not affect post-harvest contamination.

3. Global Burden of Salmonellosis

Stanaway et al. [71], while reporting on the global burden of non-typhoidal *Salmonella* invasive disease, asserted that non-typhoidal *Salmonella* remains a major cause of disease and death worldwide. Malnourished young children, the elderly, immunocompromised adults (such as HIV patients), sufferers of acute malaria and those with pre-existing debilitating sickness have greater risks. This infection can attack healthy hosts and in addition to diarrhoea, causes bacteraemia, meningitis and infections in the tonsils. In 2017, *Salmonella* enterocolitis caused 95.1 million disease conditions, 3.1 million disability-adjusted life-years and 50,771 fatalities according to The Global Burden of Diseases, Injuries, and Risk Factors Study (GBD) [71]. The Foodborne Disease Burden Epidemiology Reference Group (FERG) of the WHO in 2010 reported that *Salmonella* was responsible for a total of 180M illnesses and 298,496 deaths (Table 4).

Table 4. Global Burden of salmonellosis.

Salmonella Serovars	Illnesses	Deaths	References
S. enterica, non-typhoidal	153,097,991	56,969	[72]
Invasive non-typhoidal *S. enterica*	596,824	63,312	[72]
Invasive non-typhoidal *S. enterica*	535,000	77,500	[71]
S. enterica Paratyphi A	4,826,477	33,325	[73]
S. enterica Typhi	20,984,683	144,890	[73]

Food illnesses from invasive non-typhoidal *S. enterica* presented the highest disease burden. This is due to the pervasive nature of this organism, the acute diarrhoea it causes and frequent infection of children [74]. Kirk et al. [73] evaluated the health impact of all the serotypes of *Salmonella* and concluded that it presents the greatest foodborne burden. Combining data associated with *S. enterica* from both the invasive Non- Typhoidal *Salmonella* (iNTS), *Salmonella* Typhi and *Salmonella* Paratyphi A and diarrheal infections, a total of 8.76 million Disability-Adjusted Life Year (DALY) from all transmission sources and 6.43 million attributed to infected foods.

In France, between 2008 and 2013, disease pathogens caused between 1.28–2.23 million illnesses, 16,500–20,800 hospitalizations, and 250 deaths. *Campylobacter* spp., non-typhoidal *Salmonella* spp., and norovirus were responsible for >70% of all foodborne pathogen-associated illnesses and hospitalizations while non-typhoidal *Salmonella* spp. and *Listeria monocytogenes* were the main causes of foodborne pathogen–associated deaths. *Salmonella* spp. ranked third as the cause of foodborne illnesses (12%), second as a cause for hospitalization (24%), and first as a cause of death (27%) [75]. Furthermore, Simpson et al. [24] stated that salmonellosis is the second main cause of gastroenteritis in Australia and the most common cause of food-related deaths in the world.

In the EU, there are more than 91,000 reported *Salmonella* infections each year [76]. In 2016, there were 94,530 human cases of salmonellosis reported in the EU with S. Enteritidis accounting for 59% of all cases [50]. There was an increase of 11.5% in the trend of reported food outbreaks compared with that of 2015 and S. Enteriditis was responsible for one in six outbreaks in 2016. *Salmonella* was responsible for the highest health burden with 1766 hospitalizations (45.6%) and 50% of all deaths in outbreak cases [50]. In Australia, gastroenteritis was responsible for about $811 million annually in costs associated with treatments, deaths, loss of productive hours and government surveillance [24].

From 2009 to 2015, there was a drastic increase in hospitalizations due to salmonellosis among the EU/EEA Member States. Concerted efforts by the European Commission and stakeholders tried to level case numbers in 2015 at 12,510 hospitalizations. However, recent data show the trend is rising again with 16,816 recorded hospitalizations in 2018. The USDA ERS [77] estimated the economic cost of *Salmonella* (non-typhoidal) as $3.66B for 2014 to account for lost wages, medical costs, premature deaths, number of cases and productivity losses. In the EU, these costs are estimated to exceed €3 billion a year [3]. Other studies as shown in (Table 5) recorded the cost of illness caused by salmonellosis.

Table 5. Cost of illness studies on salmonellosis.

Country	Year (S)	Cost	Reference
UK	2018	£0.21 billion	[78]
Sweden	2018	€25.6 million	[79]
Australia	2015	AUD 146.8 million	[80]
Canada	2000–2015	CAD 287.78 million	[81]
Netherlands	2012	€6.8 million	[82]
USA	2011	USD 394 million	[41]

4. Control of Salmonellosis

The coordinated *Salmonella* control programs implemented by the EU are one of the most celebrated milestones for the fight against zoonotic diseases. Before 2004, there were over 200,000 reported human salmonellosis cases in 15 EU Member States but control programs put in place reduced this number to 90,000 cases annually in the whole 28 Member States [83]. This led to a reduction by half of the usual cases between 2005 and 2009. The amended EU Regulation 2073/2005 requires the absence of *Salmonella* in 25 g of pooled neck skin samples for broiler carcasses, turkey carcasses and most food types.

However, as evidenced by the Eurobarometer, Europeans are increasingly worried about food safety due to contaminations from pathogenic bacteria. The rising trend of reported cases makes activities aimed at increasing consumer awareness of these foodborne illnesses a requisite [3]. The European Union established an integrated approach to control *Salmonella* in the food chain. This approach involved players at the top government level of the EU Member States, the European Commission, the European Parliament, EFSA and ECDC [76]. The EU took a drastic step to curtail the spread of *Salmonella* by applying extended control programs and legislation that cover the routes of *Salmonella* exposure (Table 6). Under this regulation, an absence of *Salmonella* is required in ready-to-eat foods. Industrially, proof of its absence is a part of buying specifications for raw and finished products. Its absence is taken as evidence of microbiological examination done to support both HACCP control and due diligence. A microbiological criterion for *Salmonella* has been written into law for diverse foods such as poultry products, molluscs, dairy, meat and meat products, ready-to-eat foods, etc. [84].

Table 6. Legislations and Policies against Salmonellosis.

Organization	Regulations/Policies	Objective
European Commission	Regulation (EC) No 1177/2006	Overall implement acts on application of antimicrobial agents and vaccines for poultry birds
	Regulation (EC) No 2008/798/EC	Overall implement acts for importing live birds and eggs
	Regulation (EC) No 517/2011	Reduction in flocks of laying hens
	Regulation (EC) No 200/2010	Standard sampling and monitoring of *Gallus gallus* to reduce *Salmonella* among breeding stocks
	Decision (EC) No 1237/2007	Strict requirement mandating all eggs meant for trade must follow national control programs across the chain
	Regulation (EC) No 200/2012	Standard sampling and monitoring for reduction of *Salmonella* in broilers
	Regulation (EC) No 1190/2012	Standard sampling and monitoring for reduction of *Salmonella* in fattening and breeding turkeys
World Health Organization	Global Foodborne Infections Network (GFN)	Ensuring efficient oversight of antimicrobial-resistant *Salmonella* strains across the food chain; acquiring and testing samples along with data analysis
	WHO Advisory Group on Integrated Surveillance of Antimicrobial Resistance (AGISAR)	Working with FAO in prompt detection and response to food outbreaks by supporting national competent authorities at such periods
	International Network of Food Safety Authorities (INFOSAN)	Provides risk assessment data that serve as guidelines for international standards and recommendations through the Codex Alimentarius Commission

Regulation (EC) No 2160/2003 sets a Union target for each Member State to reduce *Salmonella* in their poultry flocks from 10 to 40% based on their number in the previous year. Every country must achieve at least a 2% reduction annually. However, Regulation (EC) 270 No 517/2011 (Table 6) as amended sets a Union target of 1% or less for *Gallus gallus* breeding flocks positive for *Salmonella enteritidis, Salmonella infantis, Salmonella hadar, Salmonella typhimurium*, monophasic *Salmonella typhimurium* with the antigenic formula 1,4, [5],12:i:-, and *Salmonella* Virchow. Regulation 517/2011 requires sampling to be at least once every 16 weeks compared to 200/2010 which required once every 15 weeks. Commission Regulation (EU) No 1190/2012 (Table 6) which repealed 584/2008 requires that the maximum percentage of *Salmonella* Enteritidis and *Salmonella* Typhimurium should be less than or equal to 1% in both breeding and fattening turkeys.

Curtailing the spread of *Salmonella* involves controls that start from poultry production on the farm until products get to the table of consumers. These controls have to be a farm to fork systematic set of processes [85]. The WHO in 2018 gave recommendations for control of *Salmonella* that cover the whole food chain. These efforts are aimed at strengthening food safety standards that enhance *Salmonella* surveillance efforts, educating consumers and training food handlers on best practices in preventing *Salmonella* and other foodborne diseases (Table 7). It further stressed the importance of national and regional surveillance networks in identifying and monitoring this disease to forestall its detrimental activities and halt its spread. The contact points between children and domesticated animals such as cats, dogs and pet reptiles are mentioned as requiring supervision. The WHO works in improving the effectiveness of national and regional laboratories in tackling salmonellosis.

Table 7. Control measures recommended by the WHO.

Recommendations	Objectives
Prevention methods	Prevention steps should be applied at all stages of the food chain: from primary production, processing, distribution, sales and consumption.
	Salmonella prevention steps recommended in the food handlers handbook should be followed.
	The contact between children and domesticated animals require supervision.
	The public is advised to follow national and regional surveillance systems on foodborne diseases to be aware, detect and respond rapidly to salmonellosis outbreaks early and halt the spread.
Recommendations for the public and travellers	Food must always be cooked properly and served hot
	Only pasteurized milk and its products should be consumed
	Fruits and vegetables should be washed adequately before consumption
	Hands should be washed adequately after contacting animals or using the restroom.
	Ice meant for consumption must be made from potable water
Recommendations for food handlers	Food handlers should observe ingredients and follow hygienic food preparation rules.
	Provision of Five keys to safer food which provides a basis for food safety training courses both for professionals and consumers. They centre on: keeping clean, separating raw from cooked foods, cooking adequately, storing at correct temperatures and use of potable water
Recommendations for producers of fruits and vegetables	Practice good personal hygiene.
	Faecal pollution should be avoided
	Only treated faecal waste is permitted
	Irrigation water should be treated and well managed.
Recommendations for producers of aquaculture products	Practice good personal hygiene.
	Pond environment should be clean
	Water quality should be managed.
	Harvest equipment should be hygienic
	Ensure fish is healthy.

4.1. Food Hygiene Practices

Food hygiene refers to the encompassing conditions and measures that prevent food contamination from production to consumption. Poor hygiene practices along the food chain from slaughtering or harvesting, processing, storage, distribution, transportation to preparation can expose the consumer to foodborne infections that may be fatal [86]. Proper food hygiene practices centre on cleanliness, separating raw meat from other raw/cooked foods, cooking at correct temperatures and chilling (storing) foods before and after cooking [87]. The USFDA [39] reported that poor hygiene during food handling can lead to the spread of *Salmonella* in foods.

Numerous foodborne outbreaks are associated with restaurants [88]. According to CDC estimates, 59% of these outbreaks in the United States happened in the foodservice industry [89]. The CDC estimates that 48 million people suffer from food-related illness, 128,000 are hospitalized and about 3000 subsequently die each year [48]. About 75% of these cases are caused by poor food handling practices in restaurants [90,91].

The catering industry is expanding massively; from 2010 it had increased by 26.5% and this trend is not abating [92]. In 2017 alone, the industry had a revenue of USD800 billion [93]. With this level of growth due to changing societal eating habits, there arises a higher chance for outbreaks of foodborne disease. Food handlers have access to food products when they are unwrapped, the equipment used in making them and places where these unwrapped products are stored or displayed, and therefore can be potential sources of contamination. Poor handling practice at this level is a high-risk factor for foodborne outbreaks. It is therefore very important that workers have adequate food safety training to sustain the industry [94].

4.2. Food Handler Effects

The Codex Alimentarius defines a food handler as "any person who directly handles packaged or unpackaged food, food equipment and utensils, or food contact surfaces and is therefore expected to comply with food hygiene requirements" [95]. Food handlers play a major role in food production and serving. They are responsible for preparing the food and this means they have more direct contact with food systems and can invariably be agents of contamination. The chance for contamination largely depends on how healthy the food handlers are, their personal hygiene, knowledge and application of food hygiene rules [96]. Solomon et al. [97] reported on a study carried out involving 387 food handlers in a meal-serving facility. A total of 159 (41%) of the food handlers had one or more intestinal parasites and 35 *Salmonella* species were isolated from them. Another study was done in Arba Minch University students' cafeteria in Ethiopia involving 345 participants. Stool cultures revealed that 6.9% were positive for *Salmonella* and 3% for *Shigella* [96]. The prevalence of salmonellosis amongst people and food handlers, in this case, increases the risk of food contamination by physical contact (i.e., touching the food with unwashed hands). A food handler can directly cross-contaminate food during preparation by allowing raw foods to come in contact with cooked or ready-to-eat foods or allowing blood or juices to flow from raw to the cooked foods [95]. FSAI further stressed that handlers can indirectly contaminate foods by touching cooked foods after preparing raw foods without prior washing of hands, using the same equipment and utensils meant for raw foods for cooked foods, displaying cooked foods in places meant for raw foods or by poor personal hygiene.

Hygienic Meat Handling Practices

Salmonella has been isolated from meat products more than any other foodstuff. Poultry and its products present the highest statistics on salmonellosis. Adequate meat handling practices start from the farm where these animals are raised. EC 853/2004 prohibits the transport of animals suspected to be sick, which come from herds known to be diseased, to the slaughterhouse without the permission of the competent authority. It also gives specific requirements for slaughterhouses to combat the spread of *Salmonella*. These include having hygienic and sufficient lairage facilities, lock rooms for diseased or suspected animals, separate rooms for evisceration and cutting, etc. The regulation aims at preventing contamination of meat, ensuring disinfectants are present, focuses a lot on slaughter hygiene, and mandates conditions in which the meat must be in during storage and transport [98]. The Hygiene rating of slaughterhouses is highly dependent on technical issues such as slaughter line speed, efficient work routines and the number of carcasses each operator has to deal with. Inadequacies in these factors raise the risks of food infections (Table 8).

Table 8. Report on food handling practices.

Region	Study Type	Issues	References
South Africa (Hospital)	Interview using questionnaire	29% of all food handlers never had a food safety training course. More than 60% of the hospital staff had either good or satisfactory Food Safety Knowledge (FSK) but these did not contribute to better Food Safety Outcomes.	[99]
South Africa (Hospices)	Semi-structured questionnaire	68% had not taken basic food safety training. There was no knowledge of appropriate temperatures for refrigeration and hot RTE foods.	[100]
Ireland (Public)	Survey	Knowledge of food handling was below 10.8% and food poisoning below 20.1%—both were critically low.	[101]
Ethiopia	Survey	Unsatisfactory meet handling practice especially after smoking, sneezing, and coughing.	[102]
Norway, Denmark, Germany, Spain and the UK	Microbiological testing and Hygiene Performance Rating audits	Hygiene is a major issue in Slaughter Operational issues	[103]
Pakistan	Cross-tabulations, chi-square, and correlation tests.	Unhygienic vending practices for ready-to-eat foods	[104]
Global	Analysis of 81 full-text articles	Internalisation of food products across several countries increases risks for poor handling and food safety	[105]

Despite the stringent controls used on farms and slaughterhouses, *Salmonella* is still present in the meat. The handling processes are not aimed at sterilizing the meat but instead at slowing down their activities. The moment these products are exposed to favourable conditions, the bacteria start to grow and multiply to dangerous levels. Hence, hygienic meat handling practices are crucial both domestically and in catering services. The proper handling of meat starts from purchasing raw meats from reputable vendors. If it is pre-packed, then the use-by dates must always be checked.

Raw meat should be kept in separate bags apart from ready-to-eat foods to avoid cross-contamination. Storing of meat is a crucial step. Raw meat/poultry should be stored in sealed bags at the bottom of the fridge as early as possible [58]. This limits the time for *Salmonella* to grow and avoids the dripping of fluids to other foods. Freezing meats before the use-by dates halt the growth of bacteria. Defrosting can be done in a tray at the bottom of the fridge. It is recommended to defrost 2.5 kg/5 lbs of meat or chicken for 24 h. However, when defrosting is done in a microwave, it should be consumed right away [106]. Hands should be washed before and after handling raw meat. All meat types need to be properly cooked before consumption to avoid the intake of bacteria. For whole chicken, cooking should be at 180 °C for 20 min. The same weight for pork and rolled meats should be cooked at the same temperature but for 35 min. Verifying all parts of the meat have received adequate heating is essential. Cutting into the thickest part of the meat to see if the juice runs clear indicates adequate cooking ensuring no part is pink [106]. A thermometer or probe should be used domestically and in catering services for checking temperatures in different parts of food. Areas where meat is handled, and utensils should be colour coded.

4.3. Ready-to-Eat (RTE) Foods and Processed Foods with Needed Control

Processed food is defined as any food that has changed in its preparation. This alteration can be freezing, canning, heating, baking, etc. [107]. *Salmonella* has been isolated from processed foods such as nut butter, frozen pot pies, chicken nuggets, and stuffed chicken entrees [25]. Huang and Hwang [108] defined RTE foods "as a group of food products that are pre-cleaned, precooked, mostly packaged and ready for consumption without prior preparation or cooking". The fact that RTE foods need no further heating step means the consumers have a heavy reliance on the control programs put in place by processors. RTE foods have a shorter shelf life compared to other processed foods. The shelf life is usually a maximum of three weeks after manufacture because they have not been subjected to lethal temperatures to conserve organoleptic properties. These foods depend on hurdle preservative steps such as acidic environment, packaging used, isotonic medium, refrigeration, etc. RTE foods have been linked to several salmonellosis outbreaks such as *Salmonella* Coelin in ready-to-eat salad mix [109], *Salmonella enterica* in chill ready-to-eat poultry meat products [110]. Due to the nature of RTE foods, the risk for contamination and cross-contamination leading to illness is quite high. Finished process testing is only valid for the verification process because the results could be coming in too late [9]. Moreover, the fact that a few samples taken from a batch of products pass microbiological criteria does not guarantee that all products are safe especially when heterogeneous and local contamination may occur [111]. However, food safety management programs based on prerequisite programs and HACCP covering all stages of production will ensure hygiene and microbiological criteria is met. There is a necessity for all food handlers to be trained and retrained periodically on food safety especially when dealing with RTE foods to improve knowledge of food handling and food poisoning (Table 9).

Table 9. A comparison of food safety training efficacies.

Country	Training Method	Study Type	Behaviour	Conclusion	Reference
USA	Knowledge and behaviour-based online training video	Seven question quiz from Servy Safe coursebook	Observation by researcher	Behaviour-based training improves handwashing better than knowledge-based training especially during peak hours	[91]
Malaysia	Food safety training course based on regulations and behaviour training	31 questions	Self-reported questionnaire and researcher observations	Behaviour-based training performed better in certain areas than the control group	[112]
USA	Two hours ServSave training	Questionnaire	Self-reported	Volunteers reported a significant increase in food safety knowledge, but behaviour is unchanged. Self-reported data is unreliable	[36]
USA	Customized lessons using ServSafe	Questionnaire	Researcher Observation	Significant improvement in Food safety knowledge	[113]
Korea	Lecture and demonstrations	Questionnaire	Self-reported questionnaire and researcher observations	Increase in knowledge was statistically significant Intervention did not produce a change in behaviour	[64]
USA	Four hours ServSafe class and behaviour training	Questionnaire	Researcher Observation	Hand washing knowledge and behaviour significantly Improved but these did not improve general compliance behaviour	[114]

4.4. Knowledge vs. Behavioural Training Models

Well-trained food handlers with adequate knowledge of food safety can reduce the risk of food hazards [91]. The fact that many restaurants use different means of ensuring food safety, but outbreaks still occur frequently and are related to poor handlings, raises the question of the efficacy of such training [92]. It is often believed that increased knowledge would directly translate to best practices, but this is not always the case [88]. Training is usually focused on passing information, assessment, and certification. All these are done in a brief period without the opportunity to see it work in real practice and assess if it is translated into behaviour [92]. Yu et al. [91] note that translating knowledge to behaviour is not an easy task just as it was shown that knowledge of proper food handling and behaviour are different things [115].

McFarland et al., [92] reviewed six studies as reported in (Table 3). Results from five of the studies indicated that an increase in the knowledge of an employee on food safety does not necessarily transfer into proper food safety behaviour. Yu et al. [91] showed that knowledge-based training is good, but behaviour training is better. The best results come from a combination of both methods. Knowledge-based training influenced behaviour in some ways, but this effect did not last if used alone. It failed during peak periods in the restaurant. Participants in the behaviour-based training still carried on good practices after the training for longer periods. Husain et al. [112] focused their study on three factors that can influence behaviour: attitude, normative beliefs, and perceived behavioural control. This study centred on food handler having a clear understanding of the importance of food safety in preventing foodborne illness. If they do not understand why they do what they do, then the behaviour would not change. Results showed that there was an improvement in personal hygiene and safe preparation of food for 12 weeks but did not translate to technical procedures such as time-temperature abuse, proper sanitation, etc. [92]. It is also very important to tailor training based on the role the employee takes and their background. The language is spoken and the level of education becomes very important. Type of training material is also important such as videos instead of text, pictures instead of just words and other languages instead of English [113].

5. Future Perspective and Conclusions

Efforts to control salmonellosis should involve both the public and private sectors. Government regulations and stricter measures being put in place can provide a framework that guides both domestic production and international importation requirements. However, this has to be infused into periodic training for food handlers. Industrially, stricter control systems need to be put in place. There should be more focus on production and process controls than on testing finished products. Consumers need to be educated both formally and informally on the basic steps of food safety. There is a need for studies that identify the most suitable means of communicating scientific information and raising awareness on salmonellosis to all strata of the population.

Author Contributions: Conceptualization, O.E., S.J. and A.K.J.; writing—original draft preparation, O.E., writing—review and editing, S.J. and A.K.J.; supervision, S.J. and A.K.J.; All authors have read and agreed to the published version of the manuscript.

Funding: This research received no external funding.

Institutional Review Board Statement: Not applicable.

Informed Consent Statement: Not applicable.

Data Availability Statement: Data sharing not applicable.

Conflicts of Interest: The authors declare no conflict of interest.

References

1. Callejón, R.M.; Rodríguez-Naranjo, M.I.; Ubeda, C.; Hornedo-Ortega, R.; Garcia-Parrilla, M.C.; Troncoso, A.M. Reported foodborne outbreaks due to fresh produce in the United States and European Union: Trends and causes. *Foodborne Pathog. Dis.* **2015**, *12*, 32–38. [CrossRef]
2. Myintzaw, P.; Moran, F.; Jaiswal, A.K. Campylobacteriosis, consumer's risk perception, and knowledge associated with domestic poultry handling in Ireland. *J. Food Saf.* **2020**, *40*, e12799. [CrossRef]
3. ECDC. *Salmonella* the Most Common Cause of Foodborne Outbreaks in the European Union. 2020. Available online: https://www.ecdc.europa.eu/en/news-events/salmonella-most-common-cause-foodborne-outbreaks-european-union (accessed on 2 March 2020).
4. Munck, N.; Smith, J.; Bates, J.; Glass, K.; Hald, T.; Kirk, M.D. Source attribution of *Salmonella* in Macadamia nuts to animal and environmental reservoirs in Queensland, Australia. *Foodborne Pathog. Dis.* **2020**, *17*, 357–364. [CrossRef] [PubMed]
5. Zweifel, C.; Stephan, R. Spices and Herbs as Source of *Salmonella*-Related Foodborne Diseases. *Food Res. Int.* **2012**, *45*, 765–769. [CrossRef]

6. HPSC. Annual Epidemiological Report. Salmonella Infection in Ireland (2018). 2019. Available online: https://www.hpsc.ie/a-z/gastroenteric/salmonellosis/publications/annualreportsonsalmonellosisinireland/Salmonella%202018%20Annual%20report.pdf (accessed on 31 March 2020).
7. Eng, S.-K.; Pusparajah, P.; Ab Mutalib, N.-S.; Ser, H.-L.; Chan, K.-G.; Lee, L.-H. *Salmonella*: A review on pathogenesis, epidemiology and antibiotic resistance. *Front. Life Sci.* **2015**, *8*, 284–293. [CrossRef]
8. Podolak, R.; Enache, E.; Stone, W.; Black, D.G.; Elliott, P.H. Sources and risk factors for contamination, survival, persistence, and heat resistance of *Salmonella* in low-moisture foods. *J. Food Prot.* **2010**, *73*, 1919–1936. [CrossRef]
9. Zwietering, M.H.; Jacxsens, L.; Membré, J.-M.; Nauta, M.; Peterz, M. Relevance of microbial finished product testing in food safety management. *Food Cont.* **2016**, *60*, 31–43. [CrossRef]
10. Ray, B.; Jezeski, J.; Busta, F. Isolation of *Salmonellae* from naturally contaminated dried milk products: Iii. Influence of pre-enrichment conditions. *J. Milk Food Technol.* **1972**, *35*, 607–614. [CrossRef]
11. Gruzdev, N.; Pinto, R.; Sela, S. Persistence of *Salmonella* enterica during dehydration and subsequent cold storage. *Food Microbiol.* **2012**, *32*, 415–422. [CrossRef]
12. Rayman, M.; D'AOUST, J.-Y.; Aris, B.; Maishment, C.; Wasik, R. Survival of microorganisms in stored pasta. *J. Food Prot.* **1979**, *42*, 330–334. [CrossRef]
13. Tamminga, S.; Beumer, R.; Kampelmacher, E.; Van Leusden, F. Survival of *Salmonella* east bourne and *Salmonella* typhimurium in chocolate. *Epidemiol. Infect.* **1976**, *76*, 41–47. [CrossRef]
14. Kotzekidou, P. Microbial stability and fate of *Salmonella* Enteritidis in halva, a low-moisture confection. *J. Food Prot.* **1998**, *61*, 181–185. [CrossRef] [PubMed]
15. Kozak, G.; MacDonald, D.; Landry, L.; Farber, J. Foodborne outbreaks in Canada linked to produce: 2001 through 2009. *J. Food Prot.* **2013**, *76*, 173–183. [CrossRef] [PubMed]
16. Burnett, S.; Gehm, E.; Weissinger, W.; Beuchat, L. Survival of *Salmonella* in peanut butter and peanut butter spread. *J. Appl. Microbiol.* **2000**, *89*, 472–477. [CrossRef] [PubMed]
17. Ray, B.; Jezeski, J.; Busta, F. Isolation of *Salmonellae* from naturally contaminated dried milk products: Ii. Influence of storage time on the isolation of *Salmonellae*. *J. Milk Food Technol.* **1971**, *34*, 423–427. [CrossRef]
18. De Jong, B.; Ekdahl, K. The comparative burden of salmonellosis in the European Union member states, associated and candidate countries. *Bmc Public Health* **2006**, *6*, 1–9. [CrossRef]
19. Obukhovska, O. The natural reservoirs of *Salmonella* Enteritidis in populations of wild birds. *Online J. Public Health Inf.* **2013**, *5*. [CrossRef]
20. PFMA.org.uk. Pet Population 2019. Available online: https://www.pfma.org.uk/pet-population-2019 (accessed on 14 March 2020).
21. Finlay, F.; Furnell, C.; Ridley, P. *Salmonella* in pets the risk to children. *Community Pract.* **2015**, *88*, 27–29.
22. Bjelland, A.M.; Sandvik, L.M.; Skarstein, M.M.; Svendal, L.; Debenham, J.J. Prevalence of *Salmonella* serovars isolated from reptiles in Norwegian zoos. *Acta Vet. Scand.* **2020**, *62*, 1–9. [CrossRef]
23. CFSPH. Reptile-Associated Salmonellosis. 2013. Available online: http://www.cfsph.iastate.edu/Factsheets/pdfs/reptile_associated_salmonellosis.pdf (accessed on 19 August 2020).
24. Simpson, K.M.; Hill-Cawthorne, G.A.; Ward, M.P.; Mor, S.M. Diversity of *Salmonella* serotypes from humans, food, domestic animals and wildlife in New South Wales, Australia. *BMC Infect. Dis.* **2018**, *18*, 1–11. [CrossRef]
25. CDC. Outbreaks Involving Salmonella. 2020. Available online: https://www.cdc.gov/Salmonella/outbreaks.html (accessed on 10 March 2020).
26. CDC. Outbreak of Salmonella Infections Linked to Small Pet Turtles. 2020. Available online: https://www.cdc.gov/Salmonella/typhimurium-1-20/index.html (accessed on 12 August 2020).
27. CDC. Outbreak of Multidrug-Resistant Salmonella Infections Linked to Contact with Pig Ear Pet Treats. 2019. Available online: https://www.cdc.gov/Salmonella/pet-treats-07-19/index.html (accessed on 12 August 2020).
28. CDC. Outbreaks of Salmonella Infections Linked to Backyard Poultry. 2019. Available online: https://www.cdc.gov/Salmonella/backyardpoultry-05-19/index.html (accessed on 12 August 2020).
29. CDC. Multistate Outbreaks of Human Salmonella Infections Linked to live Poultry in Backyard Flocks. 2017. Available online: https://www.cdc.gov/Salmonella/live-poultry-06-17/index.html (accessed on 12 August 2020).
30. FDA. FDA Investigates Outbreak of Salmonella Infections Linked To Pet Food. 2018. Available online: https://www.fda.gov/animal-veterinary/outbreaks-and-advisories/fda-investigates-outbreak-Salmonella-infections-linked-raws-paws-ground-turkey-food-pets (accessed on 12 August 2020).
31. HPSC. Reptiles and Risks of Infectious Diseases-Health Protection Surveillance Centre. 2013. Available online: https://www.hpsc.ie/a-z/zoonotic/reptilesandrisksofinfectiousdiseases/ (accessed on 13 March 2021).
32. Turgeon, P.; Ng, V.; Murray, R.; Nesbitt, A. Forecasting the incidence of salmonellosis in seniors in Canada: A trend analysis and the potential impact of the demographic shift. *PLoS ONE* **2018**, *13*, e0208124. [CrossRef]
33. Kurtz, J.R.; Goggins, J.A.; McLachlan, J.B. *Salmonella* infection: Interplay between the bacteria and host immune system. *Immunol. Let.* **2017**, *190*, 42–50. [CrossRef]
34. Schempp, C.M.; Schauer, F.; Huhn, C.K.; Venhoff, N.; Finzel, S. Skin inflammation associated with arthritis, synovitis and enthesitis. Part 2: Rheumatoid arthritis, reactive arthritis, Reiter's syndrome, Lyme borreliosis, dermatomyositis and lupus erythematosus. *J. Dtsch. Dermaltol.* **2019**, *17*, 167–181. [CrossRef]

35. Ajene, A.N.; Walker, C.L.F.; Black, R.E. Enteric pathogens and reactive arthritis: A systematic review of *Campylobacter, Salmonella* and *Shigella*-associated reactive arthritis. *J. Health Popul. Nutr.* **2013**, *31*, 299–307. [CrossRef]
36. Townes, J.M.; Deodhar, A.A.; Laine, E.S.; Smith, K.; Krug, H.E.; Barkhuizen, A.; Thompson, M.E.; Cieslak, P.R.; Sobel, J. Reactive arthritis following culture-confirmed infections with bacterial enteric pathogens in Minnesota and Oregon: A population-based study. *Ann. Rheum. Dis.* **2008**, *67*, 1689–1696. [CrossRef]
37. EFSA. Multi-Country Outbreak of Salmonella Agona Infections Possibly Linked to Ready-to-Eat Food. 2018. Available online: https://www.efsa.europa.eu/en/supporting/pub/en-1465 (accessed on 27 August 2020).
38. FDA. Salmonella (Salmonellosis). 2019. Available online: https://www.fda.gov/food/foodborne-pathogens/Salmonella-salmonellosis (accessed on 11 April 2021).
39. FDA. Outbreaks of Foodborne Illness. 2021. Available online: https://www.fda.gov/food/recalls-outbreaks-emergencies/outbreaks-foodborne-illness (accessed on 11 April 2021).
40. CDC. Outbreak of Salmonella Infections Linked to Cavi Brand Whole, Fresh Papayas. 2019. Available online: https://www.cdc.gov/Salmonella/uganda-06-19/index.html (accessed on 11 August 2020).
41. CDC. Cost of Illness. 2021. Available online: https://www.cdc.gov/policy/polaris/economics/cost-of-illness.html (accessed on 11 April 2021).
42. CDC. Reports of Selected Salmonella Outbreak Investigations. 2021. Available online: https://www.cdc.gov/Salmonella/outbreaks.html (accessed on 11 April 2021).
43. FDA. Outbreak Investigation of Salmonella Uganda: Fresh Papayas (June 2019). 2020. Available online: https://www.fda.gov/food/outbreaks-foodborne-illness/outbreak-investigation-Salmonella-uganda-fresh-papayas-june-2019 (accessed on 11 August 2020).
44. Ung, A.; Baidjoe, A.Y.; Van Cauteren, D.; Fawal, N.; Fabre, L.; Guerrisi, C.; Danis, K.; Morand, A.; Donguy, M.-P.; Lucas, E.; et al. Disentangling a complex nationwide *Salmonella* Dublin outbreak associated with raw-milk cheese consumption, France, 2015 to 2016. *Eurosurveillance* **2019**, *24*, 1700703. [CrossRef]
45. Jourdan-da Silva, N.; Fabre, L.; Robinson, E.; Fournet, N.; Nisavanh, A.; Bruyand, M.; Mailles, A.; Serre, E.; Ravel, M.; Guibert, V.; et al. Ongoing nationwide outbreak of *Salmonella* Agona associated with internationally distributed infant milk products, France, December 2017. *Eurosurveillance* **2018**, *23*, 17–00852. [CrossRef] [PubMed]
46. CDC. Outbreak of Multidrug-Resistant Salmonella Infections Linked to Raw Chicken Products. 2019. Available online: https://www.cdc.gov/Salmonella/infantis-10-18/index.html (accessed on 11 August 2020).
47. Brandwagt, D.; van den Wijngaard, C.; Tulen, A.D.; Mulder, A.C.; Hofhuis, A.; Jacobs, R.; Heck, M.; Verbruggen, A.; van den Kerkhof, H.; Slegers-Fitz-James, I.; et al. Outbreak of *Salmonella* Bovismorbificans associated with the consumption of uncooked ham products, the Netherlands, 2016 to 2017. *Eurosurveillance* **2018**, *23*, 17–00335. [CrossRef] [PubMed]
48. CDC. Food Safety. 2021. Available online: https://www.cdc.gov/foodsafety/ (accessed on 14 March 2021).
49. England, P.H. Outbreak Report Outbreak of Salmonella Enteritidis Phage Type 14B, May-September 2014. Available online: https://assets.publishing.service.gov.uk/government/uploads/system/uploads/attachment_data/file/432637/S_Enteritidis_PT14B_outbreak_report_gateway__2_pdf (accessed on 11 August 2020).
50. EFSA. Salmonella Cases No Longer Falling in the EU. 2017. Available online: https://www.efsa.europa.eu/en/press/news/171212 (accessed on 18 June 2020).
51. Bennington-Castro, J. What Is Salmonella? *Symptoms, Causes, Diagnosis, Treatment, and Prevention.* Available online: https://www.everydayhealth.com/Salmonella/guide/ (accessed on 10 March 2021).
52. Lenzi, A.; Marvasi, M.; Baldi, A. Agronomic practices to limit pre-and post-harvest contamination and proliferation of human pathogenic Enterobacteriaceae in vegetable produce. *Food Cont.* **2020**, 107486.
53. Abbott, D.W.; Boraston, A.B. Structural biology of pectin degradation by Enterobacteriaceae. *Microbiol. Mole. Biol. Rev.* **2008**, *72*, 301–316. [CrossRef] [PubMed]
54. Jacob, C.; Melotto, M. Human pathogen colonization of lettuce dependent upon plant genotype and defense response activation. *Front. Plant Sci.* **2020**, *10*, 1769. [CrossRef] [PubMed]
55. Rai, P.K.; Tripathi, B. Microbial contamination in vegetables due to irrigation with partially treated municipal wastewater in a tropical city. *Int. J. Environ. Health Res.* **2007**, *17*, 389–395. [CrossRef]
56. Critzer, F.J.; Doyle, M.P. Microbial ecology of foodborne pathogens associated with produce. *Cur. Opi. Biotechnol.* **2010**, *21*, 125–130. [CrossRef] [PubMed]
57. Barak, J.D.; Schroeder, B.K. Interrelationships of food safety and plant pathology: The life cycle of human pathogens on plants. *Ann. Rev. Phytopathol.* **2012**, *50*, 241–266. [CrossRef] [PubMed]
58. FoodSafety.gov. 4 Steps to Food Safety. 2020. Available online: https://www.foodsafety.gov/keep-food-safe/4-steps-to-food-safety (accessed on 28 August 2020).
59. Bartz, J.A.; Marvasi, M.; Teplitski, M. Salmonella and tomatoes. In *The Produce Contamination Problem: Causes and Solution Matthews, K.R., Sapers, G.M., Gerba, C.P., Eds.*; Elsevier: Amsterdam, The Netherlands, 2014; pp. 269–289.
60. Holden, N.J. Plants as reservoirs for human enteric pathogens. *Ani. Sci. Rev.* **2010**, *5*, 185. [CrossRef]
61. Ibenyassine, K.; Mhand, R.A.; Karamoko, Y.; Anajjar, B.; Chouibani, M.; Ennaji, M. Bacterial pathogens recovered from vegetables irrigated by wastewater in Morocco. *J. Environ.Environ. Health* **2007**, *69*, 47–51.

62. Iwu, C.D.; Okoh, A.I. Preharvest transmission routes of fresh produce associated bacterial pathogens with outbreak potentials: A review. *Int. J. Environ. Res. Public Health* **2019**, *16*, 4407. [CrossRef]
63. Luna-Guevara, J.J.; Arenas-Hernandez, M.M.; Martínez de la Peña, C.; Silva, J.L.; Luna-Guevara, M.L. The role of pathogenic *E. coli* in fresh vegetables: Behavior, contamination factors, and preventive measures. *Int. J. Microb.* **2019**, *2019*. [CrossRef]
64. Park, S.; Navratil, S.; Gregory, A.; Bauer, A.; Srinath, I.; Jun, M.; Szonyi, B.; Nightingale, K.; Anciso, J.; Ivanek, R. Generic *Escherichia coli* contamination of spinach at the preharvest stage: Effects of farm management and environmental factors. *Appl. Environ. Microbiol.* **2013**, *79*, 4347–4358. [CrossRef]
65. Jung, Y.; Jang, H.; Matthews, K.R. Effect of the food production chain from farm practices to vegetable processing on outbreak incidence. *Microb. Biotechnol.* **2014**, *7*, 517–527. [CrossRef]
66. Dos Santos, S.F.; Cardoso, R.D.C.V.; Borges, Í.M.P.; Almeida, A.C.; Andrade, E.S.; Ferreira, I.O.; do Carmo Ramos, L. Post-harvest losses of fruits and vegetables in supply centers in Salvador, Brazil: Analysis of determinants, volumes and reduction strategies. *Waste Manag.* **2020**, *101*, 161–170. [CrossRef]
67. Gould, L.H.; Walsh, K.A.; Vieira, A.R.; Herman, K.; Williams, I.T.; Hall, A.J.; Cole, D. Surveillance for foodborne disease outbreaks—United States, 1998–2008. *Morb. Mortal. Wkly. Rep. Surveill. Summ.* **2013**, *62*, 1–34.
68. Mandrell, R.E. Enteric human pathogens associated with fresh produce: Sources, transport and ecology. In *Microbial Safety of Fresh Produce*; Blackwell Publishing: Hoboken, NJ, USA, 2009; pp. 5–41.
69. Bennett, S.; Littrell, K.; Hill, T.; Mahovic, M.; Behravesh, C.B. Multistate foodborne disease outbreaks associated with raw tomatoes, United States, 1990–2010: A recurring public health problem. *Epidemiol. Inf.* **2015**, *143*, 1352–1359. [CrossRef] [PubMed]
70. Devleesschauwer, B.; Marvasi, M.; Giurcanu, M.C.; Hochmuth, G.J.; Speybroeck, N.; Havelaar, A.H.; Teplitski, M. High relative humidity pre-harvest reduces post-harvest proliferation of *Salmonella* in tomatoes. *Food Microbiol.* **2017**, *66*, 55–63. [CrossRef] [PubMed]
71. Stanaway, J.D.; Parisi, A.; Sarkar, K.; Blacker, B.F.; Reiner, R.C.; Hay, S.I.; Nixon, M.R.; Dolecek, C.; James, S.L.; Mokdad, A.H.; et al. The global burden of non-typhoidal *Salmonella* invasive disease: A systematic analysis for the Global Burden of Disease Study 2017. *Lancet Infect. Dis.* **2019**, *19*, 1312–1324. [CrossRef]
72. WHO. *WHO Estimates of the Global Burden of Foodborne Diseases: Foodborne Disease Burden Epidemiology Reference Group 2007–2015*; World Health Organization: Geneva, Switzerland, 2015; Available online: https://apps.who.int/iris/handle/10665/199350 (accessed on 21 June 2020).
73. Kirk, M.D.; Pires, S.M.; Black, R.E.; Caipo, M.; Crump, J.A.; Devleesschauwer, B.; Döpfer, D.; Fazil, A.; Fischer-Walker, C.L.; Hald, T. World Health Organization estimates of the global and regional disease burden of 22 foodborne bacterial, protozoal, and viral diseases, 2010: A data synthesis. *PLoS Med.* **2015**, *12*, e1001921.
74. Majowicz, S.E.; Scallan, E.; Jones-Bitton, A.; Sargeant, J.M.; Stapleton, J.; Angulo, F.J.; Yeung, D.H.; Kirk, M.D. Global incidence of human Shiga toxin–producing *Escherichia coli* infections and deaths: A systematic review and knowledge synthesis. *Foodborne Pathog. Dis.* **2014**, *11*, 447–455. [CrossRef] [PubMed]
75. Van Cauteren, D.; Le Strat, Y.; Sommen, C.; Bruyand, M.; Tourdjman, M.; Da Silva, N.J.; Couturier, E.; Fournet, N.; de Valk, H.; Desenclos, J.-C. Estimated annual numbers of foodborne pathogen–associated illnesses, hospitalizations, and deaths, France, 2008–2013. *Emerg. Infect. Dis.* **2017**, *23*, 1486. [CrossRef]
76. EFSA. *Salmonella* . 2021. Available online: https://www.efsa.europa.eu/en/topics/topic/Salmonella (accessed on 22 March 2021).
77. USDA. Cost Estimates of Foodborne Illnesses. 2021. Available online: https://www.ers.usda.gov/data-products/cost-estimates-of-foodborne-illnesses.aspx (accessed on 21 June 2020).
78. Agency, F.S. The Burden of Foodborne Disease in the UK 2018. Available online: https://www.food.gov.uk/research/research-projects/the-burden-of-foodborne-disease-in-the-uk-2018 (accessed on 19 June 2020).
79. Sundström, K. Cost of illness for five major foodborne illnesses and sequelae in Sweden. *Appl. Health Econ. Health Policy* **2018**, *16*, 243–257. [CrossRef]
80. Ford, L.; Haywood, P.; Kirk, M.D.; Lancsar, E.; Williamson, D.A.; Glass, K. Cost of *Salmonella* Infections in Australia, 2015. *J. Food Prot.* **2019**, *82*, 1607–1614. [CrossRef]
81. Jain, S.; Mukhopadhyay, K.; Thomassin, P.J. An economic analysis of *Salmonella* detection in fresh produce, poultry, and eggs using whole genome sequencing technology in Canada. *Food Res. Int.* **2019**, *116*, 802–809. [CrossRef]
82. Suijkerbuijk, A.W.; Bouwknegt, M.; Mangen, M.-J.J.; de Wit, G.A.; van Pelt, W.; Bijkerk, P.; Friesema, I.H. The economic burden of a *Salmonella* Thompson outbreak caused by smoked salmon in the Netherlands, 2012–2013. *Eur. J. Public Health* **2017**, *27*, 325–330.
83. EC. Control of *Salmonella*. 2020. Available online: https://ec.europa.eu/food/safety/biosafety/food_borne_diseases/Salmonella_en (accessed on 14 August 2020).
84. Food.gov.uk. Meat Industry Guide- Microbiological Criteria. Available online: https://www.food.gov.uk/sites/default/files/media/document/Chapter13-Microbiological-criteria.pdf (accessed on 25 August 2020).
85. WHO. *Salmonella* (non-typhoidal).2020. Available online: https://www.who.int/news-room/fact-sheets/detail/Salmonella-(non-typhoidal) (accessed on 21 March 2020).
86. WHO. Promoting Safe Food Handling. 2019. Available online: https://www.who.int/foodsafety/areas_work/food-hygiene/en/ (accessed on 27 August 2020).

87. CCOHS. Food and Kitchen Hygiene: OSH Answers. 2017. Available online: https://www.ccohs.ca/oshanswers/prevention/kitchen_hygiene.html (accessed on 27 August 2020).
88. Adesokan, H.K.; Akinseye, V.O.; Adesokan, G.A. Food safety training is associated with improved knowledge and behaviours among foodservice establishments' workers. *Int. J. Food Sci.* **2015**, *2015*. [CrossRef]
89. CDC. Vital signs: Incidence and trends of infection with pathogens transmitted commonly through food–foodborne diseases active surveillance network, 10 US sites, 1996–2010. *Mmwr. Morb. Mortal. Wkly. Rep.* **2011**, *60*, 749–755.
90. Almanza, B.A.; Nesmith, M.S. Food safety certification regulations in the United States. *J. Environ. Health* **2004**, *66*, 10.
91. Yu, H.; Neal, J.; Dawson, M.; Madera, J.M. Implementation of behavior-based training can improve food service employees' handwashing frequencies, duration, and effectiveness. *Cornell Hosp. Q.* **2018**, *59*, 70–77. [CrossRef]
92. McFarland, P.; Checinska Sielaff, A.; Rasco, B.; Smith, S. Efficacy of food safety training in commercial food service. *J. Food Sci.* **2019**, *84*, 1239–1246. [CrossRef] [PubMed]
93. National Restaurant Association. Restaurant Industry Facts at a Glance. 2019. Available online: https://www.restaurant.org/research/restaurant-statistics/restaurant-industry-facts-at-a-glance (accessed on 17 August 2020).
94. CIoF. Why Food Safety Training Is Important. 2020. Available online: https://www.foodsafety.ca/blog/why-food-safety-training-important (accessed on 27 August 2020).
95. FAO. Food Handlers: Manual • Instructor. 2017. Available online: http://www.fao.org/3/i5896e/i5896e.pdf (accessed on 21 June 2020).
96. Mama, M.; Alemu, G. Prevalence, antimicrobial susceptibility patterns and associated risk factors of *Shigella* and *Salmonella* among food handlers in Arba Minch University, South Ethiopia. *BMC Infect. Dis.* **2016**, *16*, 1–7. [CrossRef]
97. Solomon, F.B.; Wada, F.W.; Anjulo, A.A.; Koyra, H.C.; Tufa, E.G. Burden of intestinal pathogens and associated factors among asymptomatic food handlers in South Ethiopia: Emphasis on salmonellosis. *Bmc Res. Notes* **2018**, *11*, 1–6. [CrossRef] [PubMed]
98. Europe Commission. Regulation (EC) No 853/2004. Specific Hygiene Rules for Food of Animal Origin. Available online: https://www.fsai.ie/uploadedFiles/Consol_Reg853_2004(1).pdf (accessed on 28 March 2020).
99. Teffo, L.A.; Tabit, F.T. An assessment of the food safety knowledge and attitudes of food handlers in hospitals. *BMC Public Health* **2020**, *20*, 1–12. [CrossRef] [PubMed]
100. Nkhebenyane, J.S.; Lues, R. The knowledge, attitude, and practices of food handlers in central South African hospices. *Food Sci. Nut.* **2020**, *8*, 2598–2607. [CrossRef] [PubMed]
101. Moreb, N.A.; Priyadarshini, A.; Jaiswal, A.K. Knowledge of food safety and food handling practices amongst food handlers in the Republic of Ireland. *Food Cont.* **2017**, *80*, 341–349. [CrossRef]
102. Yenealem, D.G.; Yallew, W.W.; Abdulmajid, S. Food Safety Practice and Associated Factors among Meat Handlers in Gondar Town: A Cross-Sectional Study. *J. Environ. Public Health* **2020**, *2020*. [CrossRef] [PubMed]
103. Alvseike, O.; Røssvoll, E.; Røtterud, O.-J.; Nesbakken, T.; Skjerve, E.; Prieto, M.; Sandberg, M.; Johannessen, G.; Økland, M.; Urdahl, A.M. Slaughter hygiene in European cattle and sheep abattoirs assessed by microbiological testing and Hygiene Performance Rating. *Food Cont.* **2019**, *101*, 233–240. [CrossRef]
104. Afreen, A.; Ahmed, Z.; Ahmad, H.; Khalid, N. Estimates and burden of foodborne pathogens in RTE beverages in relation to vending practices. *Food Qual. Saf.* **2019**, *3*, 107–115. [CrossRef]
105. Gizaw, Z. Public health risks related to food safety issues in the food market: A systematic literature review. *Environ. Health Prev. Med.* **2019**, *24*, 1–21. [CrossRef] [PubMed]
106. SafeFood. How to Cook Meat Safely. Available online: https://www.safefood.net/food-safety/cooking-meat (accessed on 28 March 2020).
107. NHS. Eating Processed Foods. 2020. Available online: https://www.nhs.uk/live-well/eat-well/what-are-processed-foods/ (accessed on 17 August 2020).
108. Huang, L.; Hwang, C.A. In-package pasteurization of ready-to-eat meat and poultry products. In *Advances in Meat, Poultry and Seafood Packaging*; Elsevier: Amsterdam, The Netherlands, 2012; pp. 437–450.
109. Vestrheim, D.; Lange, H.; Nygård, K.; Borgen, K.; Wester, A.; Kvarme, M.; Vold, L. Are ready-to-eat salads ready to eat? An outbreak of *Salmonella* Coeln linked to imported, mixed, pre-washed and bagged salad, Norway, November 2013. *Epidemiol. Infect.* **2016**, *144*, 1756–1760. [CrossRef] [PubMed]
110. Akbar, A.; Anal, A.K. Isolation of *Salmonella* from ready-to-eat poultry meat and evaluation of its survival at low temperature, microwaving and simulated gastric fluids. *J. Food Sci. Technol.* **2015**, *52*, 3051–3057. [CrossRef]
111. Jongenburger, I.; Den Besten, H.; Zwietering, M. Statistical aspects of food safety sampling. *Annu. Rev. Food Sci. Technol.* **2015**, *6*, 479–503. [CrossRef]
112. Husain, N.R.N.; Muda, W.M.W.; Jamil, N.I.N.; Hanafi, N.N.N.; Rahman, R.A. Effect of food safety training on food handlers' knowledge and practices. *Br. Food J.* **2016**, *118*, 795–808. [CrossRef]
113. Richard, A.; Brown, J.; Radhakrishna, R.; Yoder, E.; Nieto-Montenegro, S.; Cutter, C. Development and Implementation of a "Counter-Top". *Food Prot. Trends* **2013**, *33*, 10–19.
114. Schiellerup, P.; Krogfelt, K.A.; Locht, H. A comparison of self-reported joint symptoms following infection with different enteric pathogens: Effect of HLA-B27. *J. Rheumatol.* **2008**, *35*, 480–487. [PubMed]
115. Tokuç, B.; Ekuklu, G.; Berberoğlu, U.; Bilge, E.; Dedeler, H. Knowledge, attitudes and self-reported practices of food service staff regarding food hygiene in Edirne, Turkey. *Food Cont.* **2009**, *20*, 565–568. [CrossRef]

foods

Review

Antimicrobial Blue Light versus Pathogenic Bacteria: Mechanism, Application in the Food Industry, Hurdle Technologies and Potential Resistance

Joshua Hadi [1], Shuyan Wu [1] and Gale Brightwell [1,2,*]

1 AgResearch Ltd., Hopkirk Research Institute, Cnr University and Library Road, Massey University, Palmerston North 4442, New Zealand; Joshua.Hadi@agresearch.co.nz (J.H.); Shuyan.Wu@agresearch.co.nz (S.W.)

2 New Zealand Food Safety Science and Research Centre, Tennent Drive, Massey University, Palmerston North 4474, New Zealand

* Correspondence: Gale.Brightwell@agresearch.co.nz

Received: 26 November 2020; Accepted: 16 December 2020; Published: 18 December 2020

Abstract: Blue light primarily exhibits antimicrobial activity through the activation of endogenous photosensitizers, which leads to the formation of reactive oxygen species that attack components of bacterial cells. Current data show that blue light is innocuous on the skin, but may inflict photo-damage to the eyes. Laboratory measurements indicate that antimicrobial blue light has minimal effects on the sensorial and nutritional properties of foods, although future research using human panels is required to ascertain these findings. Food properties also affect the efficacy of antimicrobial blue light, with attenuation or enhancement of the bactericidal activity observed in the presence of absorptive materials (for example, proteins on meats) or photosensitizers (for example, riboflavin in milk), respectively. Blue light can also be coupled with other treatments, such as polyphenols, essential oils and organic acids. While complete resistance to blue light has not been reported, isolated evidence suggests that bacterial tolerance to blue light may occur over time, especially through gene mutations, although at a slower rate than antibiotic resistance. Future studies can aim at characterizing the amount and type of intracellular photosensitizers across bacterial species and at assessing the oxygen-independent mechanism of blue light—for example, the inactivation of spoilage bacteria in vacuum-packed meats.

Keywords: antimicrobial blue light; pathogenic bacteria; food-borne bacteria; endogenous photosensitizers; porphyrins

1. Introduction

Annually, there are 600 million cases and 420,000 deaths associated with food-borne pathogens, with the majority of the disease burdens (550 million cases and 230,000 deaths yearly) attributed to diarrheal diseases [1]. Bacterial pathogenic agents are major contributors to these diarrheal infections, particularly *Salmonella enterica*, *Camplyobacter* spp. and *Escherichia coli* [1], and can linger in food-processing environments and food products (for example, minimally-processed foods, such as fresh-cut fruits and vegetables or raw seafood). These findings highlight the importance of robust sanitization systems in the food industry.

While heat is a potent germicidal agent, thermal processing of foods may lead to undesirable organoleptic properties and the loss of nutrients. Consumer perception of food safety has also been associated with aversion to chemical hazards, which include food preservatives, pesticides and drug residues [2,3]. Thus, there is a need for non-thermal sanitization technologies that are also free of chemicals.

The emerging non-thermal food-processing technologies include high-pressure processing (HPP) and pulsed electric field (PEF) [4–6]. However, the current forms of these technologies are more costly and less energy efficient—and thus less environmentally friendly—than thermal processing. For example, when used to pasteurize orange juice, HPP and PEF were estimated to consume 24–27 times more electricity (kW/year), incur 5–7 times higher total cost (cents/L) and emit 7–9 times more carbon dioxide than thermal pasteurization [7].

Alternatively, light-based technologies, particularly light-emitting diodes (LED), can be used as a cheap and sustainable non-thermal sanitization system [8]. It is known that ultraviolet-C (UV-C; particularly at 254 nm) exhibits bactericidal activities by inducing the formation of pyrimidine dimers in the bacterial genome and thus can be used within the food industry to sanitize food products or the processing environments [9,10]. However, health issues may arise from the use of ultraviolet (UV) radiation in the food industry, especially as constant exposure of workers to UV may increase the risks for skin cancer (for example, basal cell carcinoma, squamous cell carcinoma and malignant melanoma) [11,12]. A study also reported that accidental exposure of two healthcare workers to UV-C germicidal lamps (254 nm) led to bilateral keratoconjunctivitis and face erythema after 12–24 h, followed by other complications to the skin, eyes, nail and hair after 24 months [13].

In this review, we provide discussions on another emerging light-based sanitization technology derived from the blue region of the visible light spectrum, which is less detrimental to mammalian cells than UV [14] and thus allows for a wider application within the food supply chain due to its safety. We focus on studies that assessed the bactericidal efficacy of blue light-mediated technology on surfaces and in different food matrixes. Further, we also discuss the antimicrobial mechanisms of blue light, available technologies, safety aspects, the combination of blue light with other treatments (hurdle technology) and the potential development of bacterial tolerance to blue light. Additionally, a brief discussion on the inactivation of fish pathogenic bacteria (non-human pathogens) is also provided.

2. Pathogenic Bacteria in Food

Food-borne diseases are mainly caused by the consumption of contaminated food or water, with contamination possibly occurring at any point of production or distribution. Globally, the major food-borne pathogenic bacteria include *Salmonella* spp., *Campylobacter* spp., enterohemorrhagic *E. coli* (EHEC), *Listeria monocytogenes* and *Vibrio cholerae* [15]—the distribution of these bacteria across the globe, among other food-borne pathogenic agents, is summarized in a report by the World Health Organization [1].

In food-processing environments, Gram-negative bacteria are pre-dominant, particularly *Pseudomonas* spp., *Enterobacteriaceae* (especially *Serratia* spp.) and *Acinetobacter* spp. Among Gram-positive bacteria, lactic acid bacteria (LAB), *Staphylococcus* spp. and *Bacillus* spp. are the most commonly identified residential bacteria. While some of these tend to be innocuous, several pathogenic strains, such as *Staphylococcus aureus*, *Pseudomonas aeruginosa* and *Bacillus cereus*, are also known to linger on surfaces, especially due to their ability to form spores or biofilms [16].

The main pathogenic bacteria associated with dairy products are *L. monocytogenes*, *Salmonella* spp., *S. aureus*, *Cronobacter* spp. and Shiga toxin-producing *E. coli* (STEC) [17,18]. Dairy farm environments are a common habitat for *L. monocytogenes*, *Salmonella* spp. and STEC [19–25], whereas *S. aureus* is less prevalent in the environments and its transmissions are more likely to occur through contaminated animals (for example, those that have mastitis) [26,27]. There are no conclusive data on the natural environments of *Cronobacter* spp., especially *Cronobacter sakazakii* that is commonly associated with contaminated infant formulas [28,29], albeit these bacteria have been associated with plants [30–32] and animal feed [27,33]. The prevalence and type of dairy-associated pathogenic bacteria may also vary with animal source of the milk and geographical location [34,35]. Based upon these data, pathogens are mainly transferred to dairy products or processing environments from farm environments (soil, animal feed, etc.). For instance, two outbreaks (STEC O26:H11 in Italy and *L. monocytogenes* in Canada) were associated with contaminated dairy processing plants (cheese and milk plants) [36,37]—one study found that contaminated cheeses from a dairy plant had been distributed to approximately 300 retailers, which caused extensive cross-contamination [36]. Contamination at the retail level has also been reported, with *L. monocytogenes*, *Salmonella* spp., *Shigella* spp. and *E. coli* O157:H7 found in cheeses and raw milk [38–40].

Outbreaks related to farm-based meat products, such as beef and pork, are primarily caused by *Salmonella* spp. and EHEC (particularly *E. coli* O157:H7) [41–43]. Other bacteria have also been identified as causes of meat product-related outbreaks, namely *L. monocytogenes*, *S. aureus*, *Campylobacter* spp., *Clostridium* spp. and *B. cereus* [41,42,44]. While poultry products may also carry all of these pathogens, previous outbreaks were mostly caused by *Salmonella* spp., *Campylobacter* spp. or *Clostridium perfringens* [45–51]. In food-processing plants, fecal matters (for example, in hides or poultry skin) and aerosols generated during processing (for example, dehiding or evisceration processes) could facilitate the spread of pathogenic bacteria [44,52]. Several studies have also reported on the prevalence of pathogenic bacteria, including *E. coli* O157:H7, *Salmonella* spp., *Shigella* spp. and antibiotic-resistant *S. aureus*, in retail shops across countries—these bacteria were found on the products (raw or cooked beef, mutton, pork, chicken and turkey) and in the environments [39,53–56].

In seafood, the major pathogenic bacteria include *Vibrio* spp., *Salmonella* spp., *L. monocytogenes*, *Campylobacter* spp., EHEC, *Clostridium* spp. and *Shigella* spp., which could cause diseases ranging from mild gastroenteritis to life-threatening infections [57–60]. *Vibrio* spp. is ubiquitous in aquatic ecosystems, with infections in humans commonly associated with *Vibrio parahaemolyticus*, *Vibrio vulnificus* and *V. cholerae* [57,61,62]. Other bacteria, such as *Salmonella* spp. and *E. coli*, may also proliferate in bodies of water, particularly when contaminated with sewage effluents [57,63]. Cross-contamination during food production is the primary route of transmission for *L. monocytogenes* within the seafood industry and thus presents a major concern due to its ability to persist in the environment and to multiply during refrigeration [64]. Similarly, a study identified the contamination of a cutting board by *V. parahaemolyticus* from raw squid as a cause of a gastroenteritis outbreak at a food bazaar in South Korea [65]. Further, there is a heightened health risk in consuming raw seafood, as evident from previous outbreaks associated with uncooked (or undercooked) fish, oysters, abalone or sea squirt [66–69].

Fruits and vegetables may harbor a myriad of pathogenic bacteria, such as *Shigella* spp., *B. cereus*, *Campylobacter* spp., *Yersenia enterocolitica* and *Clostridium botulinum*, albeit previous outbreaks were mostly associated with STEC (particularly *E. coli* O157:H7), *Salmonella* spp. and *L. monocytogenes* [70–72]. Irrigation water that comes from contaminated sources is a major reservoir for these pathogens [73–75] and may occasionally carry *Vibrio* spp., for example, two studies identified *Vibrio* spp. on vegetables irrigated with untreated water from streams [76] and wastewater [77]. Other sources of contamination include pre-harvest factors, such as compost, insects, soil and wildlife animals, along with harvesting equipment or post-harvest vectors, including human (during packing), transport vehicles and processing equipment [70,71]. Pathogenic bacteria have also been detected in different horticultural products at the retail level across the globe: *L. monocytogenes* and *E. coli* isolated from frozen fruits or vegetables (England) [78]; *L. monocytogenes*, *S. enterica* or *E. coli* from ready-to-eat raw vegetables (UK, Malaysia or Nigeria) [79–81]; and *Salmonella enterica* subsp. *enterica* serovar Typhimurium, *C. perfringens*, *Campylobacter* spp. or *L. monocytogenes* from fresh produce (Mexico, Canada or New Zealand) [82–84]. In addition, a meta-analysis of 53 studies identified 453 cumulative incidences of STEC, *L. monocytogenes* and *Salmonella* spp. in fruits/vegetables from retail establishments across Europe between the years 2001 and 2017, with *L. monocytogenes* dominating in vegetables and STEC in fruits [85].

Bacteria may form biofilms to resist physical, mechanical and/or chemical stresses, including chemical disinfectants used in food-processing environments. For instance, several pathogenic staphylococci isolated from food or food equipment, namely *Staphylococcus capitis*, *Staphylococcus cohini*, *Staphylococcus saprophyticus* and *Staphylococcus epidermidis*, had shown abilities to form biofilms on polystyrene and stainless steel [86]. The stability of these biofilms against disinfectants (benzalkonium chloride) or denaturation enzymes (dispersin B, proteinase K or trypsin) is dictated by their compositions, which are determined by the presence of genes encoding either cell wall anchored proteins (CWA) or polysaccharide intracellular adhesin (PIA) [87]. This finding presents an alternative mechanism of biofilm formation in *Staphylococcus* spp., which is predominantly attributed to the presence of *ica* operon (*icaADBC* locus and *icaR* regulatory gene) that encodes PIA [88]—for example, *icaA* gene was found to be correlated to strong biofilm formation in food-related staphylococci isolates [86]. Other major biofilm-forming pathogenic bacteria include *L. monocytogenes* (poultry, red meat, seafood and dairy), *Salmonella* spp. (poultry, red meat, seafood and horticulture), *E. coli* O157:H7 (red meat and horticulture), *B. cereus* (dairy, seafood and horticulture), *Vibrio* spp. (seafood) and *Campylobacter* spp. (poultry) [89–92].

In addition, biofilms are composed of bacterial aggregates enclosed in extracellular polymeric matrix, which constitutes polysaccharides, proteins, lipids and exogenous deoxyribonucleic acids (DNA), and can function as a platform for physical/social interactions (for example, microbial consortia) that enhance gene transfers [93]. Several bacteria, such as *B. cereus* and *E. coli* O157:H7, also form multispecies biofilms to enhance their survival in food-processing lines [94,95]. The formation of biofilms is also dependent on bacterial structures that are responsible for initial surface attachment, such as flagella and/or fimbriae (for example, curli) in *L. monocytogenes* [96,97], *S.* Typhimurium [98,99], *E. coli* O157:H7 [100] and *V. cholerae* [101].

3. Antimicrobial Blue Light

3.1. Mechanism

Bactericidal effects of blue light are mostly attributed to the wavelength range of 400 to 450 nm [102], although several reports have demonstrated the antimicrobial efficacy of blue light at longer wavelengths (460, 465 or 470 nm) [103–106]. Blue light-mediated inactivation of bacteria is associated with the generation of reactive oxygen species (ROS) when the light is absorbed by endogenous photosensitizers, which can be found in different types of bacteria (Gram positive and Gram negative; aerobic and anaerobic) [107]. Given that these photosensitizers, such as protoporphyrin, coproporphyrin and uroporphyrin, are intermediate species in the heme biosynthesis, it is likely that they are accumulated in the cytoplasmic matrix [108,109], although their precise locations within the bacterial cell are not fully understood.

The blue light-mediated photosensitization process is dependent on the presence of oxygen and mainly induces cytotoxicity (apoptosis or necrosis) through oxidative stresses caused by singlet oxygen species ($^1O_2$) [110]. Upon illumination, photosensitizers at a ground state (lowest energy level) are converted into their excited singlet state (short-lived) or triplet state (long-lived), which, in the presence oxygen, can undergo two types of energy transfer: (1) type I that produces toxic oxygen species, such as hydrogen peroxide (H_2O_2), superoxide or hydroxyl radicals; (2) type II that generates $^1O_2$ [111]. Subsequently, these ROS can induce damages to different parts of the bacterial cells, including the cell membrane, cell wall and genome (Figure 1).

An increase in blue light-induced membrane permeability was observed across several studies [112–115], although the precise mechanism is not fully elucidated. A study found that blue light illumination (405 nm) did not affect the lipid membrane of *Salmonella* spp.—there was an absence of malondialdehyde, which is a product of lipid peroxidation [116]. In contrast, two studies demonstrated that blue light inactivation (415 nm) of methicillin-resistant *S. aureus* (MRSA) or *C. sakazakii* involved lipid peroxidation, as determined by the detection of malondialdehyde and reduction in post-treatment unsaturated fatty acids ($C_{16:1}$ in both bacteria, $C_{20:1}$ and $C_{20:4}$ in MRSA, and $C_{18:1}$ and $C_{18:2}$ in *C. sakazakii*) [113,115]. Further, while one study observed the presence of blue light-induced oxidation of guanine residues in the bacterial DNA of *Salmonella* spp. (presence of 8-hydroxydeoxyguanosine) [116], others reported no DNA breakage in blue light-treated (405 nm) *E. coli* O157:H7, *Shigella sonnei* and *S.* Typhimurium [112]. These discrepancies are potentially due to the fact that the type and amount of endogenous photosensitizers vary across different bacterial species, although further investigations are needed to explain the different susceptibilities of bacteria to blue light [117].

In addition to lipids and nucleic acids, blue light can also attack proteins, carbohydrates (polysaccharide) and peptidoglycan (polymers of amino acids and sugars in bacterial cell walls). Blue light treatments, in the presence of exogenous cationic photosensitizers, induced the loss of cell membrane-associated proteins in *S. aureus* [118] and the reduction of 81% in the polysaccharide content within *P. aeruginosa* biofilms [119]. In two studies, images taken by transmission electron microscopy revealed blue light-induced breakages of bacterial cell walls in MRSA [120] and *Acinetobacter baumannii* [121]. Further, *E. coli* lipopolysaccharide coated on titanium disc was inactivated upon illumination by blue light (405 nm), as evident from the reduced activities of mouse macrophages post-treatment [122]. However, the current literature lacks data on the effect of blue light on lipopolysaccharide (endotoxin) contained within intact outer membranes of Gram-negative bacteria and thus it is a subject of future studies.

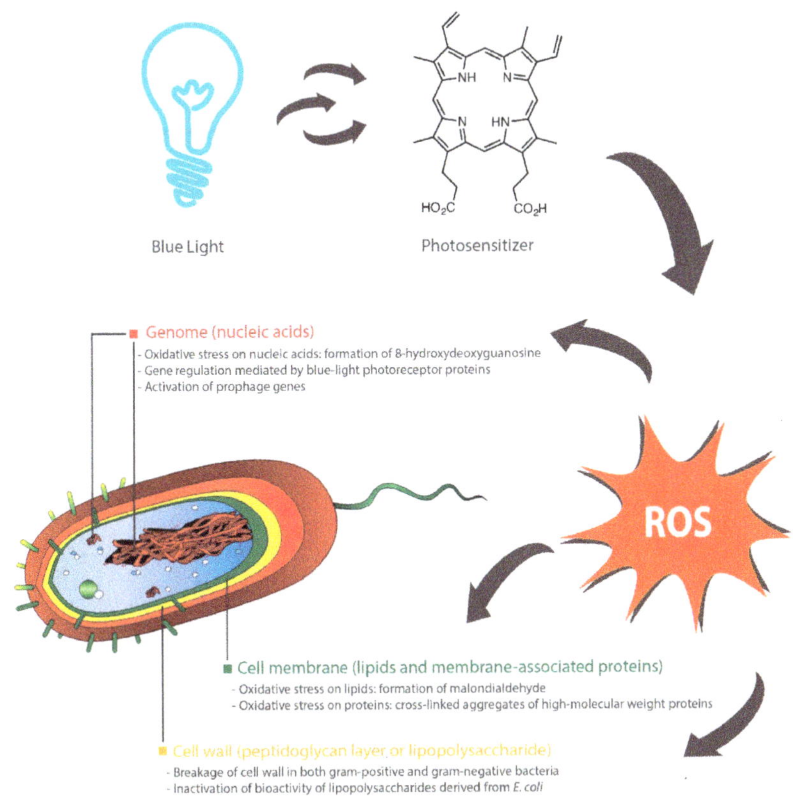

Figure 1. Bactericidal activities of blue light rely on activation of endogenous photosensitizers, such as porphyrins, which subsequently induces the production of reactive oxygen species (ROS). These ROS inflict oxidative damages to nucleic acids [116], lipids [113,115] and proteins [118]. Inhibition of biofilm formation can also occur through blue light-regulated transcriptional pathways [123–125] or bacterial inactivation through the activation of prophage genes [126]. Breakages of cell walls [120,121] and inactivation of lipopolysaccharides (outer membrane of Gram-negative bacteria) [122] have been reported, although the precise effects of antimicrobial blue light on peptidoglycan and lipopolysaccharide are not fully elucidated.

Blue light can act as a transcriptional regulator in bacteria [127,128], especially due to the presence of blue light receptor proteins [128]. These photoreceptors include the blue light-sensing flavin adenine diphosphate (BLUF) proteins that can undergo conformational changes upon illumination by blue light and subsequently elicit downstream effects on bacterial surface attachments, biofilm formation and motility [129]. For instance, two studies found that a BLUF-associated protein, namely YcgF, downregulated the synthesis of curli fibres, but upregulated biofilm formation in *E. coli* [123,124]. In contrast, others reported that *A. baumannii*-harboring *blsA* gene, which encodes BLUF-containing photoreceptor proteins, did not form biofilms under blue light (462 nm), whereas biofilms were observed in a mutant strain with no functional *blsA* [125]. The viability of these bacteria was not affected by blue light in both wild and mutated strains, although blue light had a negative effect on bacterial motility and pellicle formation [125]. These findings indicate that there is a variety of blue light-sensing pathways in bacteria, which could be further explored as an alternative method for controlling the growth of bacteria. Another study also observed an alternative molecular mechanism of bactericidal blue light (460 nm) that involved the activation of prophage genes in MRSA, which subsequently led to the killing of the bacteria [126]. Future studies could aim at investigating the presence of similar genes (light-sensing and prophage genes) in food-borne bacteria and subsequently at designing targeted blue light-mediated interventions for controlling the persistence of these bacteria in food or food-processing environments.

In summary, antimicrobial blue light can act upon different parts of the bacterial cell, primarily through the action of ROS. These ROS can induce oxidative damages to a range of macromolecules, such as lipids (cell membrane), proteins (cell wall-associated proteins), nucleic acids (DNA, RNA or plasmids) and polysaccharides (extracellular matrix of biofilms). Additionally, several bacterial species, such as *E. coli* and *A. baumannii*, possess blue light receptors that control biofilm formation and motility, and thus can be targeted to reduce their persistence in the environments. Further, several prophage genes may be activated by blue light and induce inactivation of the carrying bacteria (Figure 1).

3.2. *Available Technologies*

The majority of studies on antimicrobial blue light have used light-emitting diodes (LED) as a light source. LED is commonly comprised of semiconductor materials that are doped with impurities, which create free electrons on the *n* side and holes (absence of electrons) on the *p* side—also known as the *p–n* junction. When electrical voltage is applied, current flows from the positively-charged end (*p* side; anode) to the negatively-charged end (*n* side; cathode), with electrons moving in the opposite direction. Subsequently, as an electron interacts with a hole, it falls to a lower energy state through the release of a photon. In this process, the resulting color emitted corresponds to the band gap energy within the *p–n* junction, which depends on the semiconductor materials and impurities used [130,131]. Currently, a typical blue LED is made of indium/gallium nitride (InGaN) layers grown on sapphire or silicon substrates, which can theoretically cover the entire visible light spectrum—365 nm (GaN) to 1771 nm (InN)—albeit the quality of materials deposited within the LED structure declines continuously beyond 480 nm due to a range of inherent material challenges [132].

Laser diode, which emits light with a higher coherence and narrower emission band than LED, is another source of blue light that has been used in clinical settings. The photomodulative effects of these two light sources on biological systems have been a subject of debates, especially due to their differences in light coherence and wavelength bandwidth. However, accumulating evidence suggests that these parameters have little effect on the biological efficacy of light-based technologies—for example, two studies found similar effects of red LED and laser diode upon tissue repair in rats [133,134]. Others also proposed that biological effects of light were dependent on dosage and wavelength, but not on light sources, with similar healing effects of LEDs and laser diodes on skin wounds reported across different studies [135]. Similarly, the antimicrobial potency of blue light is independent of the light source used, as one in vitro study revealed that LED (405 nm; non-coherent light) and laser diode

(405 nm; coherent light) were equally efficient in inactivating MRSA across four light dosages (40, 54, 81 or 121 J/cm^2) [136]. Although LED is relatively cheaper than laser diodes, it remains unclear which technology is more efficient based upon their germicidal output per unit electrical power input. In addition, superluminous diode (SLD; 405 or 470 nm), which is an intermediate between LED and laser diode in its light intensity, coherence and emission bandwidth [137–139], has also shown bactericidal activities against *P. aeruginosa*, MRSA and *S. aureus* in vitro [140–142].

A femtosecond laser, which emits ultrasecond pulses at approximately 10^{-15} s per pulse, is another technology that can be used to deliver antimicrobial blue light. At light dosages of 18.9–75.6 J/cm^2 (5–20 min), a blue femtosecond laser (400, 410 and 420 nm) inhibited the growth of *S. aureus* and *P. aeruginosa* on agar plates (inhibition zones observed), possibly due to DNA damages induced by ROS [143]. In agreement, a femtosecond laser (425 nm; 800 J/cm^2; 1 h) reduced a mutant *S.* Typhimurium lacking RecA proteins (reponsible for damaged DNA repair) by 5 log colony forming units (CFU), whereas only 0.5-log reduction (CFU) was observed for the wild-type bacteria and thus this finding enhanced the view that DNA damage was a predominant inactivation mechanism of a bactericidal femtosecond laser [144]. However, the two studies used different methods for measuring bactericidal activity (qualitative or quantitative), and also differed in their light dosages (max. 75.6 J/cm^2 or 800 J/cm^2) and treatment times (max. 20 min or 60 min) [143,144]. Therefore, the potential use of a femtosecond laser as an antimicrobial technology depends on future investigations into its energy efficiency and also its efficacy against different types of bacteria.

3.3. Blue Light Regimes

Antimicrobial blue light may be delivered at high irradiance with short duration times (HI-SD) or low irradiance with long duration times (LI-SD). A study demonstrated that the bactericidal activity of blue light (405 nm) was dependent on light dosage: the highest inactivation of four bacteria, namely *S. aureus*, *Streptococcus pneumoniae*, *E. coli* and *P. aeruginosa*, was achieved at the highest irradiance (approximately 9 mW/cm^2) for a constant treatment time (120 min) or in the longest illumination time (250 min) at a constant irradiance (approximately 9 mW/cm^2) [145].

In the same study, HI-SD treatment (approximately 9 mW/cm^2 for 250 min) was also less effective than LI-SD (approximately 2.25 mW/cm^2 for 1000 min) in inactivating pathogenic bacteria. Isolated colonies were observed on the perimeter of plates exposed to HI-SD, whereas confluent border present on LI-SD plates, indicating post-treatment migration of bacteria to the nutrient-rich and non-treated areas on HI-SD plates. Thus, LI-SD seemed to exhibit higher bactericidal and bacteriostatic effects on a qualitative level [145]. A similar finding was reported for *L. monocytogenes*, with LI-SD treatment of 10 mW/cm^2 for 180 min yielding a 5.18-log reduction (CFU/mL), whereas HI-SD treatments of 20 mW/cm^2 for 90 or 30 mW/cm^2 for 60 min produced bacterial inactivation of approximately 5 log CFU/mL—the differences were not statistically significant [146]. However, neither study assessed the germicidal efficiency of HI-SD or LI-SD treatments per unit energy [145,146] and thus it remains inconclusive whether either regime is more suitable for practical applications in the food industry.

Alternatively, blue light can be delivered as pulses to increase its bactericidal efficiency. Pulsed blue light technology (450 nm; 33% duty cycle; three times a day for 3 days at 30 min intervals between each treatment) was reported to inactivate planktonic MRSA and *Propionibacterium acnes* (7 log CFU/mL) at light dosages of 7.6 and 5 J/cm^2, respectively [147]. The same technology (7.6 J/cm^2) also disrupted the biofilm networks of both bacteria and reduced the number of viable bacteria within the biofilm structures by approximately 1.89 and 1.56 log CFU/mL for MRSA and *P. acnes*, respectively [147]. In support of this view, pulsed blue LED (450 nm; 33% duty cycle) had a higher bactericidal efficiency against *P. acnes* than two other regimes (20% or 100% duty cycle), with a 7-log reduction (CFU/mL) achieved at a light dosage of 5 J/cm^2 (2 mW/cm^2 repeated nine times at 3-h intervals) [148].

For *S. aureus*, pulsed blue LED (405 nm; 25, 50 or 75% duty cycle) and continuous blue light (405 nm; 100% duty cycle) had similar inactivation efficiency (95–98%), albeit the pulsed blue light had approximately 83% higher optical efficiency (bacterial reduction in CFU/mL per J/cm^2) [149].

Based upon these findings [147–149], pulsed blue light is preferred than continuous blue light in vitro, but its utilization in food settings is a subject of further investigations into its ability to remain energy efficient during scale up.

3.4. Safety of Blue Light

Safety assessments of blue light have mostly been conducted in clinical settings. Generally, exposure of skin to blue light is safe, albeit high fluences at certain wavelengths could induce cytotoxic effects. In one study, eight volunteers were exposed to blue light (380–480 nm; peak at 420 nm; 100 J/cm^2 per day) for five consecutive days and the subsequent results of their skin biopsies were reported as follows: (1) no significant change in the expression of p53, i.e., no DNA damage; (2) no inflammatory cells and sunburn before and after treatment; (3) transient melanogenesis and vacuolization of keratinocytes observed, although these changes did not result in cell apoptosis [150]. Similarly, an in vitro study demonstrated that blue light (415 nm) could be used to inactivate *P. aeruginosa* on skin burns without inflicting any damage on the mouse skin at an effective antimicrobial dosage of 55.8 J/cm^2 [151]. Exposure to the same blue light at a dosage of 109.9 J/cm^2 inactivated human keratinocytes and *P. aeruginosa* by 0.16 log cell/mL and 7.48 log CFU/mL, respectively. However, cytotoxic effects of blue light on human endothelial and keratinocyte cells were observed at wavelengths of 412, 419 and 426 nm (66–100 J/cm^2) or 453 nm (>500 J/cm^2) [152].

In an in vitro study, damages on human corneal and conjuctival epithelial cells were observed after prolonged (17 h) exposure to blue light (420 and 430 nm at 1.13 and 1.16 W/cm^2, respectively), with the authors reporting decreased cellular viabilities, morphological changes of the cells, accumulation of ROS and altered mRNA expression of biomarkers associated with cellular inflammatory response and antioxidant defense system [153]. A review article presented evidence of the adverse effects that blue light (415–455 nm) inflicted on retina (oxidative stress), lens (cataract due to accumulating ROS) and blood-retinal barrier functions [154]. Another group of researchers also reported the suppression of plasma melatonin in eight human subjects exposed to blue light (469 nm; corneal irradiance 0.1–600 W/cm^2 for 90 min)—the extent of suppression was significantly higher at higher irradiances ($p < 0.0001$)—which suggests that blue light has the potential to disrupt circadian rhythm [155].

Widespread implementation of blue light-based technologies requires robust safety standards. According to the American Conference of Governmental Industrial Hygienists, daily exposure of workers to blue light is recommended to follow these rules: (1) for an exposure of 10,000 s (2.8 h) or more, the maximum intensity of the light source is ≤0.01 W/cm^2.sr; (2) for light intensity above 0.01 W/cm^2.sr, the maximum light dosage is 100 J/cm^2.sr, where light dosage (J/cm^2.sr) = light intensity (W/cm^2.sr) × time of exposure (s); (3) for a light source subtending an angle less than 0.011 radian, the maximum light intensity is 10^{-4} W/cm^2 for viewing durations greater than 100 s [156]. In accordance with these recommendations, a study analyzed blue light-related hazards through optical radiation measurements of several light sources [157]—the methodology in this study can be applied within the food industry for assessing the safety of different antimicrobial blue light technologies.

Based upon the findings presented in this section, blue light is innocuous on the skin, but deleterious to the eyes. Thus, safety glasses can be prescribed for personnel working within the proximity of high-intensity blue light sources. A study reported that several glasses and light filters significantly reduced ($p < 0.001$) the transmission of blue light from two LEDs (389–500 nm at 1625 mW/cm^2 or 410–510 nm at 1680 mW/cm^2; 10 s) by at least 97% [158]. Others reported that the use of blue light-blocking amber glasses improved the sleep quality of people with sleep disorders (self-reported or clinically diagnosed) [159,160], with an earlier endogenous dim-light melatonin onset observed when patients wore amber glasses [160]. However, there are no available data on whether anti-blue light glasses or filters can prevent damages to ocular cells. In addition, there is a need for a universal safety standard that governs the use of antimicrobial blue light within the food industry and thus the scientific community should aim at establishing the effective antimicrobial light dosages for different food-borne bacteria.

4. Application of Antimicrobial Blue Light on Surfaces and in Food Matrixes

4.1. Inactivation of Bacteria on Food Packaging and Work Surfaces

Heating effects induced by blue light treatment would be undesirable in industrial settings. Two studies reported that the surface temperatures of stainless steel increased to approximately 50–56 °C when treated with antimicrobial blue light (405 nm; 150–306 mW/cm^2; 180–185 J/cm^2) [161,162]. In these studies, bacterial reductions of 5 and <1 log CFU were achieved for *Campylobacter* spp. [162] and other pathogenic bacteria [161], respectively (Table 1). However, others observed a temperature increase of only 2.5 °C when stainless steel was continuously illuminated with blue light for 8 h (405 nm; 26 mW/cm^2, 748.8 J/cm^2) [163]. The discrepancies between these studies can be attributed to the different light intensities (mW/cm^2) used and thus optimization studies are required to determine the suitable combination of light intensity and treatment time, i.e., high intensity-short duration or low intensity-long duration. For example, the reduction of blue light dosage from approximately 183–186 to 89–92 J/cm^2 alleviated surface heating effects—final surface temperatures were approximately 44–56 and 31–36 °C at 183–186 and 89–92 J/cm^2, respectively—although the bacterial inactivation was also reduced from 5 log CFU (183–186 J/cm^2) to 1.1–3.1 log CFU (89–92 J/cm^2) [162].

At 4, 15, and 20 °C, the formation of biofilm was inhibited by blue light (405 nm; 748.8 J/cm^2) on stainless steel and acrylic coupons contaminated with *L. monocytogenes*-laden salmon exudates. However, the bacterial population within blue light-treated pre-formed biofilms was only significantly reduced ($p < 0.05$) at 25 °C [163]. This finding suggests that the blue light is more effective when used on cells contained in forming biofilms than in established biofilms. The efficacy of blue light in inactivating biofilms on other surfaces is a subject of future studies.

Blue light was able to traverse transparent solid surfaces, such as glass and acrylic slides, which was evident from the same inactivation rates of *E. coli* biofilms on top (direct exposure) or at the bottom (indirect exposure) of these slides, although four percent of the light irradiance was lost during transmission across both slides [164]. However, another study found that the inactivation of *L. monocytogenes* on tryptic soy agar was dependent on the ability of blue light (406–470 nm) to penetrate several packaging materials used to cover the agar—for example, no inhibition was observed when polyethylene + nylon was used, whereas maximum inhibition was obtained with polypropylene [165]. Thus, it is pertinent that packaging or surface materials are taken into considerations prior to designing blue light treatments intended to inactivate bacteria located behind these materials.

Table 1. Blue light inactivation of pathogenic bacteria on surfaces.

Bacteria	Surface	Reduction (CFU, CFU/mL or CFU/g)	Light Dosage	Light Source; Temperature; Distance; Photosensitizer [β]	Reference
B. cereus (vegetative or spores)	Food packaging (yellow trays; LINPAC)	4 log (vegetative); 2.7 log (spores)	18 J/cm^2	Blue LED (400 nm; 20 mW/cm^2); ALA [1] (3 mmol/L)	[166]
B. cereus (vegetative or spores) *L. monocytogenes* (planktonic or biofilms)	Polyolefin food packaging (yellow trays; LINPAC)	4–4.5 log (vegetative); 2–5 log (spores) 4.2 log	3600–10,800 J (vegetative); 3600 J (spores) 3600 J	Blue LED (405 nm; 12 mW/cm^2); CHL [2] (7.5×10^{-7}–1.5×10^{-6} M for vegetative cells or 7.5×10^{-6}–7.5×10^{-5} M for spores) Blue LED (405 nm; 12 mW/cm^2); CHL [2] (7.5×10^{-7} M for planktonic cells or 1.5×10^{-4} M for biofilms)	[167]
L. monocytogenes (planktonic or biofilms)	Polyolefin food packaging (yellow trays; LINPAC)	2.3–3.7 log (planktonic); 1.7–3 log (biofilm)	18 J/cm^2	Blue LED (400 nm; 20 mW/cm^2); ALA [1] (7.5 or 10 mM)	[168]
Uropathogenic *E. coli*; *E. coli* O157:H7; *Salmonella* spp.; *L. monocytogenes*; *S. aureus*	STC [2] contaminated with bacteria-laden chicken purge	0.23–1.01 log	180 J/cm^2	Blue LED (405 nm; 150 or 300 mW/cm^2); 10 °C; 23 cm	[161]
Campylobacter jejuni *Campylobacter coli*	STC [3] contaminated with bacteria-laden chicken exudate	1.1 or 4.9 log 3.1 or 5.1 log	91.7 or 183.4 J/cm^2 89.2 or 185.8 J/cm^2	Blue LED (405 nm; 151, 226 or 306 mW/cm^2); 10 °C; 20.3 cm	[162]
E. coli *L. monocytogenes* *S. aureus* *P. aeruginosa*	Glass or acrylic surfaces	7–8 log (glass); 5 log (acrylic) [α] 2.48 (glass) 2.75 (glass) 3.72 (glass)	504 J/cm^2 (E. coli) or 168 J/cm^2 (other bacterial species)	Blue LED (405 nm; 141.48 mW/cm^2); RT [4]; 5 cm	[164]
C. jejuni	Stainless steel Cutting board (polyvinylchloride)	2 log 4 log	1.20–2.10 J/cm^2	NUV–vis [5] LED (395 nm); RT [4]; 3, 12 or 23 cm	[169]
Salmonella enterica subsp. *enterica* serovar Enteritidis *L. monocytogenes*	Polyvinylchloride (PVC) or acrylic surfaces	1.90–2.19 log (PVC); 1.18–1.63 log (acrylic) 0.68–0.90 log (PVC); 0.21–0.42 log (acrylic)	15–45 J/cm^2 (PVC); 15–60 J/cm^2 (acrylic)	Blue LED (405 nm; 110 mW/cm^2)	[146]
L. monocytogenes (planktonic or biofilm)	STC [3] or AC [6] contaminated with bacteria-laden salmon exudate	Planktonic: 1.9–2.4 log (STC [3]); 2.4–2.8 log (AC [6]) Biofilm: 0.7–1.6 log	748.8 J/cm^2	Blue LED (405 nm; 26 mW/cm^2); 4, 15, 25 °C; 4.5 cm	[163]

[1] ALA = 5-aminolevulinic acid (precursor in photosensitizer synthesis); [2] CHL = sodium chlorophyllin. [3] STC = stainless steel coupon; [4] RT = room temperature; [5] NUV–vis = near ultraviolet–visible (395 ± 5 nm); [6] AC = acrylic coupon. [α] Several concentrations of *E. coli* biofilm were tested (developed for 4 h to 72 h), but bacterial reductions presented in this review were only for those developed for 72 h (glass) or 48 h (acrylic). [β] Experimental temperature or distance were not specified in several studies, whereas photosensitizers were only used in some studies.

4.2. Inactivation of Bacteria in Dairy and Liquid Foods: Milk, Cheese and Orange Juice

Current data suggest that antimicrobial blue light is effective against pathogenic and spoilage bacteria in dairy and liquid foods. At least 3-log reduction was achieved in all studies reviewed (Table 2), with the extent of bactericidal efficacy depending on temperature and light wavelength. For milk products, two studies assessed the blue light-mediated inactivation of pathogenic bacteria in skim and whole milk, but no data are available on blue light inactivation in concentrated milk. A study reported that pulsed white light (200–1100 nm) was able to inactivate *E. coli* in skim and whole milk, albeit not in concentrated milk [170]. Thus, future research is needed to ascertain whether antimicrobial blue light can retain its bactericidal potency in milk products with varying total solid contents.

Interestingly, a study found that blue light inactivation of several bacterial strains in milk was more efficient than in a clear liquid matrix (PBS), except for *Mycobacterium fortuitum*. Two explanations were proposed: (1) blue light was absorbed by riboflavin (photosensitizer) in milk, as apparent from the significant ($p < 0.05$) reduction in the amount of riboflavin post-treatment, which subsequently generated ROS; (2) milk strongly scattered light and retained the light longer within its matrix, relative to PBS [171]. In contrast, blue light inactivation of *Campylobacter* spp. was significantly higher ($p < 0.05$) in transparent Brucella broth than in opaque chicken exudate [162]. These findings suggest that the type of solid particulate in liquid matrixes determines whether bactericidal efficacy of blue light is enhanced or attenuated.

Further, the bactericidal efficacy of blue light in liquid matrix also varies across different bacterial species/strains tested. In PBS, blue- light treatment (405 nm) resulted in a 5-log reduction (CFU/mL) of *C. jejuni* (18 J/cm^2) [172] and *L. monocytogenes* (185 J/cm^2) [173], whereas *Salmonella* spp. and *E. coli* O157:H7 was only reduced by less than 1.5 log CFU/mL at light dosages of 180–185 J/cm^2 [172,173]. Thus, future studies are required to establish the effects of liquid opacity, particularly for liquid foods, on blue light inactivation of different bacteria.

Table 2. Blue light inactivation of pathogenic bacteria on dairy products.

Bacteria	Food Matrix	Reduction (CFU/mL or CFU/g)	Light Dosage	Light Source; Temperature; Distance [β]	Reference
E. coli	UHT skim milk	4.69–5.27 log (405 nm); 4.11–5.04 log (433 nm); 3.41–4.64 log (460 nm)	Approx. 250 J/cm^2 (405 nm); 313 J/cm^2 (433 nm); 376 J/cm^2 (460 nm) [α]	Blue LED (405, 433 or 460 nm; 10 W); 5–15 °C; 30 mm	[174]
S. aureus; *E. coli*; *P. aeruginosa*; S. Typhimurium; *M. fortuitum*	Whole milk	5 log	228.84–583.5 J/cm^2	Blue LED (413 nm; 100 mW/cm^2); 1 mm	[171]
P. fluorescens (spoilage bacteria)	Ricotta cheese	3–5 log	6.36 J/cm^2	Near UV–vis LED (395 nm; 16 mW/cm^2); 6 cm	[176]
L. monocytogenes *P. fluorescens* (spoilage bacteria)	Sliced cheese (packaged)	5.14 log (4 °C); 1.95 log (25 °C) 3.60 log (4 °C); 1.85 log (25 °C)	604.8 J/cm^2 (4 °C); 172.8 J/cm^2 (25 °C)	Blue LED (460–470 nm; 1 mW/cm^2); 4 or 25 °C; 10 mm	[165]
S. enterica (Gaminara, Montevideo, Newport, Typhimurium and Saintpaul)	Orange juice	2–5 log	4500 J/cm^2	Blue LED (460 nm; 92, 147.7 or 254.7 mW/cm^2); 4, 12 or 20 °C	[175]

[α] Dosage for achieving maximum bacterial reduction was approximated using the formula: light intensity (10 W) × treatment lengths (s) at each wavelength/area of application (143.75 cm^2).
[β] Experimental temperature or distance were not specified in several studies.

Blue light inactivation of *E. coli* in liquid milk was more efficient at lower wavelengths and higher temperatures, with optimal treatment (reduction of 5 log CFU/mL and minimum color change) achieved at 405 nm, 13.8 °C and for 37.83 min [174]. This view was corroborated by others, who found that *S. enterica* in orange juice was inactivated by blue light (460 min; 4500 J; 92 W/cm^2) to a higher degree at 12 or 20 °C than at 4 °C, although the inactivation rate was the same across these temperatures at higher light intensities (147.7 or 254.7 mW/cm^2) [175].

Further, major milk components, namely proteins, lipids and lactose, were retained after 2 h of blue light treatment (720 J/cm^2), albeit the loss of riboflavin (vitamin B_2) had resulted in a bleaching effect that was perceptible to naked eyes [171]. In orange juice, blue light treatment also induced a color change in a temperature- and light intensity-dependent manner, particularly when the treatment was applied at a low intensity with long duration (light dosage was constant) [175].

On cheeses, blue light treatments were effective against *L. monocytogenes* and *Pseudomonas fluorescens* (Table 2), with no color changes observed in treated ricotta [176] and packaged slice cheeses [165]. Hyun and Lee (2020) also found that the efficacy of blue light on packaged sliced cheese was higher at 4 °C than at 25 °C [165].

4.3. Inactivation of Bacteria in Horticultural Products

Several studies found that blue light sanitization of fruits and vegetables was dependent on the type of product. Glueck et al. observed that the photosensitizer-mediated inactivation of blue light (435 nm; 33.8 J/cm^2) was affected by the geometry of the food, with higher efficacy observed in flat-surfaced vegetables (cucumber, tomatoes and lettuce) than in those with complex structures (fenugreek seeds, mung bean seeds and mung bean germlings) (Table 3) [177]. In support of this view, three other studies showed varying bactericidal efficacies of blue light across different application media. Tortik et al. demonstrated that the combination of blue light (435 nm; 33.8 J/cm^2) and curcumin (50 μM) reduced the bacterial load of *S. aureus* on peppers and cucumber by 2.5–2.6 log CFU [178], whereas an identical treatment in a clear liquid matrix (PBS) resulted in a 7-log reduction (CFU) in the number of *S. aureus* [179]. Buchovec et al. also found that *S.* Typhimurium was inactivated by chlorophyllin/chitosan-mediated blue light treatment (405 nm; 38 J/cm^2) to a lower degree on strawberries (2.2 log CFU/mL) than in PBS (6.5 log CFU/mL) [180]. Possible explanations include the varying light-reflecting properties of different matrixes, the adsorption of photosensitizers onto the cuticle of vegetables/fruits and the presence of antioxidants in vegetables/fruits that reduced the efficacy of blue light [178,180].

Table 3. Blue light inactivation of pathogenic bacteria in horticultural products.

Bacteria	Food Matrix	Reduction (CFU, CFU/mL or CFU/g)	Light Dosage	Light Source; Temperature; Distance; Photosensitizer [β]	Reference
L. monocytogenes	Basil	1.6 log	9 J/cm^2	Blue LED (405 nm; 10 mW/cm^2); RT [1]; 6 cm; chlorophyllin (1.5 × 10^{-4} M)	[186]
E. coli	Grape	2.4 log	36.3 J/cm^2	Blue LED (465–470 nm; 4.5–30.2 mW/cm^2); RT [1]; 19 cm; curcumin (1.6 × 10^{-3} M)	[103]
L. monocytogenes *Salmonella* spp.	Cantaloupe rinds	At 405 nm: 2.4–2.9 log (no CHL); 2.8–3 log (CHL) At 460 nm: 2.7 log (no CHL); 2.2–2.3 log (CHL) At 405 nm: 2.3 (no CHL); 2.9 (CHL) At 460 nm: 1.1 log	1210 J/cm^2 (405 nm); 5360 J (460 nm)	Blue LED (405 or 460 nm; 7 or 31 mW/cm^2); 4 or 20 °C; CHL [2] (100 µM)	[104]
Salmonella spp.	Fresh-cut papaya	1–1.2 log (4 °C); 0.3–1.3 log (10 °C); 0.8–1.6 log (20 °C)	900–1700 J/cm^2	Blue LED (405 nm); 4, 10 or 20 °C; 2.3 or 4.5 cm	[116]
Mesophilic bacteria *B. cereus* *L. monocytogenes*	Cherry tomatoes	2.4 log 1.5 log 1.6 log	3–9 J/cm^2	Blue LED (405 nm; 10 mW/cm^2); RT [1]; 6 cm NCCHL [3] (1.5 × 10^{-4} M)	[181]
S. Typhimurium	Strawberries	2.2. log	38 J/cm^2	Blue LED (405 nm; 10–11 mW/cm^2); 37 °C; 3.5 or 6 cm; CHL-CHN [4]	[180]
S. Typhimurium	Cucumber peels	Approx. 3.9 log	18 J/cm^2	Supra-luminous diode (SLD; 464 nm; 16.6 mW/cm^2)	[187]
E. coli O157:H7 *E. coli* K-12 *S. Enteritidis* non-pathogenic S. Typhimurium	Almond kernel	1.43–2.44 log 1.64–1.84 log 0.55–0.70 log 0.64–0.96 log	2000 J [§]	Blue LED (405 nm; 3.4 W); RT [1]; 7 cm	[185]

[1] RT = room temperature; [2] CHL = sodium-copper chlorophyllin; [3] NCCHL = non-copperized sodium chlorophyllin; [4] CHL-CHN = non-copperized sodium chlorophyllin (1.5 × 10^{-5} M)-chitosan (0.1%). [5] PVP-C = Curcumin bound to polyvinylpyrrolidone. [§] Light dosage presented was used to treat three almond kernels (4 g) and approximated by multiplying the maximum treatment time (10 min) by light intensity (3.4 W). [β] Experimental temperature or distance were not specified in several studies, whereas photosensitizers were only used in some studies.

Table 3. *Cont.*

Bacteria	Food Matrix	Reduction (CFU, CFU/mL or CFU/g)	Light Dosage	Light Source; Temperature; Distance; Photosensitizer [β]	Reference
S. aureus	Cucumber Pepper (green, red or yellow)	2.6 log 2.5 log 3 log (10 µM); 4 log (50 µM); 4.5 log (100 µM)	33.8 J/cm^2	Blue LED (435 nm; 9.4 mW/cm^2); RT [1]; PVP-C [5] (50 or 100 µM)	[178]
E. coli	Cucumber Tomatoes Lettuce	Approx. 3 log (10 µM); 6 log (50 µM); 3 log (100 µM) Approx. 3 log (10 µM); 7 log (50 µM); 6 log (100 µM) Approx. 3 log (10 µM); 5 log (50 µM); 4.5 log (100 µM)	33.8 J/cm^2	Blue LED (435 nm; 9.4 mW/cm^2);15 cm; cationic curcumin derivative (10, 50 or 100 µM)	[177]
E. coli	Fenugreek seeds Mung beans Mung bean germling	Approx. 2.5 log (10 µM); 2 log (50 µM); 3.5 log (100 µM) Approx. 0.5 log (10 µM); 1 log (50 µM); 0.5 log (100 µM)	33.8 J/cm^2	Blue LED (435 nm; 9.4 mW/cm^2);15 cm; cationic curcumin derivative (10, 50 or 100 µM)	[177]
Salmonella spp.	Fresh-cut pineapple	0.61–1.72 log	Approx. 8000 J/cm^2	Blue LED (460 nm; 92–257 mW/cm^2); 7, 16 or 25 °C; 2.5–4.5 cm	[184]
E. coli O157:H7, *Salmonella* spp. or *L. monocytogenes*	Fresh-cut mangoes	1–1.6 log	1700–3500 J/cm^2	Blue LED (405 nm; 20 mW/cm^2); 4, 10 or 20 °C; 4.5 cm	[182]
L. monocytogenes Mesophilic bacteria Yeasts and microfungi	Strawberries	1.8 log 1.7 log 0.87 log	14.4 J/cm^2	Blue LED (400 nm; 12 mW/cm^2); NCCHL [3] (1 mM)	[188]
E. coli	Fresh-cut Fuji apple	0.95 log	152 J/cm^2	Blue LED (420 nm; 298 mW/cm^2); 4 cm; curcumin (2 µM)	[183]

[1] RT = room temperature; [2] CHL = sodium-copper chlorophyllin; [3] NCCHL = non-copperized sodium chlorophyllin; [4] CHL-CHN = non-copperized sodium chlorophyllin (1.5×10^{-5} M)-chitosan (0.1%). [5] PVP-C = Curcumin bound to polyvinylpyrrolidone. [§] Light dosage presented was used to treat three almond kernels (4 g) and approximated by multiplying the maximum treatment time (10 min) by light intensity (3.4 W). [β] Experimental temperature or distance were not specified in several studies, whereas photosensitizers were only used in some studies.

In the majority of studies that we reviewed, blue light treatments resulted in no detrimental effects upon the sensorial and nutritional properties of fruits and vegetables. For instance, the antioxidant activities of cherry tomatoes, fresh-cut mangoes and papayas were retained after treatment with blue light (405 nm) [116,181,182]. Most nutrients (vitamin C, β-carotene and lycopene) were also preserved in blue light-treated papayas or fresh-cut mangoes (405 nm), although there was a significant increase ($p < 0.05$) in the amount of flavonoids during storage (4 °C or 20 °C) in the illuminated papayas [116]—flavonoid content remained stable in fresh-cut mangoes [182]. Others observed no adverse visual quality on blue light-treated strawberries (405 nm), as compared with the untreated controls [180]. In fresh-cut Fuji apples, polyphenol oxidase and peroxidase was inhibited by the combination of curcumin (2 μM) and blue light (420 nm) and thus there was significantly less browning ($p < 0.01$) in treated apples as compared with untreated controls [183]. On the contrary, blue light (460 nm) induced a bleaching effect in fresh-cut pineapples, as measured by the reduction in its yellowness index [184], although no human observers were used to determine whether this change would be perceived as undesirable.

The non-thermal nature of blue light treatments also allows for their application on low-moisture products, such as almonds, albeit improvements on its bactericidal efficiency would be required (at least 4-log reduction is needed) [185]. Additionally, blue light (405 nm) delayed the regrowth of *L. monocytogenes* on cherry tomatoes by 14 days, albeit this finding should prompt food producers to be vigilant in determining whether the blue light used is bacteriostatic or bactericidal against pathogenic bacteria [181].

4.4. *Inactivation of Bacteria in Meat Products and Seafood (Chicken, Beef and Fish)*

Generally, blue light treatment is less effective in meat and seafood products than on surfaces or in dairy and horticultural products (Tables 1–4), possibly due to the presence of absorptive materials and ROS-neutralizing substances, such as proteins. For instance, the inactivation of *S.* Enteritidis by blue light was less efficient on cooked chicken meat (0.8–0.9 log CFU/cm^2) than in the transparent PBS (1.3–2.4 log CFU/mL) [189]. However, blue light could still induce injuries on bacterial cells that render them more susceptible to subsequent stresses. On cooked chicken meat, *S.* Enteritidis lost its resistance to four antibiotics, relative to the untreated controls (details in Section 7.1.) [189]. Similarly, blue light treatment rendered *L. monocytogenes* and *Salmonella* spp. on fresh salmon significantly more susceptible ($p < 0.05$) to gastric digestion (pH 2) than untreated cells, especially at lower temperatures [190]. These findings indicate that bactericidal efficacy of blue light-mediated treatments could be improved by combining it with other treatments, such as organic acids, essential oils agents or polyphenols (details in Section 6).

Table 4. Blue light inactivation of pathogenic bacteria in meat and seafood products.

Bacteria	Food Matrix	Reduction (CFU, CFU/mL or CFU/g)	Light Dosage	Light Source; Temperature; Distance; Photosensitizer [β]	Reference
Uropathogenic *E. coli*; *E. coli* O157:H7; *Salmonella* spp.; *L. monocytogenes*; *S. aureus*	Chicken skin	0.19–0.40 log	180 J/cm^2	Blue LED (405 nm; 150 or 300 mW/cm^2); 10 °C; 23 cm	[161]
C. jejuni *C. coli*	Chicken skin	1.7 log 2.1 log	184 J/cm^2 185.8 J/cm^2	Blue LED (405 nm; 151, 226 or 306 mW/cm^2); 10 °C; 20.3 cm	[162]
L. monocytogenes	Hot dog	<1 log	120 J/cm^2	SLD (405 or 464 nm); 3–5 mm	[192]
E. coli	Hot dog	2.43 log	100 J/cm^2	SLD (405 nm; 83.3 mW/cm^2); 3–5 mm	[187]
L. monocytogenes *Salmonella* spp.	Fresh salmon	0.4 log (4 °C); 0.3 log (12 °C) 0.5 log (4 °C); 0.4 log (12 °C)	460.8 J/cm^2	Blue LED (405 nm; 16 mW/cm^2); 4 or 12 °C; 7.9 cm	[190]
C. jejuni	Skinless chicken fillet Chicken skin	1.43–2.62 log Approx. 6.7 log (3 cm); 1 log (12 cm); 0.7 log (23 cm)	1.20–2.10 J/cm^2 9 J/cm^2 (3 cm); 4.23 J/cm^2 (12 cm); 1.20 J/cm^2 (23 cm)	NUV–vis [1] LED (395 nm); RT [2]; 3, 12 or 23 cm	[169]
L. monocytogenes	Smoked salmon fillets	0.7–1.2 log	2400 J/cm^2	Blue LED (460 nm; 15, 31 or 58 mW/cm^2); 4 or 12 °C; 5.4–9 cm; riboflavin (25, 50 or 100 μM)	[191]
S. aureus	Chicken meat (with skin)	1.7 log	33.8 J/cm^2	Blue LED (435 nm; 9.4 mW/cm^2); RT [2]; curcumin (50 or 100 μM)	[178]
S. Enteritidis	Cooked chicken	0.8–0.9 log	1.58–3.80 J/cm^2	Blue LED (405 nm; 22 mW/cm^2); 4 °C; 4 cm	[189]

[1] NUV–vis = near ultraviolet–visible; [2] RT = room temperature. [β] Experimental distance was not specified in several studies, whereas photosensitizers were only used in some studies.

The efficacy of blue light treatment on chicken products is dependent on the wavelength used, possibly related to wavelength-specific activation of different endogenous photosensitizers. Two studies found that *Campylobacter* spp. in chicken (fillet or skin) were reduced by 1.7–2.4 log CFU or 0.7–6.7 log CFU/g through treatments with blue light at 405 nm [162] or 395 nm [169], respectively. Other than the type of light, these differences may also be attributable to the different bacterial species, treatment distances and treatment lengths used (Table 4) [162,169]. In agreement, light dosage was found to be inversely proportional to treatment distance, for example, the inactivation of *C. jejuni* on chicken skin was higher at a treatment distance of 3 cm (6.7 log CFU/cm^2) than at 12 cm (1 log CFU/cm^2) or 23 cm (0.7 log CFU/cm^2) (Table 4) [169].

Sensorial properties of chicken and salmon could be affected by light treatment, especially when using light at the UV–vis region or photosensitizers. A study reported on the heating effects of ultraviolet/blue light (395 nm), especially at shorter treatment distance and longer treatment length, which resulted in significant color changes ($p < 0.05$) in chicken fillet and skin [169]. Others found that while discoloration was absent in smoked salmon illuminated with blue light alone (460 nm), the whiteness index significantly increased ($p < 0.05$) in samples treated with riboflavin-mediated blue light, relative to the untreated control samples [191]. Consistently, an extended illumination (8 h) of fresh salmon with blue light alone (405 nm) did not result in color changes [190].

Further, the introduction of exogenous photosensitizers could improve the inactivation of pathogenic bacteria on cooked food. The combination of curcumin (50 or 100 μM) and blue light (435 nm; 33.8 J/cm^2) resulted in a 1.7-log reduction (CFU) of *S. aureus* in cooked chicken meat, whereas the treatment of blue light alone had no effect on the bacterial load. Albeit, the authors suggested that the lipophilicity of curcumin could make it susceptible to attenuation by the fatty regions on the chicken skin and thus modification of this photosensitizer (or alternative photosensitizers) would be required to achieve a higher bactericidal activity [178]. Another study found that blue light (460 nm) reduced the population of *L. monocytogenes* on smoked salmon fillets by up to 1.12 log from the initial concentration of 3.5 log CFU/cm^2, but only when riboflavin was present. The light dosage required to achieve the first log reduction was lower at 4 °C (1600 J/cm^2) than at 12 °C (2000 J/cm^2), although the difference was not statistically significant [191].

5. Potential Application of Blue Light in Food Supply Chain

5.1. Food Processing and Farms: Airborne and Surface Inactivation

Practical applications of any blue light technology within the food industry are dependent on its ability to inactivate pathogens over distances beyond those typically used in laboratory-scale experiments. In clinical settings, three studies found that a ceiling-mounted high-intensity narrow-spectrum light environmental decontamination system (HINS-light EDS; 405 nm) significantly reduced ($p < 0.05$) the total viable counts, including MRSA and *S. aureus*, on surfaces (for example, bed, table, chair, worktop or bins) [193–195]. The safety of HINS-light EDS also allowed it to be operated in the presence of humans, such as patients and healthcare workers, which is in contrast to ultraviolet germicidal lamps [193–195].

These findings suggest that there is a potential for HINS-light EDS (or similar technologies) to be used for environmental sanitization in food-processing plants. While blue light inactivation of planktonic and biofilm-associated bacteria has been tested on food packaging and also on work surfaces (for example, stainless steel, acrylic or glasses), there are limited studies on the sporicidal effects of blue light on food packaging (Table 1). In suspensions, antimicrobial blue light (405 nm; 1730 J/cm^2) also reduced the population of bacterial endospores, namely those of *B. cereus*, *B. subtilis*, *Bacillus megaterium* and *Clostridium difficile*, by 4 log CFU/mL [196]. Thus, future in vivo validations are required to assess the ability of blue light to inactivate different forms of bacteria on a range of surfaces and at varying distances.

Further, bacteria may be aerosolized during food processing, persist in the air and subsequently spread across indoor premises. Interestingly, a study found that aerosolized *S. epidermidis* was significantly inactivated by approx. 2 log CFU/mL ($p < 0.001$) by blue light (405 nm; 39.5 J/cm^2), with the susceptibility of the bacteria to blue light being 2–4 times higher in aerosols than in liquids or on surfaces [197]. Future studies should explore the potential of blue light for inactivating food-borne bacteria in aerosols.

Light treatments could also be used to treat veterinary diseases in farm animals, such as mastitis in cows. For example, there were significantly lower ($p < 0.05$) bacterial loads of *Streptococcus dysgalactiae* and coagulase-negative staphylococci in milk produced by cows treated with the combination red LED (635 nm; 200 J/cm^2) and toluidine blue (2%), relative to the untreated groups—this treatment was not intended for direct decontamination milk, but for alleviating incidences of mastitis in cows, with bacteriological characteristic of milk samples only used as an indicator [198]. In vitro, bacteria isolated from bovine mastitis, namely *S. dysgalactiae*, *S. aureus* and *Streptococcus agalactiae*, were inactivated by red LED (662 nm; 3–12 J/cm^2) and methylene blue (50 μM) [199]. Currently, there are no data available on the application of blue light against farm-animal pathogens. However, given the fact that blue light can act on endogenous chromophores, it may have practical advantages over the existing red LED that depends on exogenous photosensitizers to inactivate bacteria on cows.

5.2. Aquaculture

Pathogenic bacteria that attack fish include *Vibrio* spp., *Photobacterium damselae* subsp. *piscicida*, *Edwardsiella tarda* and *Edwardsiella ictaluri* [200]. Although most of these bacteria are not known to infect humans, high incidences of disease in farmed fish may inflict adverse economic consequences upon fish farmers. Thus, the availability of methods for inactivating these bacteria in aquaculture systems is paramount to sustain viable fisheries. Several studies have assessed the application of antimicrobial blue light against fish pathogens. In PBS, the blue light inactivation of several pathogenic fish bacteria was 132–543.7 and 247–2178 J/cm^2 at 405 and 465 nm, respectively. Generally, these bacteria were more susceptible to blue light at 405 nm than at 465 nm, although there were variations across different bacterial species (Figure 2) [201].

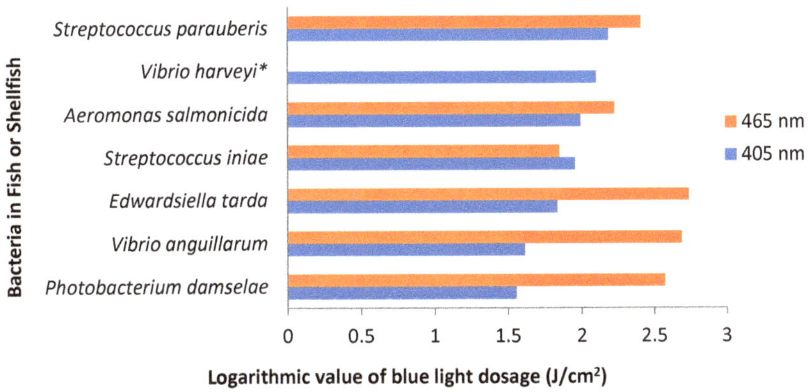

Figure 2. Blue light dosage required to achieve 1-log reduction of pathogenic bacteria in fish or shellfish. Light dosage was converted to logarithmic values (log) and thus an increase of one unit on the x axis represents a tenfold increase in the light dosage. This graph was created using data taken from Roh et al., which was published under the Creative Commons Attribution 4.0. International License (http://creativecommons.org/licenses/by/4.0/) [201]. * *Vibrio harveyi* was not inactivated by blue light at 465 nm.

The presence of particulates in aquaculture water reduced the bactericidal efficacy of artificial white light (380–700 nm), in combination with a cationic porphyrin (Tri-Py$^+$-Me-PF), against *Vibrio fischeri*: (1) in unfiltered water, 50 μM of porphyrin and 43.2–64.8 J/cm^2 of light were required to achieve a 7-log reduction (CFU/mL); (2) in filtered water, the combination of porphyrin (at least 10 μM) and light (64.8 J/cm^2) led to a bacterial reduction of 7 log CFU/mL. When tested in PBS, a complete inactivation of *V. fischeri* (7 log CFU/mL) was achieved, regardless of variations in acidity (pH of 6.5–8.5), salinity (20–40 g/L), temperature (10–25 °C) and oxygen concentration (5.3–5.9 mg/L), although the rate of inactivation was highest at the physiological pH (7.4) and ambient temperature (25 °C) [202]. However, when similar treatments (white light at 380–700 nm; Tri-Py$^+$-Me-PF at 5–50 μM) were used against heterotrophic bacteria cultivated from aquaculture water samples, the bactericidal efficacies varied across different water samples, ranging from 1.2 to 2 log CFU/mL. Nevertheless, in PBS, a complete inactivation (8 log CFU/mL) was observed for several pathogenic bacteria isolated from aquaculture water, namely *Vibrio* spp., *P. damselae*, *Enterococcus faecalis*, *E. coli* and *S. aureus*, after exposure to a combined treatment of artificial white light (380–700 nm; up to 648 J/cm^2) and Tri-Py$^+$-Me-PF (5 μM) [203].

In the absence of photosensitizer, blue LED (405 or 465 nm) was able to significantly reduce ($p < 0.05$ or 0.01) the bacterial loads of *Edwardsiella piscida* in rearing water, which subsequently also decreased the number of bacterial infections in Fancy carps (*Cyprinus caprio*) [204]. The light treatment did not induce damages on the fish eyes and skin, with also no increase in the production of heat-shock proteins or unusual feeding behaviour observed in the treated fish, relative to the untreated controls [204]. Albeit, precautions are needed as continuous exposure of some fish, such as sea bass and sole, to blue light (435–500 nm) could result in increased malformations and poor survival of the fish larvae [205]. Thus, additional studies are needed to assess the effects of antimicrobial blue light on live seafood, including fish, oysters and mussels.

5.3. Retail: Prolonging Shelf-Life

As previously discussed, food contamination at retail establishments could lead to outbreaks (Section 2), particularly as inactivation treatments are not usually present at this stage within the food supply chain. In Section 4, we have reviewed studies on the use of antimicrobial blue light against pathogenic bacteria in an array of food products and also on surfaces. However, the use of blue light at the retail level also requires it to inhibit spoilage microorganisms and thus extends the shelf-life of foods. There are also concerns about spoilage bacteria surviving cleaning regimes and persist on surfaces in food-processing plants, including *Pseudomonas* spp., *Serratia* spp., *Hafnia* spp. and LAB [16], and thus additional control measures are required. Further, several review articles have identified major spoilage microorganisms in dairy [206,207], horticultural [208], meat [209] and seafood [210] products, which may be a subject of future studies on antimicrobial blue light.

Although still limited in number, there are several studies that assessed blue light-mediated inactivation of spoilage microorganisms in food products and subsequently the shelf-life of these treated foods. For example, blue light significantly reduced ($p < 0.05$) the initial loads of mesophilic bacteria, yeasts or other microfungi on strawberries and cherry tomatoes, which delayed the spoilage onset by 2 and 4 days, respectively (Table 3) [181,188]. In Hami melon (cantaloupe), the combination of curcumin (50 μM) and blue LED (470 nm) significantly reduced ($p < 0.05$) the initial amount of total aerobic microorganisms by 1.38 log CFU/g, with the treated melons also having 1.8-log lower bacterial load (CFU/g) than the untreated controls after 9 days of storage at 4 °C. The soluble solid content, color, water content and firmness of Hami melon were also better preserved in the blue light-treated group than the untreated controls [211].

The combination of curcumin (10 μM) and blue light (470 nm; 5.4 J/cm^2) extended the shelf-life of fresh oysters from 8 to 12 days at 4 °C, as determined by total aerobic plate count (shelf-life limit of 10^7 CFU/g) and total volatile basic nitrogen analysis (shelf-life limit of 30 mg *n*/100 g oyster) [212]. Sensorial properties of the treated oyster were also improved at the shelf-life terminal point of the untreated group (8th day): (1) human panels rated treated oyster more favorably than untreated control in terms of smell, body color, mucus appearance and texture; (2) electronic nose indicated that the generation of spoilage metabolites was reduced in the treated group, relative to untreated controls. Retention of flavorful free amino acids and reduction in the oxidation of lipids and fatty acids were also observed in the blue light-treated group [212]. Similarly, the treatment of fresh sturgeons with curcumin (30 μM) and blue light (470 nm) significantly reduced ($p < 0.05$) the prevalence of spoilage *Pseudomonas* spp. during storage at 4 °C for 6–9 days [213].

The number of spoilage *P. fluorescens* was lowered by 1.85–3.60 log CFU/g in blue light-treated (460–470 nm) packaged sliced cheese, relative to untreated controls, after 2 and 7 d of storage at 25 and 4 °C, respectively (Table 2) [165].

6. Hurdle Technology

6.1. Photosensitizers

In all studies that combined exogenous photosensitizers with blue light treatments, the addition of photosensitizing agents improved the bactericidal efficacy of blue light, relative to when blue light was used alone (Tables 1, 3 and 4). These photosensitizers facilitate the production of ROS, which subsequently induces bacterial inactivation. However, the bactericidal efficacy may also vary with the type of photosensitizer used. For instance, the combination of blue LED (427–470 nm; 30 J/cm^2) and rose bengal (160 μg/mL) inactivated 7.1 log CFU/mL of *Porphyromonas gingivalis*, whereas when the same blue light was combined with erythrosine or phloxine, the bacterial reductions were only 0.9 or 1 log (CFU/mL), respectively [214].

Similarly, the efficacy of photosensitizers also depends on the delivery method within different application matrixes. For example, curcumin bound to polyvinylpyrrolidone did not improve the blue light inactivation on chicken skin, but was effective on vegetables. The lipophilic curcumin may be readily released by the hydrophilic polyvinylpyrrolidone to the fatty regions of the chicken skin. In contrast, micellar formulation of curcumin (NovaSol®-C) was effective on chicken skin, potentially due to the retention of curcumin in micellar form prior to contact with the bacterial cells [178].

Non-toxic inorganic salts can be used to potentiate photosensitization in photodynamic treatments, particularly through the production of non-oxygen reactive species, such as azide radicals from sodium azide salts or reactive iodine species from potassium iodide [215]. In an in vitro study, the combination of blue light (415 nm; 10 J/cm^2), Photofrin (10 μM) and potassium iodide (100 mM) inactivated 6 log of five Gram-negative bacterial species, namely *E. coli*, *P. aeruginosa*, *Klebsiella pneumoniae*, *Proteus mirabilis* and *A. baumannii*, whereas in the absence of potassium iodide, photosensitization treatments resulted in no inactivation [216].

Precursors of endogenous photosensitizers, such as 5-aminolevulinic acid (ALA), can also be added to facilitate photo-inactivation of bacteria [110]. A study found that ALA and its derivatives induced the formation of photo-active porphyrins in Gram-negative (*E. coli* K-12, *E. coli* Ti05 and *P. aeruginosa)* and Gram-positive (*S. aureus*) bacteria, although the amount or type of porphyrin produced, and also the extent of bacterial photo-inactivation (white light; 120 J/cm^2), depended on three factors: (1) type of precursor used; (2) bacterial species/strain tested; (3) concentration of precursor added [217]. In support of this view, another study demonstrated that ALA-mediated inactivation of bacteria by blue light (407–420 nm; 50–100 mJ/cm^2) was more profound in Gram-positive (5–7 log CFU/mL; except for *B. cereus* and *Streptococcus faecalis*) than in Gram-negative bacteria (1–2 log CFU/mL)—this difference was possibly due to the higher amount of coproporphyrin present in Gram-positive bacteria tested (*B. cereus* produced 37–45% lower coproporphyrin than the other Gram-positive bacteria and was reduced by only 1–2 log CFU/mL; porphyrin production was not observed in *S. faecalis* and no reduction was reported) [218].

6.2. Acidity and Temperature

In clear liquid suspensions, the susceptibility of *E. coli* and *L. monocytogenes* to blue light (405 nm) was enhanced in the presence of environmental stresses: (1) the light dosages required to inactivate both bacteria (5 log CFU/mL) at stressful temperatures (4 °C or 45 °C) were significantly lower ($p < 0.05$) than at a non-stressful temperature (22 °C); (2) the light dosages required to inactivate *E. coli* and *L. monocytogenes* at pH 3 were reduced by 77% and 50%, respectively, relative to that required at pH 7; (3) the bacterial inactivation was significantly higher ($p < 0.05$) under osmotically-stressful conditions (salt concentrations of 10% and 15% for *E. coli* or 10% for *L. monocytogenes*) than under non-stressful conditions (salt concentrations of 0 or 0.8%) [219]. On nitrocellulose surface, both bacteria were also inactivated by blue light (405 nm; 36 J/cm^2) to a higher extent at pH 3 (reduction of 95–99% reduction) than at pH 7 (reduction of 13–26%) [219]. In addition, the type of acid present determined the extent of bactericidal inactivation of blue LED (461 nm; 596.7 J/cm^2) against *E. coli* O157:H7, *S.* Typhimurium, *L. monocytogenes* and *S. aureus*, with the highest inactivation rates at pH 4.5 achieved using lactic acid, followed by citric and malic acids [220].

However, there has been no consensus on how temperature affects the efficacy of antimicrobial blue light, with current data suggesting that it depends on the bacterial species and light wavelength used. The growth of *S.* Enteritidis on cooked chicken meat was only delayed when treated with blue light (405 nm) at 10 and 20 °C, but it was inactivated at 4 °C [189]. Similarly, five *S. enterica* serovars on fresh-cut pineapple, namely Typhimurium, Newport, Gaminara, Montevideo and Saintpaul, were inactivated by blue light (460 nm) at 7 and 16 °C (bactericidal), but only inhibited at 25 °C (bacteriostatic) [184]. On the contrary, *S. aureus*, *Lactobacillus plantarum* and *V. parahaemolyticus* in PBS were inactivated by blue light (405 or 460 nm) at all experimental temperatures (4, 10 or 25 °C), albeit the extent of bactericidal effect varied across bacterial species and also light wavelengths [221]. Another study also demonstrated that *E. coli* O157:H7, *L. monocytogenes* and *Salmonella* spp. on fresh-cut mangoes were inactivated by blue light (405 nm) at 4 and 10 °C, but only inhibited at 20 °C, with both bactericidal and bacteriostatic effects varying with bacterial species [182]. Interestingly, the population of *L. monocytogenes* in pre-biofilms on stainless steel and acrylic coupons was significantly reduced ($p < 0.05$) at 25 °C, but not at 4 °C [163].

6.3. Nanoparticle

Silver nanoparticles (AgNPs) are able to interact with negatively-charged molecules in bacterial cells, such as proteins or nucleic acids and subsequently induce damages to the cells, for example, by increasing cell membrane permeability, causing DNA damage or inhibiting protein synthesis [222]. A review by Carbone et al. summarized available data on the potential use of AgNPs as an antimicrobial agent in food packaging [223]. In photodynamic treatments, the combination of blue light (460 nm) with AgNPs was significantly more effective ($p < 0.001$) against MRSA and *P. aeruginosa* than each treatment alone [224,225]. The formation of *P. aeruginosa* biofilm on gelatin-based discs was also significantly inhibited ($p < 0.001$) by the combined treatment, relative to treatments with blue light or AgNPs alone [225].

Similarly, metal oxides can be used in photocatalytic processes to generate superoxides or hydroxyl radicals for inactivating microorganisms or oxidizing organic substances. In an in vitro study, zinc oxide nanoparticles (ZnO-NPs; 0.5 mg/mL) were combined with blue light (462 nm; 5.4 J/cm^2) to inactivate 5 log CFU/mL of *A. baumannii*, with ZnO-NPs and blue light alone resulted in zero and less than 1-log reduction (CFU/mL), respectively. The combination of ZnO-NPs (0.1 mg/mL) and blue light (462 nm; 10.8 J/cm^2) also significantly reduced the number of antibiotic-resistant (colistin or imipenem) *A. baumannii* ($p < 0.005$), *K. pneumoniae* ($p < 0.005$) or *Candida albicans* (fungus; $p < 0.05$) in culture medium. Further, transmission electron microscopical image revealed that the cell membrane was damaged in *A. baumannii*, but an analysis using gel electrophoresis showed no fragmentations of plasmid DNA post-treatment and thus these findings indicate that ZnO-NPs/blue light photocatalysis only attacks cytoplasmic membrane and not the bacterial genome [226].

6.4. Plant Extracts: Polyphenols and Essential Oils

Polyphenols are known to exhibit antimicrobial potency through their ability to bind to the cell membrane, cell wall and their associated proteins and thus compromising the structural integrity of the bacterial cell [227,228]. Inhibition of bacterial adhesins and quorum sensing by polyphenols also resulted in failures to form biofilms [228]. Interestingly, several studies have reported on the synergistic antimicrobial effects of blue light and plant-derived polyphenols, particularly at the blue light wavelength range of 385–400 nm. For example, gallic acid (4 mM) was combined with blue light (400 nm; 72 J/cm^2) to inactivate 7.5 log CFU/mL of *S. aureus*, with lipid peroxidation observed through the detection of malondialdehyde. This lipid peroxidation was likely to be caused by the formation of hydroxyl radicals, which were detected by electron spin resonance [229]. In a follow-up study, the combination of blue light (400 nm; 75–150 J/cm^2) and several polyphenols (1 mg/mL; caffeic acid, gallic acid, chlorogenic acid, epigallocatechin, epigallocatechin gallate and proanthocyanidin) induced significant inactivation ($p < 0.05$ or $p < 0.01$) of *E. faecalis*, *S. aureus*, *Streptococcus mutans*, *Aggregatibacter actinomycetemcomitans*, *E. coli* and *P. aeruginosa*. Damage to the DNA was also reported, which suggested that the polyphenols were incorporated into the bacterial cells, probably facilitated by the high affinity of polyphenols to the cell membrane [230].

Similarly, inactivation of *S. mutans* within biofilms was reported after exposure to the combination of caffeic acid (0–2 mg/mL) and light (365, 385 and 400 nm; 120–480 J/cm^2), with the highest inactivation (5 log CFU/mL) achieved at caffeic concentration of 2 mg/mL and blue light dosage of 480 J/cm^2 (385 nm) [231]. Other authors also reported the synergistic antimicrobial activities of blue light (400 nm) and wine grape-derived polyphenols (for example, catechin and its isotopic ingredients) against *S. aureus* or *P. aeruginosa* (5 log CFU/mL) [232,233].

In addition, essential oils possess antimicrobial activity by inducing membrane breakage and permeability, albeit Gram-negative bacteria are known to be more resistant than Gram-positive bacteria due to the presence of hydrophilic outer membrane [234,235]. Membrane leakages can lead to loss of cellular constituents, such as ions, genetic materials or adenosine triphosphate (ATP), and subsequently cell death [236,237]. One study found that essential oils derived from eucalyptus (5%), clove (0.5%) and thyme (0.5%) improved the blue light (469–470 nm) inactivation of *S. epidermidis* and *P. aeruginosa* by 2–7 and 3–8 log CFU/mL, respectively, as compared with inactivation achieved by light alone. Relative to treatments with essential oils alone, samples treated with the combination of essential oils and blue light had 3–6 and 3–5 log CFU/mL lower counts of *S. epidermidis* and *P. aeruginosa*, respectively [238].

7. Blue Light versus Antimicrobial Resistance and Consequences of Sub-Lethal Light Exposures

Antimicrobial resistance presents a challenge to the food industry, with multiroute transmissions of resistant bacteria occurring through the contamination of food-processing environments, transfer of genes originating from microorganisms intentionally added to foods (starter cultures, bio-preservative bacteria or bacteriophage) and cross-contamination of foods [239–242]. Throughout the years, antimicrobial-resistant pathogenic bacteria have emerged across different food industries: horticultural (vegetables and fruits) [240], seafood [241] and meat (food and livestock) [242]. In previous sections, we have reviewed several studies that successfully used blue light against drug-resistant bacteria, such as MRSA. However, in this section, we focus on the ability of blue light to sensitize multidrug-resistant bacteria to antibiotics and subsequently on the inactivation of biofilms (monomicrobial or polymicrobial) and also on the potential development of bacterial tolerance/resistance to blue light.

7.1. Resistant Bacteria: Improved Sensitivity to Antibiotics

Generally, bacteria achieve antimicrobial resistance through three mechanisms: (1) preventing the drug from reaching the target (limiting uptake or active efflux); (2) modifying the target sites, such as alterations of penicillin-binding proteins in Gram-positive bacteria; (3) inactivating the drug through degradation or chemical modulation [243]. Theoretically, blue light could induce damages that disrupt the ability of bacteria to limit drug uptakes or to perform active efflux (mechanism 1). As mentioned previously (Section 3.1), blue light could induce breakage of cell walls, increase the permeability of the cell membrane (lipid peroxidation) and inactivate lipopolysaccharides (constituents of outer membrane in Gram-negative bacteria)—all of these structures play roles in conferring barriers against antibiotics [243]. In addition, one study showed that blue light-treated MRSA suffered from potassium ion leakages, which suggested that several transmembrane proteins (for example, Na^+/K^+ ion pumps) may have been denatured [113] and thus potentially limiting the role of these transmembrane proteins in pumping drugs out of the bacterial cells (active efflux) [243].

Several studies also found that blue light rendered bacteria more susceptible to antibiotics. In an in vitro study, blue light-treated (411 nm; 150 J/cm^2 per cycle; 15 cycles) *S. aureus* (methicillin-sensitive and resistant) had a higher susceptibility to gentamycin and doxycycline, but not vancomycin, ciprofloxacin, chloramphenicol and rifampicin, than untreated controls [244]. Another report showed that the minimum inhibitory concentration (MIC) of gentamycin, ceftazidime and meropenem against drug-resistant *P. aeruginosa* was reduced by up to 8-fold in the blue light-treated groups (405 nm; 10 or 12 J/cm^2), relative to the untreated controls [245]. On cooked chicken meat, blue light-treated (405 nm; 1700 J/cm^2) *S.* Enteritidis became more susceptible to ampicillin, chloramphenicol, nalidixic acid and rifampicin, relative to freshly-cultured or non-illuminated controls [189]. On the contrary, one study reported no change in the susceptibility of *S. aureus* (methicillin sensitive and resistant) to antibiotics (panel of 10, including gentamycin) after fifteen cycles of sub-lethal exposures to blue light (405 nm; 108 J/cm^2 per cycle) [246].

Further, He et al. found that blue light could be combined with tetracycline-class antibiotics to inactivate drug-resistant *E. coli* and MRSA: (1) demeclocycline (DMCT; 10–50 μM) could be activated as a photosensitizer by blue light (415 nm; 10 J/cm^2) and reduced the bacterial load of resistant *E. coli* and MRSA by 6 log CFU/mL; (2) sub-lethally injured *E. coli* underwent inactivation during the sub-culturing of these bacteria post-treatment. This indicated that the antibiotic was still active, even in the absence of light, and continued to inactivate the sensitized bacteria by inhibiting their ribosomes; (3) minimum inhibitory concentrations of DMCT, doxycycline, minocycline and tetracycline against drug-resistant *E. coli* and MRSA were reduced by up to 8-fold, when applied in the presence of blue light (415 nm), relative to the dark controls [247].

7.2. Inactivation of Biofilms

Biofilms facilitate horizontal gene transfers between individual bacteria within the matrix, especially in those that contain more than one bacterial species. These exchanges of mobile genetic elements may lead to an increase in antimicrobial resistance and environmental persistence [248]. To mitigate this issue, a group of researchers deployed antimicrobial blue light (405 nm; 500 J/cm^2) against polymicrobial biofilms and achieved log reductions of 2.37 and 3.40 CFU/mL for MRSA and *P. aeruginosa*, respectively, within a dual-species biofilm. The same blue light treatment on another dual-species biofilm inactivated *P. aeruginosa* and *C. albicans* by 6.34 and 3.11 log CFU/mL, respectively. As expected, monomicrobial biofilms were more susceptible to blue light, with damages to the exopolysaccharide matrix also observed across different types of biofilm [249].

The efficacy of blue light against biofilms also varies across bacterial species. For example, blue light (405 nm; 108–206 J/cm^2) significantly inactivated monomicrobial biofilms of drug-resistant *A. baumannii*, *P. aeruginosa* and *Neisseria gonorrhoeae* (4–8 log CFU/mL; $p < 0.01$ or $p < 0.0001$), whereas the same blue light treatment did not significantly affect the biofilms of *E. coli*, *E. faecalis* and *Proteus mirabilis*. Further, blue light (405 nm; 216 J/cm^2) significantly reduced ($p < 0.01$) the number of MRSA in biofilms grown for 24 h, but not in biofilms grown for 48 h [250].

7.3. Sub-Lethal Exposures Induce Cellular Processes Potentially Leading to Tolerance

A study found that fifteen cycles of sub-lethal exposures of *S. aureus* to blue light (411 nm; 150 J/cm^2 per cycle) resulted in the development of tolerance due to genetic alterations, which was stable after five successive sub-culturing. There was an increase in the expression of *recA* and *umuC* genes in the blue light-tolerant *S. aureus*, whereas mutant strains with non-functional *recA* and *umuC* genes did not develop tolerance. This finding confirmed that SOS-dependent mechanism played a role in the development of blue light-tolerant phenotype, although direct mutations from DNA damage were also possible [244]. In contrast, another study showed that *S. aureus* did not develop tolerance to blue light (405 nm; 108 J/cm^2 per cycle) after fifteen cycles of sub-lethal treatment [246]. Exposures of *P. aeruginosa*, *A. baumannii* and *E. coli* to twenty cycles of blue light (405 nm; dosages enough to induce bacterial reduction of 4 log CFU) also did not result in the development of tolerance [251]. Possible explanations for the discrepancies between these studies include different wavelengths and blue light dosages used, albeit further investigations are needed to ascertain the effects of sub-lethal exposures of bacteria to blue light.

Two reports found that exposure of *S. aureus* to blue light had reduced its susceptibility to H_2O_2 [244,246]. In response to oxidative stress induced by the blue light, bacteria may up-regulate the expression of *katA* that encodes the production of H_2O_2-scavenging catalase protein and thus exhibit higher tolerance to H_2O_2 [118]. Adair and Drum identified thirty-two other genes in *S. aureus* regulated (up- or downregulated) by blue light (465 nm; 250 J/cm^2), which included those responsible for the production of cell envelope components and heat-shock proteins [252]. Similarly, light-mediated gene regulations occurred in other major food-borne pathogens, namely *V. cholerae* and *C. sazakii*. In response to ROS generated by blue light (fluorescent black light; UV-filtered), *V. cholerae* showed differential expression of 222 genes (6.3%), relative to untreated cells, especially those encoding enzymes that protect or repair lipids and nucleic acids (genome)—these transcriptional responses to blue light were regulated by ChrR and MerR-like proteins [253]. Blue light (415 nm; up to 20.04 J/cm^2) was also found to upregulate the expression of genes in *C. sakazakii* that encode oxidative stress-resistance chaperone, an adhesin and a capsule biosynthesis protein (CapC) [115].

These findings indicate that sub-lethal exposures of bacteria to blue light induce cell responses that may lead to the development of tolerance over time. However, complete resistance has not been reported, as even when tolerance was observed at a particular blue light dosage, increasing the light dosage was sufficient to eliminate the tolerant bacteria [244]. Nevertheless, other strategies are required to antagonize any development of bacterial tolerance to blue light. For example, the combination of blue light at 460 and 405 nm was reported to be effective against blue light-tolerant *S. aureus*. Leanse et al. [254] revealed that blue light at 460 nm (90–360 J/cm^2) inactivated staphyloxanthin (a ROS scavenger; antioxidant) through photolysis and thus disrupted the ability of *S. aureus* to resist blue light treatment. Subsequent treatment with blue light at 405 nm (90–180 J/cm^2) inactivated the bacteria (planktonic or in biofilm) at a higher rate than when single wavelengths were used, albeit inactivation was dependent on the dosage of both 405- and 460-nm blue light. In addition, as blue light attacks multiple targets in the bacterial cells, the development of resistance to blue light is likely to be slower than resistance to antibiotics.

8. Research Gap and Future Outlook

While it is a well-established fact that bacterial cells contain photo-active endogenous photosensitizers, the amount and type of these intracellular chromophores—and thus the susceptibility to blue light—may vary across bacterial species. For example, spectroscopic measurements revealed the presence of flavins and porphyrins in the cell lysates of *A. actinomycetecomitans*, although these compounds were not detected in *E. coli*. When illuminated by blue light (460 nm; 150 J/cm^2), *A. actinomycetecomitans* (serotype b; ATCC 43718) were reduced by 5 log CFU, whereas *E. coli* (ATCC 25922) remained unaffected [255]. Another group of researchers also used spectroscopic technique to identify protoporphyrin IX and coproporphyrin as the main intracellular photosensitizers in *Helicobacter pylori*, which could exist in monomeric, dimeric or aggregated forms [256]. Others utilized high-performance liquid chromatography to characterize the endogenous photosensitizers in *P. aeruginosa* and *A. baumannii*, including coproporphyrin (I or III) and protporphyrin IX [257,258]. These studies provide technical foundation that future researches could build upon, particularly for characterizing endogenous photosensitizers in food-borne pathogenic bacteria.

There are limited data on the bactericidal activity of blue light against spoilage microorganisms, especially bacteria that are capable of growing in anaerobic conditions, such as *Clostridium estertheticum* in vacuum-packed meats. A study showed that while *E. coli*, *S. aureus* and *E. faecalis* in liquid media were unaffected by blue light (405 nm; 5.73 J/cm^2) under anaerobic environments, the bacterial loads of *Prevotella intermedia* and *Prevotella nigrescenes* were significantly reduced by 1 and 2 log CFU/mL ($p < 0.05$), respectively [259]. Others observed 1-log reduction of *P. gingivalis* after illumination with blue light (405 nm; 3.42 J/cm^2) [260]. The authors of both studies attributed blue light-mediated anaerobic inactivation of bacteria to the generation of organic radicals directly from the triplet state of endogenous photosensitizers [259,260]. These findings indicate that there is a potential for anaerobic application of blue light, although there is a need for improvements in the bactericidal efficacy of blue light under oxygen-scarce conditions. Azide salts can be used to facilitate anaerobic photodynamic treatments [216]. In addition, future studies are required to assess the effects of blue light on spore germination or the production of bacterial toxins.

Sensorial properties of blue light-treated foods need to be assessed beyond the quantitative measurements conducted within laboratory settings. For example, the quality of blue light-treated oysters were evaluated using both chemical/microbiological analyses and human panels (sensory evaluation), which provided the researchers with a comprehensive information to determine the shelf-life of seafood [212]. Future research can be designed to assess the sensorial properties of other types of blue light-treated food, including fruits, vegetables, meat and dairy products.

Lastly, spectral readings from Fourier-transform infrared (FTIR) indicated that blue light (470 nm) and UV (254 nm) primarily attacked the DNA during inactivation of MRSA, although the two lights targeted different DNA conformations: blue light induced damage on A-DNA, whereas UV predominantly inactivated B-DNA—these are two of the three conformations of double helical DNA, with the other one being Z-DNA. These findings suggest that blue and UV lights may be used as complementary treatments against microbes [261]. For safety, far ultraviolet-C (UV-C; 207 or 222 nm) can be used as an alternative to the conventional UV-C (254 nm). Accumulating evidence indicates that unlike the conventional UV-C, far UV-C exhibits bactericidal activity, but only has minimal effects on mammalian cells, such as the eye and skin of mice or human skin cells [262–267].

Author Contributions: Conceptualization, G.B.; investigation, J.H. and S.W.; writing—original draft preparation, J.H.; writing—review and editing, J.H., S.W. and G.B.; visualization, J.H.; project administration, G.B.; funding acquisition, G.B. All authors have read and agreed to the published version of the manuscript.

Funding: This work is funded through the AgResearch Strategic Science Investment Fund (SSIF) Food Integrity (contract A25768).

Acknowledgments: The authors wish to thank Aswathi Soni and Delphine Rapp for their technical inputs. Elisabeth Belianty and Erinna Hadi are also acknowledged for their assistance with the illustrations.

Conflicts of Interest: The authors declare no conflict of interest. The funders had no role in the design of the study; in the collection, analyses, or interpretation of data; in the writing of the manuscript, or in the decision to publish the work.

References

1. World Health Organization. WHO|WHO Estimates of the Global Burden of Foodborne Diseases. Available online: http://www.who.int/foodsafety/publications/foodborne_disease/fergreport/en/ (accessed on 10 October 2020).
2. Dickson-Spillmann, M.; Siegrist, M.; Keller, C. Attitudes toward Chemicals Are Associated with Preference for Natural Food. *Food Qual. Prefer.* **2011**, *22*, 149–156. [CrossRef]

3. Ha, T.M.; Shakur, S.; Pham Do, K.H. Consumer Concern about Food Safety in Hanoi, Vietnam. *Food Control* **2019**, *98*, 238–244. [CrossRef]
4. Soni, A.; Oey, I.; Silcock, P.; Bremer, P. Bacillus Spores in the Food Industry: A Review on Resistance and Response to Novel Inactivation Technologies. *Compr. Rev. Food Sci. Food Saf.* **2016**, *15*, 1139–1148. [CrossRef]
5. Wang, M.S.; Wang, L.H.; Bekhit, A.E.D.A.; Yang, J.; Hou, Z.P.; Wang, Y.Z.; Dai, Q.Z.; Zeng, X.A. A Review of Sublethal Effects of Pulsed Electric Field on Cells in Food Processing. *J. Food Eng.* **2018**, *223*, 32–41. [CrossRef]
6. Huang, H.W.; Wu, S.J.; Lu, J.K.; Shyu, Y.T.; Wang, C.Y. Current Status and Future Trends of High-Pressure Processing in Food Industry. *Food Control* **2017**, *72*, 1–8. [CrossRef]
7. Sampedro, F.; McAloon, A.; Yee, W.; Fan, X.; Geveke, D.J. Cost Analysis and Environmental Impact of Pulsed Electric Fields and High Pressure Processing in Comparison with Thermal Pasteurization. *Food Bioprocess Technol.* **2014**, *7*, 1928–1937. [CrossRef]
8. D'Souza, C.; Yuk, H.G.; Khoo, G.H.; Zhou, W. Application of Light-Emitting Diodes in Food Production, Postharvest Preservation, and Microbiological Food Safety. *Compr. Rev. Food Sci. Food Saf.* **2015**, *14*, 719–740. [CrossRef]
9. Bintsis, T.; Litopoulou-Tzanetaki, E.; Robinson, R.K. Existing and Potential Applications of Ultraviolet Light in the Food Industry—A Critical Review. *J. Sci. Food Agric.* **2000**, *80*, 637–645. [CrossRef]
10. Koutchma, T. UV Light Microbial Inactivation in Foods. In *Ultraviolet Light in Food Technology: Principles and Applications*; CRC Press: Boca Raton, FL, USA, 2019; pp. 1–344.
11. D'Orazio, J.; Jarrett, S.; Amaro-Ortiz, A.; Scott, T. UV Radiation and the Skin. *Int. J. Mol. Sci.* **2013**, *14*, 12222–12248. [CrossRef]
12. Kütting, B.; Drexler, H. UV-Induced Skin Cancer at Workplace and Evidence-Based Prevention. *Int. Arch. Occup. Environ. Health* **2010**, *83*, 843–854. [CrossRef]
13. Zaffina, S.; Camisa, V.; Lembo, M.; Vinci, M.R.; Tucci, M.G.; Borra, M.; Napolitano, A.; Cannatã, V. Accidental Exposure to UV Radiation Produced by Germicidal Lamp: Case Report and Risk Assessment. *Photochem. Photobiol.* **2012**, *88*, 1001–1004. [CrossRef] [PubMed]
14. Yin, R.; Dai, T.; Avci, P.; Jorge, A.E.S.; De Melo, W.C.M.A.; Vecchio, D.; Huang, Y.Y.; Gupta, A.; Hamblin, M.R. Light Based Anti-Infectives: Ultraviolet C Irradiation, Photodynamic Therapy, Blue Light, and Beyond. *Curr. Opin. Pharmacol.* **2013**, *13*, 731–762. [CrossRef] [PubMed]
15. World Health Organization. Food Safety. Available online: https://www.who.int/news-room/fact-sheets/detail/food-safety (accessed on 17 December 2020).
16. Møretrø, T.; Langsrud, S. Residential Bacteria on Surfaces in the Food Industry and Their Implications for Food Safety and Quality. *Compr. Rev. Food Sci. Food Saf.* **2017**, *16*, 1022–1041. [CrossRef]
17. Fox, E.M.; Jiang, Y.; Gobius, K.S. Key Pathogenic Bacteria Associated with Dairy Foods: On-Farm Ecology and Products Associated with Foodborne Pathogen Transmission. *Int. Dairy J.* **2018**, *84*, 28–35. [CrossRef]
18. Cancino-Padilla, N.; Fellenberg, M.A.; Franco, W.; Ibáñez, R.A.; Vargas-Bello-Pérez, E. Foodborne Bacteria in Dairy Products: Detection by Molecular Techniques. *Cien. Inv. Agr.* **2017**, *44*, 215–229. [CrossRef]
19. Castro, H.; Jaakkonen, A.; Hakkinen, M.; Korkeala, H.; Lindström, M. Occurrence, Persistence, and Contamination Routes of Listeria Monocytogenes Genotypes on Three Finnish Dairy Cattle Farms: A Longitudinal Study. *Appl. Environ. Microbiol.* **2018**, *84*, 2000–2017. [CrossRef]
20. Haley, B.J.; Sonnier, J.; Schukken, Y.H.; Karns, J.S.; Van Kessel, J.A.S. Diversity of Listeria Monocytogenes Within a U.S. Dairy Herd, 2004–2010. *Foodborne Pathog. Dis.* **2015**, *12*, 844–850. [CrossRef]
21. Browne, A.S.; Midwinter, A.C.; Withers, H.; Cookson, A.L.; Biggs, P.J.; Marshall, J.C.; Benschop, J.; Hathaway, S.; Haack, N.A.; Akhter, R.N.; et al. Molecular Epidemiology of Shiga Toxinproducing Escherichia Coli (STEC) on New Zealand Dairy Farms: Application of a Culture-Independent Assay and Whole-Genome Sequencing. *Appl. Environ. Microbiol.* **2018**, *84*, 481–499. [CrossRef]
22. Venegas-Vargas, C.; Henderson, S.; Khare, A.; Mosci, R.E.; Lehnert, J.D.; Singh, P.; Ouellette, L.M.; Norby, B.; Funk, J.A.; Rust, S.; et al. Factors Associated with Shiga Toxin-Producing Escherichia Coli Shedding by Dairy and Beef Cattle. *Appl. Environ. Microbiol.* **2016**, *82*, 5049–5056. [CrossRef]
23. Ross, C.M.; Rapp, D.; Cave, V.M.; Brightwell, G. Prevalence of Shiga Toxin-Producing Escherichia Coli in Pasture-Based Dairy Herds. *Lett. Appl. Microbiol.* **2019**, *68*, 112–119. [CrossRef]
24. McAuley, C.M.; McMillan, K.E.; Moore, S.C.; Fegan, N.; Fox, E.M. Characterization of Escherichia Coli and Salmonella from Victoria, Australia, Dairy Farm Environments. *J. Food Prot.* **2017**, *80*, 2078–2082. [CrossRef]

25. Rodriguez-Rivera, L.D.; Wright, E.M.; Siler, J.D.; Elton, M.; Cummings, K.J.; Warnick, L.D.; Wiedmann, M. Subtype Analysis of Salmonella Isolated from Subclinically Infected Dairy Cattle and Dairy Farm Environments Reveals the Presence of Both Human- and Bovine-Associated Subtypes. *Vet. Microbiol.* **2014**, *170*, 307–316. [CrossRef] [PubMed]
26. Leuenberger, A.; Sartori, C.; Boss, R.; Resch, G.; Oechslin, F.; Steiner, A.; Moreillon, P.; Graber, H.U. Genotypes of Staphylococcus Aureus: On-Farm Epidemiology and the Consequences for Prevention of Intramammary Infections. *J. Dairy Sci.* **2019**, *102*, 3295–3309. [CrossRef] [PubMed]
27. McAuley, C.M.; McMillan, K.; Moore, S.C.; Fegan, N.; Fox, E.M. Prevalence and Characterization of Foodborne Pathogens from Australian Dairy Farm Environments. *J. Dairy Sci.* **2014**, *97*, 7402–7412. [CrossRef]
28. Hunter, C.J.; Bean, J.F. Cronobacter: An Emerging Opportunistic Pathogen Associated with Neonatal Meningitis, Sepsis and Necrotizing Enterocolitis. *J. Perinatol.* **2013**, *33*, 581–585. [CrossRef]
29. Norberg, S.; Stanton, C.; Ross, R.P.; Hill, C.; Fitzgerald, G.F.; Cotter, P.D. *Cronobacter* spp. in Powdered Infant Formula. *J. Food Prot.* **2012**, *75*, 607–620. [CrossRef]
30. Vojkovska, H.; Karpiskova, R.; Orieskova, M.; Drahovska, H. Characterization of Cronobacter Spp. Isolated from Food of Plant Origin and Environmental Samples Collected from Farms and from Supermarkets in the Czech Republic. *Int. J. Food Microbiol.* **2016**, *217*, 130–136. [CrossRef]
31. Jaradat, Z.W.; Ababneh, Q.O.; Saadoun, I.M.; Samara, N.A.; Rashdan, A.M. Isolation of Cronobacter Spp. (Formerly Enterobacter Sakazakii) from Infant Food, Herbs and Environmental Samples and the Subsequent Identification and Confirmation of the Isolates Using Biochemical, Chromogenic Assays, PCR and 16S RRNA Sequencing. *BMC Microbiol.* **2009**, *9*, 225. [CrossRef]
32. Schmid, M.; Iversen, C.; Gontia, I.; Stephan, R.; Hofmann, A.; Hartmann, A.; Jha, B.; Eberl, L.; Riedel, K.; Lehner, A. Evidence for a Plant-Associated Natural Habitat for Cronobacter spp. *Res. Microbiol.* **2009**, *160*, 608–614. [CrossRef]
33. Molloy, C.; Cagney, C.; O'Brien, S.; Iversen, C.; Fanning, S.; Duffy, G. Surveillance and Characterisation by Pulsed-Field Gel Electrophoresis of Cronobacter spp. in Farming and Domestic Environments, Food Production Animals and Retail Foods. *Int. J. Food Microbiol.* **2009**, *136*, 198–203. [CrossRef]
34. Quigley, L.; O'Sullivan, O.; Stanton, C.; Beresford, T.P.; Ross, R.P.; Fitzgerald, G.F.; Cotter, P.D. The Complex Microbiota of Raw Milk. *FEMS Microbiol. Rev.* **2013**, *37*, 664–698. [CrossRef]
35. Oliver, S.P.; Murinda, S.E. Milk and Raw Milk Consumption as a Vector for Human Disease. In *Zoonotic Pathogens in the Food Chain*; Krause, D., Hendrick, S., Eds.; CABI: Oxfordshire, UK, 2010; pp. 99–118.
36. Gaulin, C.; Ramsay, D.; Bekal, S. Widespread Listeriosis Outbreak Attributable to Pasteurized Cheese, Which Led to Extensive Cross-Contamination Affecting Cheese Retailers, Quebec, Canada, 2008. *J. Food Prot.* **2012**, *75*, 71–78. [CrossRef]
37. Germinario, C.; Caprioli, A.; Giordano, M.; Chironna, M.; Gallone, M.S.; Tafuri, S.; Minelli, F.; Maugliani, A.; Michelacci, V.; Santangelo, L.; et al. Community-Wide Outbreak of Haemolytic Uraemic Syndrome Associated with Shiga Toxin 2-Producing Escherichia Coli O26: H11 in Southern Italy, Summer 2013. *Eurosurveillance* **2016**, *21*, 30343. [CrossRef]
38. Oxaran, V.; Lee, S.H.I.; Chaul, L.T.; Corassin, C.H.; Barancelli, G.V.; Alves, V.F.; de Oliveira, C.A.F.; Gram, L.; De Martinis, E.C.P. Listeria Monocytogenes Incidence Changes and Diversity in Some Brazilian Dairy Industries and Retail Products. *Food Microbiol.* **2017**, *68*, 16–23. [CrossRef]
39. Ahmed, A.M.; Shimamoto, T. Isolation and Molecular Characterization of Salmonella Enterica, Escherichia Coli O157: H7 and Shigella Spp. from Meat and Dairy Products in Egypt. *Int. J. Food Microbiol.* **2014**, *168–169*, 57–62. [CrossRef]
40. Derra, F.A.; Karlsmose, S.; Monga, D.P.; Mache, A.; Svendsen, C.A.; Félix, B.; Granier, S.A.; Geyid, A.; Taye, G.; Hendriksen, R.S. Occurrence of Listeria Spp. in Retail Meat and Dairy Products in the Area of Addis Ababa, Ethiopia. *Foodborne Pathog. Dis.* **2013**, *10*, 577–579. [CrossRef]
41. Omer, M.K.; Álvarez-Ordoñez, A.; Prieto, M.; Skjerve, E.; Asehun, T.; Alvseike, O.A. A Systematic Review of Bacterial Foodborne Outbreaks Related to Red Meat and Meat Products. *Foodborne Pathog. Dis.* **2018**, *15*, 598–611. [CrossRef]
42. Tesson, V.; Federighi, M.; Cummins, E.; de Oliveira Mota, J.; Guillou, S.; Boué, G. A Systematic Review of Beef Meat Quantitative Microbial Risk Assessment Models. *Int. J. Environ. Res. Public Health* **2020**, *17*, 688. [CrossRef]

43. Ferrari, R.G.; Rosario, D.K.A.; Cunha-Neto, A.; Mano, S.B.; Figueiredo, E.E.S.; Conte-Junior, C.A. Worldwide Epidemiology of Salmonella Serovars in Animal-Based Foods: A Meta-Analysis. *Am. Soc. Microbiol.* **2019**. [CrossRef]
44. Mor-Mur, M.; Yuste, J. Emerging Bacterial Pathogens in Meat and Poultry: An Overview. *Food Bioprocess Technol.* **2010**, *3*, 24–35.
45. Gieraltowski, L.; Higa, J.; Peralta, V.; Green, A.; Schwensohn, C.; Rosen, H.; Libby, T.; Kissler, B.; Marsden-Haug, N.; Booth, H.; et al. National Outbreak of Multidrug Resistant Salmonella Heidelberg Infections Linked to a Single Poultry Company. *PLoS ONE* **2016**, *11*, e0162369. [CrossRef] [PubMed]
46. Grass, J.E.; Gould, L.H.; Mahon, B.E. Epidemiology of Foodborne Disease Outbreaks Caused by Clostridium Perfringens, United States, 1998–2010. *Foodborne Pathog. Dis.* **2013**, *10*, 131–136. [CrossRef] [PubMed]
47. Abid, M.; Wimalarathna, H.; Mills, J.; Saldana, L.; Pang, W.; Richardson, J.F.; Maiden, M.C.J.; McCarthy, N.D. Duck Liver–Associated Outbreak of Campylobacteriosis among Humans, United Kingdom, 2011. *Emerg. Infect. Dis.* **2013**, *19*, 1310–1313. [CrossRef] [PubMed]
48. O'Leary, M.C.; Harding, O.; Fisher, L.; Cowden, J. A Continuous Common-Source Outbreak of Campylobacteriosis Associated with Changes to the Preparation of Chicken Liver Pâté. *Epidemiol. Infect.* **2009**, *137*, 383–388. [CrossRef]
49. Chai, S.J.; Cole, D.; Nisler, A.; Mahon, B.E. Poultry: The Most Common Food in Outbreaks with Known Pathogens, United States, 1998–2012. *Epidemiol. Infect.* **2017**, *145*, 316–325. [CrossRef]
50. Centers for Disease Control and Prevention (CDC). Outbreak of Salmonella Heidelberg Infections Linked to a Single Poultry Producer—13 States, 2012-2013. *MMWR Morb. Mortal. Wkly. Rep.* **2013**, *62*, 553–556.
51. Little, C.L.; Gormley, F.J.; Rawal, N.; Richardson, J.F. A Recipe for Disaster: Outbreaks of Campylobacteriosis Associated with Poultry Liver Pâté in England and Wales. *Epidemiol. Infect.* **2010**, *138*, 1691–1694. [CrossRef]
52. Rouger, A.; Tresse, O.; Zagorec, M. Bacterial Contaminants of Poultry Meat: Sources, Species, and Dynamics. *Microorganisms* **2017**, *5*, 50. [CrossRef]
53. Yang, B.; Qu, D.; Zhang, X.; Shen, J.; Cui, S.; Shi, Y.; Xi, M.; Sheng, M.; Zhi, S.; Meng, J. Prevalence and Characterization of Salmonella Serovars in Retail Meats of Marketplace in Shaanxi, China. *Int. J. Food Microbiol.* **2010**, *141*, 63–72. [CrossRef]
54. Wu, S.; Huang, J.; Wu, Q.; Zhang, J.; Zhang, F.; Yang, X.; Wu, H.; Zeng, H.; Chen, M.; Ding, Y.; et al. Staphylococcus Aureusisolated from Retail Meat and Meat Products in China: Incidence, Antibiotic Resistance and Genetic Diversity. *Front. Microbiol.* **2018**, *9*, 2767. [CrossRef]
55. Ali, N.H.; Farooqui, A.; Khan, A.; Khan, A.Y.; Kazmi, S.U. Microbial Contamination of Raw Meat and Its Environment in Retail Shops in Karachi, Pakistan. *J. Infect. Dev. Ctries.* **2010**, *4*, 382–388.
56. Ge, B.; Mukherjee, S.; Hsu, C.H.; Davis, J.A.; Tran, T.T.T.; Yang, Q.; Abbott, J.W.; Ayers, S.L.; Young, S.R.; Crarey, E.T.; et al. MRSA and Multidrug-Resistant Staphylococcus Aureus in U.S. Retail Meats, 2010–2011. *Food Microbiol.* **2017**, *62*, 289–297. [CrossRef] [PubMed]
57. Iwamoto, M.; Ayers, T.; Mahon, B.E.; Swerdlow, D.L. Epidemiology of Seafood-Associated Infections in the United States. *Clin. Microbiol. Rev.* **2010**, *23*, 399–411. [CrossRef] [PubMed]
58. Novoslavskij, A.; Terentjeva, M.; Eizenberga, I.; Valciņa, O.; Bartkevičs, V.; Bērziņš, A. Major Foodborne Pathogens in Fish and Fish Products: A Review. *Ann. Microbiol.* **2016**, *66*, 1–15. [CrossRef]
59. Wang, F.; Jiang, L.; Yang, Q.; Han, F.; Chen, S.; Pu, S.; Vance, A.; Ge, B. Prevalence and Antimicrobial Susceptibility of Major Foodborne Pathogens in Imported Seafood. *J. Food Prot.* **2011**, *74*, 1451–1461. [CrossRef]
60. Dumen, E.; Ekici, G.; Ergin, S.; Bayrakal, G.M. Presence of Foodborne Pathogens in Seafood and Risk Ranking for Pathogens. *Foodborne Pathog. Dis.* **2020**, *17*, 541–546. [CrossRef]
61. Novotny, L.; Dvorska, L.; Lorencova, A.; Beran, V.; Pavlik, I. Fish: A Potential Source of Bacterial Pathogens for Human Beings. *Vet. Med.* **2004**, *49*, 343–358. [CrossRef]
62. Ali, A.; Parisi, A.; Conversano, M.C.; Iannacci, A.; D'Emilio, F.; Mercurio, V.; Normanno, G. Food-Borne Bacteria Associated with Seafoods: A Brief Review. *J. Food Qual. Hazards Control.* **2020**, *7*, 4–10. [CrossRef]
63. Herwig, R. Pathogens Transmitted by Seafood. In *Foodborne Disease Handbook: Volume IV: Seafood and Environmental Toxins*; Hui, Y.H., Kitts, D., Stanfield, P.S., Eds.; CRC Press: Boca Raton, FL, USA, 2017; p. 109.
64. Jami, M.; Ghanbari, M.; Zunabovic, M.; Domig, K.J.; Kneifel, W. Listeria Monocytogenes in Aquatic Food Products—A Review. *Compr. Rev. Food Sci. Food Saf.* **2014**, *13*, 798–813. [CrossRef]

65. Jung, S.W. A Foodborne Outbreak of Gastroenteritis Caused by Vibrio Parahaemolyticus Associated with Cross-Contamination from Squid in Korea. *Epidemiol. Health* **2018**, *40*, e2018056. [CrossRef]
66. Kim, J.H.; Lee, J.; Hong, S.; Lee, S.; Na, H.Y.; Jeong, Y.-I.; Choi, E.J.; Kim, J.; Kawk, H.S.; Cho, E. Cholera Outbreak Due to Raw Seafood Consumption in South Korea, 2016. *Am. J. Trop. Med. Hyg.* **2018**, *99*, 168–170. [CrossRef]
67. Hassan, R.; Tecle, S.; Adcock, B.; Kellis, M.; Weiss, J.; Saupe, A.; Sorenson, A.; Klos, R.; Blankenship, J.; Blessington, T.; et al. Multistate Outbreak of Salmonella Paratyphi B Variant L(+) Tartrate(+) and Salmonella Weltevreden Infections Linked to Imported Frozen Raw Tuna: USA, March–July 2015. *Epidemiol. Infect.* **2018**, *146*, 1461–1467. [CrossRef]
68. Heinitz, M.L.; Ruble, R.D.; Wagner, D.E.; Tatini, S.R. Incidence of Salmonella in Fish and Seafood. *J. Food Prot.* **2000**, *63*, 579–592. [CrossRef]
69. Taylor, M.; Cheng, J.; Sharma, D.; Bitzikos, O.; Gustafson, R.; Fyfe, M.; Greve, R.; Murti, M.; Stone, J.; Honish, L.; et al. Outbreak of Vibrio Parahaemolyticus Associated with Consumption of Raw Oysters in Canada, 2015. *Foodborne Pathog. Dis.* **2018**, *15*, 554–559. [CrossRef]
70. Olaimat, A.N.; Holley, R.A. Factors Influencing the Microbial Safety of Fresh Produce: A Review. *Food Microbiol.* **2012**, *32*, 1–19. [CrossRef]
71. Alegbeleye, O.O.; Singleton, I.; Sant'Ana, A.S. Sources and Contamination Routes of Microbial Pathogens to Fresh Produce during Field Cultivation: A Review. *Food Microbiol.* **2018**, *73*, 177–208. [CrossRef]
72. Beuchat, L.R. Ecological Factors Influencing Survival and Growth of Human Pathogens on Raw Fruits and Vegetables. *Microbes Infect.* **2002**, *4*, 413–423. [CrossRef]
73. Araújo, S.; Silva, I.A.T.; Tacão, M.; Patinha, C.; Alves, A.; Henriques, I. Characterization of Antibiotic Resistant and Pathogenic Escherichia Coli in Irrigation Water and Vegetables in Household Farms. *Int. J. Food Microbiol.* **2017**, *257*, 192–200. [CrossRef]
74. Akinde, S.B.; Sunday, A.A.; Adeyemi, F.M.; Fakayode, I.B.; Oluwajide, O.O.; Adebunmi, A.A.; Oloke, J.K.; Adebooye, C.O. Microbes in Irrigation Water and Fresh Vegetables: Potential Pathogenic Bacteria Assessment and Implications for Food Safety. *Appl. Biosaf.* **2016**, *21*, 89–97. [CrossRef]
75. Steele, M.; Odumeru, J. Irrigation Water as Source of Foodborne Pathogens on Fruit and Vegetables. *J. Food Prot.* **2004**, *67*, 2839–2849. [CrossRef]
76. Okafo, C.N.; Umoh, V.J.; Galadima, M. Occurrence of Pathogens on Vegetables Harvested from Soils Irrigated with Contaminated Streams. *Sci. Total Environ.* **2003**, *311*, 49–56. [CrossRef]
77. Hounmanou, Y.M.G.; Mdegela, R.H.; Dougnon, T.V.; Mhongole, O.J.; Mayila, E.S.; Malakalinga, J.; Makingi, G.; Dalsgaard, A. Toxigenic Vibrio Cholerae O1 in Vegetables and Fish Raised in Wastewater Irrigated Fields and Stabilization Ponds during a Non-Cholera Outbreak Period in Morogoro, Tanzania: An Environmental Health Study. *BMC Res. Notes* **2016**, *9*, 466. [CrossRef]
78. Willis, C.; McLauchlin, J.; Aird, H.; Amar, C.; Barker, C.; Dallman, T.; Elviss, N.; Lai, S.; Sadler-Reeves, L. Occurrence of Listeria and Escherichia Coli in Frozen Fruit and Vegetables Collected from Retail and Catering Premises in England 2018–2019. *Int. J. Food Microbiol.* **2020**, *334*, 108849. [CrossRef]
79. Sagoo, S.K.; Little, C.L.; Ward, L.; Gillespie, I.A.; Mitchell, R.T. Microbiological Study of Ready-to-Eat Salad Vegetables from Retail Establishments Uncovers a National Outbreak of Salmonellosis. *J. Food Prot.* **2003**, *66*, 403–409. [CrossRef]
80. Ponniah, J.; Robin, T.; Paie, M.S.; Radu, S.; Ghazali, F.M.; Kqueen, C.Y.; Nishibuchi, M.; Nakaguchi, Y.; Malakar, P.K. Listeria Monocytogenes in Raw Salad Vegetables Sold at Retail Level in Malaysia. *Food Control* **2010**, *21*, 774–778. [CrossRef]
81. Ajayeoba, T.A.; Atanda, O.O.; Obadina, A.O.; Bankole, M.O.; Adelowo, O.O. The Incidence and Distribution of Listeria Monocytogenes in Ready-to-Eat Vegetables in South-Western Nigeria. *Food Sci. Nutr.* **2016**, *4*, 59–66. [CrossRef]
82. Gomez-Govea, M.; Solis, L.; Heredia, N.; García, S. Analysis of Microbial Contamination Levels of Fruits and Vegetables at Retail in Monterrey, Mexico. *J. Food Agric. Environ.* **2005**, *10*, 152–156.
83. Denis, N.; Zhang, H.; Leroux, A.; Trudel, R.; Bietlot, H. Prevalence and Trends of Bacterial Contamination in Fresh Fruits and Vegetables Sold at Retail in Canada. *Food Control* **2016**, *67*, 225–234. [CrossRef]
84. Zhu, Q.; Gooneratne, R.; Hussain, M.A. Detection of Listeria Species in Fresh Produce Samples from Different Retail Shops in Canterbury, New Zealand. *Adv. Food Technol. Nutr. Sci.Open J.* **2016**, *2*, 96–102. [CrossRef]

85. Nunes Silva, B.; Cadavez, V.; Teixeira, J.A.; Gonzales-Barron, U. Meta-Analysis of the Incidence of Foodborne Pathogens in Vegetables and Fruits from Retail Establishments in Europe. *Curr. Opin. Food Sci.* **2017**, *18*, 21–28. [CrossRef]
86. Møretrø, T.; Hermansen, L.; Holck, A.L.; Sidhu, M.S.; Rudi, K.; Langsrud, S. Biofilm Formation and the Presence of the Intercellular Adhesion Locus Ica among Staphylococi from Food and Food Processing Environments. *Appl. Environ. Microbiol.* **2003**, *69*, 5648–5655. [PubMed]
87. Fagerlund, A.; Langsrud, S.; Heir, E.; Mikkelsen, M.I.; Møretrø, T. Biofilm Matrix Composition Affects the Susceptibility of Food Associated Staphylococci to Cleaning and Disinfection Agents. *Front. Microbiol.* **2016**, *7*, 856. [PubMed]
88. Arciola, C.R.; Campoccia, D.; Ravaioli, S.; Montanaro, L. Polysaccharide Intercellular Adhesin in Biofilm: Structural and Regulatory Aspects. *Front. Cell. Infect. Microbiol.* **2015**, *5*, 7. [PubMed]
89. Fratamico, P.M.; Annous, B.A.; Guenther, N.W. *Biofilms in the Food and Beverage Industries*; Woodhead Publishing Limited: Cambridge, UK, 2009.
90. Furkanur, M.; Mizan, R.; Jahid, I.K.; Ha, S.-D. Microbial Biofilms in Seafood: A Food-Hygiene Challenge. *Food Microbiol.* **2015**, *49*, 41–55.
91. Srey, S.; Jahid, I.K.; Ha, S.-D. Biofilm Formation in Food Industries: A Food Safety Concern. *Food Control* **2013**, *31*, 572–585.
92. Galié, S.; García-Gutiérrez, C.; Miguélez, E.M.; Villar, C.J.; Lombó, F. Biofilms in the Food Industry: Health Aspects and Control Methods. *Front. Microbiol.* **2018**, *9*, 898.
93. Flemming, H.C.; Wingender, J.; Szewzyk, U.; Steinberg, P.; Rice, S.A.; Kjelleberg, S. Biofilms: An Emergent Form of Bacterial Life. *Nat. Rev. Microbiol.* **2016**, *14*, 563–575.
94. Majed, R.; Faille, C.; Kallassy, M.; Gohar, M. Bacillus Cereus Biofilms-Same, Only Different. *Front. Microbiol.* **2016**, *7*, 1054.
95. Liu, N.T.; Nou, X.; Lefcourt, A.M.; Shelton, D.R.; Lo, Y.M. Dual-Species Biofilm Formation by Escherichia Coli O157: H7 and Environmental Bacteria Isolated from Fresh-Cut Processing Facilities. *Int. J. Food Microbiol.* **2014**, *171*, 15–20.
96. Alonso, A.N.; Perry, K.J.; Regeimbal, J.M.; Regan, P.M.; Higgins, D.E. Identification of Listeria Monocytogenes Determinants Required for Biofilm Formation. *PLoS ONE* **2014**, *9*, e113696.
97. Zhang, T.; Bae, D.; Wang, C. Listeria Monocytogenes DNA Glycosylase AdlP Affects Flagellar Motility, Biofilm Formation, Virulence, and Stress Responses. *Appl. Environ. Microbiol.* **2016**, *82*, 5144–5152. [CrossRef] [PubMed]
98. Crawford, R.W.; Reeve, K.E.; Gunn, J.S. Flagellated but Not Hyperfimbriated Salmonella Enterica Serovar Typhimurium Attaches to and Forms Biofilms on Cholesterol-Coated Surfaces. *J. Bacteriol.* **2010**, *192*, 2981–2990. [CrossRef] [PubMed]
99. Jonas, K.; Tomenius, H.; Kader, A.; Normark, S.; Römling, U.; Belova, L.M.; Melefors, Ö. Roles of Curli, Cellulose and BapA in Salmonella Biofilm Morphology Studied by Atomic Force Microscopy. *BMC Microbiol.* **2007**, *7*, 70. [CrossRef] [PubMed]
100. Carter, M.Q.; Louie, J.W.; Feng, D.; Zhong, W.; Brandl, M.T. Curli Fimbriae Are Conditionally Required in Escherichia Coli O157: H7 for Initial Attachment and Biofilm Formation. *Food Microbiol.* **2016**, *57*, 81–89. [CrossRef]
101. Giacomucci, S.; Cros, C.D.N.; Perron, X.; Mathieu-Denoncourt, A.; Duperthuy, M. Flagella-Dependent Inhibition of Biofilm Formation by Sub-Inhibitory Concentration of Polymyxin B in Vibrio Cholerae. *PLoS ONE* **2019**, *14*, e0221431. [CrossRef]
102. Gwynne, P.J.; Gallagher, M.P. Light as a Broad-Spectrum Antimicrobial. *Front. Microbiol.* **2018**, *9*, 119. [CrossRef]
103. Aurum, F.S.; Nguyen, L.T. Efficacy of Photoactivated Curcumin to Decontaminate Food Surfaces under Blue Light Emitting Diode. *J. Food Process Eng.* **2019**, *42*. [CrossRef]
104. Josewin, S.W.; Kim, M.J.; Yuk, H.G. Inactivation of Listeria Monocytogenes and Salmonella spp. on Cantaloupe Rinds by Blue Light Emitting Diodes (LEDs). *Food Microbiol.* **2018**, *76*, 219–225. [CrossRef]
105. Bumah, V.V.; Whelan, H.T.; Masson-Meyers, D.S.; Quirk, B.; Buchmann, E.; Enwemeka, C.S. The Bactericidal Effect of 470-Nm Light and Hyperbaric Oxygen on Methicillin-Resistant Staphylococcus Aureus (MRSA). *Lasers Med. Sci.* **2015**, *30*, 1153–1159. [CrossRef]

106. De Lucca, A.J.; Carter-Wientjes, C.; Williams, K.A.; Bhatnagar, D. Blue Light (470 nm) Effectively Inhibits Bacterial and Fungal Growth. *Lett. Appl. Microbiol.* **2012**, *55*, 460–466. [CrossRef]
107. Dietel, W.; Pottier, R.; Pfister, W.; Schleier, P.; Zinner, K. 5-Aminolaevulinic Acid (ALA) Induced Formation of Different Fluorescent Porphyrins: A Study of the Biosynthesis of Porphyrins by Bacteria of the Human Digestive Tract. *J. Photochem. Photobiol. B Biol.* **2007**, *86*, 77–86. [CrossRef] [PubMed]
108. Choby, J.E.; Skaar, E.P. Heme Synthesis and Acquisition in Bacterial Pathogens. *J. Mol. Biol.* **2016**, *428*, 3408–3428. [CrossRef] [PubMed]
109. Dailey, H.A.; Dailey, T.A.; Gerdes, S.; Jahn, D.; Jahn, M.; O'Brian, M.R.; Warren, M.J. Prokaryotic Heme Biosynthesis: Multiple Pathways to a Common Essential Product. *Microbiol. Mol. Biol. Rev.* **2017**, *81*. [CrossRef] [PubMed]
110. Lukšienė, Ž. New Approach to Inactivation of Harmful and Pathogenic Microorganisms by Photosensitization. *Food Technol. Biotechnol.* **2005**, *43*, 411–418.
111. Hu, X.; Huang, Y.Y.; Wang, Y.; Wang, X.; Hamblin, M.R. Antimicrobial Photodynamic Therapy to Control Clinically Relevant Biofilm Infections. *Front. Microbiol.* **2018**, *9*, 1299. [CrossRef]
112. Kim, M.J.; Mikš-Krajnik, M.; Kumar, A.; Yuk, H.G. Inactivation by 405 ± 5 nm Light Emitting Diode on Escherichia Coli O157: H7, Salmonella Typhimurium, and Shigella Sonnei under Refrigerated Condition Might Be Due to the Loss of Membrane Integrity. *Food Control* **2016**, *59*, 99–107. [CrossRef]
113. Wu, J.; Chu, Z.; Ruan, Z.; Wang, X.; Dai, T.; Hu, X. Changes of Intracellular Porphyrin, Reactive Oxygen Species, and Fatty Acids Profiles During Inactivation of Methicillin-Resistant Staphylococcus Aureus by Antimicrobial Blue Light. *Front. Physiol.* **2018**, *9*, 1658. [CrossRef]
114. Bhavya, M.L.; Umesh Hebbar, H. Efficacy of Blue LED in Microbial Inactivation: Effect of Photosensitization and Process Parameters. *Int. J. Food Microbiol.* **2019**, *290*, 296–304. [CrossRef]
115. Chu, Z.; Hu, X.; Wang, X.; Wu, J.; Dai, T.; Wang, X. Inactivation of Cronobacter Sakazakii by Blue Light Illumination and the Resulting Oxidative Damage to Fatty Acids. *Can. J. Microbiol.* **2019**, *65*, 922–929. [CrossRef]
116. Kim, M.J.; Bang, W.S.; Yuk, H.G. 405 ± 5 nm Light Emitting Diode Illumination Causes Photodynamic Inactivation of Salmonella Spp. on Fresh-Cut Papaya without Deterioration. *Food Microbiol.* **2017**, *62*, 124–132. [CrossRef]
117. Wang, Y.; Wang, Y.; Wang, Y.; Murray, C.K.; Hamblin, M.R.; Hooper, D.C.; Dai, T. Antimicrobial Blue Light Inactivation of Pathogenic Microbes: State of the Art. *Drug Resist. Updat.* **2017**, *33–35*, 1–22. [CrossRef] [PubMed]
118. Dosselli, R.; Millioni, R.; Puricelli, L.; Tessari, P.; Arrigoni, G.; Franchin, C.; Segalla, A.; Teardo, E.; Reddi, E. Molecular Targets of Antimicrobial Photodynamic Therapy Identified by a Proteomic Approach. *J. Proteomics* **2012**, *77*, 329–343. [CrossRef] [PubMed]
119. Beirão, S.; Fernandes, S.; Coelho, J.; Faustino, M.A.F.; Tomé, J.P.C.; Neves, M.G.P.M.S.; Tomé, A.C.; Almeida, A.; Cunha, A. Photodynamic Inactivation of Bacterial and Yeast Biofilms with a Cationic Porphyrin. *Photochem. Photobiol.* **2014**, *90*, 1387–1396. [CrossRef] [PubMed]
120. Dai, T.; Gupta, A.; Huang, Y.Y.; Sherwood, M.E.; Murray, C.K.; Vrahas, M.S.; Kielian, T.; Hamblin, M.R. Blue Light Eliminates Community-Acquired Methicillin-Resistant Staphylococcus Aureus in Infected Mouse Skin Abrasions. *Photomed. Laser Surg.* **2013**, *31*, 531–538. [CrossRef] [PubMed]
121. Zhang, Y.; Zhu, Y.; Gupta, A.; Huang, Y.; Murray, C.K.; Vrahas, M.S.; Sherwood, M.E.; Baer, D.G.; Hamblin, M.R.; Dai, T. Antimicrobial Blue Light Therapy for Multidrug-Resistant Acinetobacter Baumannii Infection in a Mouse Burn Model: Implications for Prophylaxis and Treatment of Combat-Related Wound Infections. *J. Infect. Dis.* **2014**, *209*, 1963–1971. [CrossRef]
122. Giannelli, M.; Landini, G.; Materassi, F.; Chellini, F.; Antonelli, A.; Tani, A.; Nosi, D.; Zecchi-Orlandini, S.; Rossolini, G.M.; Bani, D. Effects of Photodynamic Laser and Violet-Blue Led Irradiation on Staphylococcus Aureus Biofilm and Escherichia Coli Lipopolysaccharide Attached to Moderately Rough Titanium Surface: In Vitro Study. *Lasers Med. Sci.* **2017**, *32*, 857–864. [CrossRef]
123. Tschowri, N.; Lindenberg, S.; Hengge, R. Molecular Function and Potential Evolution of the Biofilm-Modulating Blue Light-Signalling Pathway of Escherichia Coli. *Mol. Microbiol.* **2012**, *85*, 893–906. [CrossRef]
124. Tschowri, N.; Busse, S.; Hengge, R. The BLUF-EAL Protein YcgF Acts as a Direct Anti-Repressor in a Blue-Light Response of Escherichia Coli. *Genes Dev.* **2009**, *23*, 522–534. [CrossRef]

125. Mussi, M.A.; Gaddy, J.A.; Cabruja, M.; Arivett, B.A.; Viale, A.M.; Rasia, R.; Actis, L.A. The Opportunistic Human Pathogen Acinetobacter Baumannii Senses and Responds to Light. *J. Bacteriol.* **2010**, *192*, 6336–6345. [CrossRef]
126. Yang, P.; Wang, N.; Wang, C.; Yao, Y.; Fu, X.; Yu, W.; Cai, R.; Yao, M. 460 Nm Visible Light Irradiation Eradicates MRSA via Inducing Prophage Activation. *J. Photochem. Photobiol. B Biol.* **2017**, *166*, 311–322. [CrossRef]
127. Jayaraman, P.; Devarajan, K.; Chua, T.K.; Zhang, H.; Gunawan, E.; Poh, C.L. Blue Light-Mediated Transcriptional Activation and Repression of Gene Expression in Bacteria. *Nucleic Acids Res.* **2016**, *44*, 6994–7005. [CrossRef] [PubMed]
128. Braatsch, S.; Klug, G. Blue Light Perception in Bacteria. *Photosynth. Res.* **2004**, *79*, 45–57. [CrossRef] [PubMed]
129. Gomelsky, M.; Hoff, W.D. Light Helps Bacteria Make Important Lifestyle Decisions. *Trends Microbiol.* **2011**, *19*, 441–448. [CrossRef] [PubMed]
130. Yeh, N.; Chung, J.P. High-Brightness LEDs-Energy Efficient Lighting Sources and Their Potential in Indoor Plant Cultivation. *Renew. Sustain. Energy Rev.* **2009**, *13*, 2175–2180. [CrossRef]
131. Prasad, A.; Du, L.; Zubair, M.; Subedi, S.; Ullah, A.; Roopesh, M.S. Applications of Light-Emitting Diodes (LEDs) in Food Processing and Water Treatment. *Food Eng. Rev.* **2020**, *12*, 268–289. [CrossRef]
132. Feezell, D.; Nakamura, S. Invention, Development, and Status of the Blue Light-Emitting Diode, the Enabler of Solid-State Lighting. *Comptes Rendus Phys.* **2018**, *19*, 113–133. [CrossRef]
133. Dall Agnol, M.A.; Nicolau, R.A.; De Lima, C.J.; Munin, E. Comparative Analysis of Coherent Light Action (Laser) versus Non-Coherent Light (Light-Emitting Diode) for Tissue Repair in Diabetic Rats. *Lasers Med. Sci.* **2009**, *24*, 909–916. [CrossRef]
134. Corazza, A.V.; Jorge, J.; Kurachi, C.; Bagnato, V.S. Photobiomodulation on the Angiogenesis of Skin Wounds in Rats Using Different Light Sources. *Photomed. Laser Surg.* **2007**, *25*, 102–106. [CrossRef]
135. de Abreu Chaves, M.E.; Piancastelli, A.C.C.; de Araújo, A.R.; Pinotti, M. Effects of Low-Power Light Therapy on Wound Healing: LASER x LED. *An. Bras. Dermatol.* **2014**, *89*, 616–623. [CrossRef]
136. Masson-Meyers, D.S.; Bumah, V.V.; Biener, G.; Raicu, V.; Enwemeka, C.S. The Relative Antimicrobial Effect of Blue 405 Nm LED and Blue 405 Nm Laser on Methicillin-Resistant Staphylococcus Aureus in Vitro. *Lasers Med. Sci.* **2015**, *30*, 2265–2271. [CrossRef]
137. Feltin, E.; Castiglia, A.; Cosendey, G.; Sulmoni, L.; Carlin, J.F.; Grandjean, N.; Rossetti, M.; Dorsaz, J.; Laino, V.; Duelk, M.; et al. Broadband Blue Superluminescent Light-Emitting Diodes Based on GaN. *Appl. Phys. Lett.* **2009**, *95*. [CrossRef]
138. Rossetti, M.; Napierala, J.; Matuschek, N.; Achatz, U.; Duelk, M.; Vélez, C.; Castiglia, A.; Grandjean, N.; Dorsaz, J.; Feltin, E. Superluminescent Light Emitting Diodes: The Best out of Two Worlds. *MOEMS Miniaturized Syst. XI* **2012**, *8252*, 825208.
139. Zhang, Z.; Hogg, R.; Lv, X.; Wang, Z. Self-Assembled Quantum-Dot Superluminescent Light-Emitting Diodes. *Adv. Opt. Photonics* **2010**, *2*, 201–228. [CrossRef]
140. Enwemeka, C.S.; Williams, D.; Hollosi, S.; Yens, D.; Enwemeka, S.K. Visible 405 Nm SLD Light Photo-Destroys Methicillin-Resistant Staphylococcus Aureus (MRSA) in Vitro. *Lasers Surg. Med.* **2008**, *40*, 734–737. [CrossRef]
141. Guffey, J.S.; Wilborn, J. Effects of Combined 405-Nm and 880-Nm Light on Staphylococcus Aureus and Pseudomonas Aeruginosa in Vitro. *Photomed. Laser Surg.* **2006**, *24*, 680–683.
142. Guffey, J.S.; Wilborn, J. In Vitro Bactericidal Effects of 405-Nm and 470-Nm Blue Light. *Photomed. Laser Surg.* **2006**, *24*, 684–688. [CrossRef]
143. Ahmed, E.; El-Gendy, A.O.; Moniem Radi, N.A.; Mohamed, T. The Bactericidal Efficacy of Femtosecond Laser-Based Therapy on the Most Common Infectious Bacterial Pathogens in Chronic Wounds: An in Vitro Study. *Lasers Med. Sci.* **2020**. [CrossRef]
144. Tsen, K.T.; Tsen, S.-W.D.; Fu, Q.; Lindsay, S.M.; Li, Z.; Cope, S.; Vaiana, S.; Kiang, J.G. Studies of Inactivation of Encephalomyocarditis Virus, M13 Bacteriophage, and Salmonella Typhimurium by Using a Visible Femtosecond Laser: Insight into the Possible Inactivation Mechanisms. *J. Biomed. Opt.* **2011**, *16*, 078003.
145. Barneck, M.D.; Rhodes, N.L.R.; de la Presa, M.; Allen, J.P.; Poursaid, A.E.; Nourian, M.M.; Firpo, M.A.; Langell, J.T. Violet 405-Nm Light: A Novel Therapeutic Agent against Common Pathogenic Bacteria. *J. Surg. Res.* **2016**, *206*, 316–324.

146. Murdoch, L.E.; MacLean, M.; Endarko, E.; MacGregor, S.J.; Anderson, J.G. Bactericidal Effects of 405nm Light Exposure Demonstrated by Inactivation of Escherichia, Salmonella, Shigella, Listeria, and Mycobacterium Species in Liquid Suspensions and on Exposed Surfaces. *Sci. World J.* **2012**, *2012*, 137805.
147. Bumah, V.V.; Masson-Meyers, D.S.; Enwemeka, C.S. Pulsed 450 nm Blue Light Suppresses MRSA and Propionibacterium Acnes in Planktonic Cultures and Bacterial Biofilms. *J. Photochem. Photobiol. B Biol.* **2020**, *202*, 111702. [CrossRef] [PubMed]
148. Masson-Meyers, D.S.; Bumah, V.V.; Castel, C.; Castel, D.; Enwemeka, C.S. Pulsed 450 Nm Blue Light Significantly Inactivates Propionibacterium Acnes More than Continuous Wave Blue Light. *J. Photochem. Photobiol. B Biol.* **2020**, *202*, 111719. [CrossRef] [PubMed]
149. Gillespie, J.B.; Maclean, M.; Given, M.J.; Wilson, M.P.; Judd, M.D.; Timoshkin, I.V.; Macgregor, S.J. Efficacy of Pulsed 405-Nm Light-Emitting Diodes for Antimicrobial Photodynamic Inactivation: Effects of Intensity, Frequency, and Duty Cycle. *Photomed. Laser Surg.* **2017**, *35*, 150–156. [CrossRef] [PubMed]
150. Kleinpenning, M.M.; Smits, T.; Frunt, M.H.A.; van Erp, P.E.J.; van de Kerkhof, P.C.M.; Gerritsen, R.M.J.P. Clinical and Histological Effects of Blue Light on Normal Skin. *Photodermatol. Photoimmunol. Photomed.* **2010**, *26*, 16–21. [CrossRef]
151. Dai, T.; Gupta, A.; Huang, Y.Y.; Yin, R.; Murray, C.K.; Vrahas, M.S.; Sherwood, M.E.; Tegos, G.P.; Hamblin, M.R. Blue Light Rescues Mice from Potentially Fatal Pseudomonas Aeruginosa Burn Infection: Efficacy, Safety, and Mechanism of Action. *Antimicrob. Agents Chemother.* **2013**, *57*, 1238–1245. [CrossRef]
152. Liebmann, J.; Born, M.; Kolb-Bachofen, V. Blue-Light Irradiation Regulates Proliferation and Differentiation in Human Skin Cells. *J. Investig. Dermatol.* **2010**, *130*, 259–269. [CrossRef]
153. Marek, V.; Mélik-Parsadaniantz, S.; Villette, T.; Montoya, F.; Baudouin, C.; Brignole-Baudouin, F.; Denoyer, A. Blue Light Phototoxicity toward Human Corneal and Conjunctival Epithelial Cells in Basal and Hyperosmolar Conditions. *Free Radic. Biol. Med.* **2018**, *126*, 27–40. [CrossRef]
154. Zhao, Z.C.; Zhou, Y.; Tan, G.; Li, J. Research Progress about the Effect and Prevention of Blue Light on Eyes. *Int. J. Ophthalmol.* **2018**, *11*, 1999–2003.
155. West, K.E.; Jablonski, M.R.; Warfield, B.; Cecil, K.S.; James, M.; Ayers, M.A.; Maida, J.; Bowen, C.; Sliney, D.H.; Rollag, M.D.; et al. Blue Light from Light-Emitting Diodes Elicits a Dose-Dependent Suppression of Melatonin in Humans. *J. Appl. Physiol.* **2011**, *110*, 619–626. [CrossRef]
156. American Conference of Governmental Industrial Hygienists. ACGIH—Threshold Limit Values (TLVs) and Biological Exposure Indices (BEIs). Available online: https://www.nsc.org/Portals/0/Documents/facultyportal/Documents/fih-6e-appendix-b.pdf (accessed on 21 October 2020).
157. Leccese, F.; Vandelanotte, V.; Salvadori, G.; Rocca, M. Blue Light Hazard and Risk Group Classification of 8 W LED Tubes, Replacing Fluorescent Tubes, through Optical Radiation Measurements. *Sustainability* **2015**, *7*, 13454–13468. [CrossRef]
158. Soares, C.J.; de Paula Rodrigues, M.; Vilela, A.B.F.; Rizo, E.R.C.; Ferreira, L.B.; Giannini, M.; Price, R.B. Evaluation of Eye Protection Filters Used with Broad-Spectrum and Conventional LED Curing Lights. *Braz. Dent. J.* **2017**, *28*, 9–15. [CrossRef] [PubMed]
159. Kimberly, B.; James, R.P. Amber Lenses to Block Blue Light and Improve Sleep: A Randomized Trial. *Chronobiol. Int.* **2009**, *26*, 1602–1612. [CrossRef] [PubMed]
160. Esaki, Y.; Kitajima, T.; Ito, Y.; Koike, S.; Nakao, Y.; Tsuchiya, A.; Hirose, M.; Iwata, N. Wearing Blue Light-Blocking Glasses in the Evening Advances Circadian Rhythms in the Patients with Delayed Sleep Phase Disorder: An Open-Label Trial. *Chronobiol. Int.* **2016**, *33*, 1037–1044. [CrossRef] [PubMed]
161. Sommers, C.; Gunther, N.W.; Sheen, S. Inactivation of Salmonella Spp., Pathogenic Escherichia Coli, Staphylococcus Spp., or Listeria Monocytogenes in Chicken Purge or Skin Using a 405-nm LED Array. *Food Microbiol.* **2017**, *64*, 135–138. [CrossRef] [PubMed]
162. Gunther, N.W.; Phillips, J.G.; Sommers, C. The Effects of 405-Nm Visible Light on the Survival of Campylobacter on Chicken Skin and Stainless Steel. *Foodborne Pathog. Dis.* **2016**, *13*, 245–250. [CrossRef] [PubMed]
163. Li, X.; Kim, M.J.; Bang, W.S.; Yuk, H.G. Anti-Biofilm Effect of 405-Nm LEDs against Listeria Monocytogenes in Simulated Ready-to-Eat Fresh Salmon Storage Conditions. *Food Control* **2018**, *84*, 513–521. [CrossRef]
164. McKenzie, K.; Maclean, M.; Timoshkin, I.V.; Endarko, E.; Macgregor, S.J.; Anderson, J.G. Photoinactivation of Bacteria Attached to Glass and Acrylic Surfaces by 405 Nm Light: Potential Application for Biofilm Decontamination. *Photochem. Photobiol.* **2013**, *89*, 927–935. [CrossRef]

165. Hyun, J.E.; Lee, S.Y. Antibacterial Effect and Mechanisms of Action of 460–470 nm Light-Emitting Diode against Listeria Monocytogenes and Pseudomonas Fluorescens on the Surface of Packaged Sliced Cheese. *Food Microbiol.* **2020**, *86*, 103314. [CrossRef]
166. Luksiene, Z.; Buchovec, I.; Paskeviciute, E. Inactivation of Food Pathogen Bacillus Cereus by Photosensitization in Vitro and on the Surface of Packaging Material. *J. Appl. Microbiol.* **2009**, *107*, 2037–2046. [CrossRef]
167. Luksiene, Z.; Paskeviciute, E. Reprint of: Novel Approach to Decontaminate Food-Packaging from Pathogens in Non-Thermal and Not Chemical Way: Chlorophyllin-Based Photosensitization Q. *J. Food Eng.* **2012**, *110*, 317–323. [CrossRef]
168. Buchovec, I.; Paskeviciute, E.; Luksiene, Z. Photosensitization-Based Inactivation of Food Pathogen Listeria Monocytogenes in Vitro and on the Surface of Packaging Material. *J. Photochem. Photobiol. B Biol.* **2010**, *99*, 9–14. [CrossRef] [PubMed]
169. Haughton, P.N.; Grau, E.G.; Lyng, J.; Cronin, D.; Fanning, S.; Whyte, P. Susceptibility of Campylobacter to High Intensity near Ultraviolet/Visible 395 ± 5 nm Light and Its Effectiveness for the Decontamination of Raw Chicken and Contact Surfaces. *Int. J. Food Microbiol.* **2012**, *159*, 267–273. [CrossRef] [PubMed]
170. Miller, B.M.; Sauer, A.; Moraru, C.I. Inactivation of Escherichia Coli in Milk and Concentrated Milk Using Pulsed-Light Treatment. *J. Dairy Sci.* **2012**, *95*, 5597–5603. [CrossRef] [PubMed]
171. dos Anjos, C.; Sellera, F.P.; de Freitas, L.M.; Gargano, R.G.; Telles, E.O.; Freitas, R.O.; Baptista, M.S.; Ribeiro, M.S.; Lincopan, N.; Pogliani, F.C.; et al. Inactivation of Milk-Borne Pathogens by Blue Light Exposure. *J. Dairy Sci.* **2020**, *103*, 1261–1268. [CrossRef] [PubMed]
172. Murdoch, L.E.; MacLean, M.; MacGregor, S.J.; Anderson, J.G. Inactivation of Campylobacter Jejuni by Exposure to High-Intensity 405-Nm Visible Light. *Foodborne Pathog. Dis.* **2010**, *7*, 1211–1216. [CrossRef] [PubMed]
173. Endarko, E.; Maclean, M.; Timoshkin, I.V.; MacGregor, S.J.; Anderson, J.G. High-Intensity 405 Nm Light Inactivation of Listeria Monocytogenes. *Photochem. Photobiol.* **2012**, *88*, 1280–1286. [CrossRef] [PubMed]
174. Srimagal, A.; Ramesh, T.; Sahu, J.K. Effect of Light Emitting Diode Treatment on Inactivation of Escherichia Coli in Milk. *LWT-Food Sci. Technol.* **2016**, *71*, 378–385. [CrossRef]
175. Ghate, V.; Kumar, A.; Zhou, W.; Yuk, H.G. Irradiance and Temperature Influence the Bactericidal Effect of 460-Nanometer Light-Emitting Diodes on Salmonella in Orange Juice. *J. Food Prot.* **2016**, *79*, 553–560. [CrossRef]
176. Ricciardi, E.F.; Pedros-Garrido, S.; Papoutsis, K.; Lyng, J.G.; Conte, A.; Del Nobile, M.A. Novel Technologies for Preserving Ricotta Cheese: Effects of Ultraviolet and near-Ultraviolet–Visible Light. *Foods* **2020**, *9*, 580. [CrossRef]
177. Glueck, M.; Schamberger, B.; Eckl, P.; Plaetzer, K. New Horizons in Microbiological Food Safety: Photodynamic Decontamination Based on a Curcumin Derivative. *Photochem. Photobiol. Sci.* **2017**, *16*, 1784–1791. [CrossRef]
178. Tortik, N.; Spaeth, A.; Plaetzer, K. Photodynamic Decontamination of Foodstuff from Staphylococcus Aureus Based on Novel Formulations of Curcumin. *Photochem. Photobiol. Sci.* **2014**, *13*, 1402–1409. [CrossRef]
179. Winter, S.; Tortik, N.; Kubin, A.; Krammer, B.; Plaetzer, K. Back to the Roots: Photodynamic Inactivation of Bacteria Based on Water-Soluble Curcumin Bound to Polyvinylpyrrolidone as a Photosensitizer. *Photochem. Photobiol. Sci.* **2013**, *12*, 1795–1802. [CrossRef] [PubMed]
180. Buchovec, I.; Lukseviciute, V.; Marsalka, A.; Reklaitis, I.; Luksiene, Z. Effective Photosensitization-Based Inactivation of Gram (−) Food Pathogens and Molds Using the Chlorophyllin-Chitosan Complex: Towards Photoactive Edible Coatings to Preserve Strawberries. *Photochem. Photobiol. Sci.* **2016**, *15*, 506–516. [CrossRef] [PubMed]
181. Paskeviciute, E.; Zudyte, B.; Luksiene, Z. Towards Better Microbial Safety of Fresh Produce: Chlorophyllin-Based Photosensitization for Microbial Control of Foodborne Pathogens on Cherry Tomatoes. *J. Photochem. Photobiol. B Biol.* **2018**, *182*, 130–136. [CrossRef] [PubMed]
182. MinJeong, K.; CheeHwa, T.; WooSuk, B.; HyunGyun, Y. Antibacterial Effect of 405±5 Nm Light Emitting Diode Illumination against Escherichia Coli O157:H7, Listeria Monocytogenes, and Salmonella on the Surface of Fresh-Cut Mango and Its Influence on Fruit Quality. *Int. J. Food Microbiol.* **2017**, *244*, 82–89.
183. Tao, R.; Zhang, F.; Tang, Q.-J.; Xu, C.-S.; Ni, Z.J.; Meng, X.-H. Effects of Curcumin-Based Photodynamic Treatment on the Storage Quality of Fresh-Cut Apples. *Food Chem.* **2019**, *274*, 415–421. [CrossRef]

184. Ghate, V.; Kumar, A.; Kim, M.J.; Bang, W.S.; Zhou, W.; Yuk, H.G. Effect of 460 Nm Light Emitting Diode Illumination on Survival of Salmonella Spp. on Fresh-Cut Pineapples at Different Irradiances and Temperatures. *J. Food Eng.* **2017**, *196*, 130–138. [CrossRef]
185. Lacombe, A.; Niemira, B.A.; Sites, J.; Boyd, G.; Gurtler, J.B.; Tyrell, B.; Fleck, M. Reduction of Bacterial Pathogens and Potential Surrogates on the Surface of Almonds Using High-Intensity 405-Nanometer Light. *J. Food Prot.* **2016**, *79*, 1840–1845. [CrossRef]
186. Paskeviciute, E.; Zudyte, B.; Luksiene, Z. Innovative Nonthermal Technologies: Chlorophyllin and Visible Light Significantly Reduce Microbial Load on Basil. *Food Technol. Biotechnol.* **2019**, *57*, 126–132. [CrossRef]
187. Guffey, J.S.; Payne, W.C.; Motts, S.D.; Towery, P.; Hobson, T.; Harrell, G.; Meurer, L.; Lancaster, K. Inactivation of Salmonella on Tainted Foods: Using Blue Light to Disinfect Cucumbers and Processed Meat Products. *Food Sci. Nutr.* **2016**, *4*, 878–887. [CrossRef]
188. Luksiene, Z.; Paskeviciute, E. Novel Approach to the Microbial Decontamination of Strawberries: Chlorophyllin-Based Photosensitization. *J. Appl. Microbiol.* **2011**, *110*, 1274–1283. [CrossRef] [PubMed]
189. Kim, M.J.; Adeline Ng, B.X.; Zwe, Y.H.; Yuk, H.G. Photodynamic Inactivation of Salmonella Enterica Enteritidis by 405 ± 5-Nm Light-Emitting Diode and Its Application to Control Salmonellosis on Cooked Chicken. *Food Control* **2017**, *82*, 305–315. [CrossRef]
190. Li, X.; Kim, M.J.; Yuk, H.G. Influence of 405 nm Light-Emitting Diode Illumination on the Inactivation of Listeria Monocytogenes and Salmonella Spp. on Ready-to-Eat Fresh Salmon Surface at Chilling Storage for 8 h and Their Susceptibility to Simulated Gastric Fluid. *Food Control* **2018**, *88*, 61–68. [CrossRef]
191. Josewin, S.W.; Ghate, V.; Kim, M.J.; Yuk, H.G. Antibacterial Effect of 460 Nm Light-Emitting Diode in Combination with Riboflavin against Listeria Monocytogenes on Smoked Salmon. *Food Control* **2018**, *84*, 354–361. [CrossRef]
192. Stephen Guffey, J.; Stephen, J.; Professor, A. The Use of 405nm and 464nm Blue Light to Inhibit Listeria Monocytogenes in Ready-to-Eat (RTE) Meat. *Eur. J. Acad. Essays ISSN* **2016**, *3*, 76–80.
193. Maclean, M.; MacGregor, S.J.; Anderson, J.G.; Woolsey, G.A.; Coia, J.E.; Hamilton, K.; Taggart, I.; Watson, S.B.; Thakker, B.; Gettinby, G. Environmental Decontamination of a Hospital Isolation Room Using High-Intensity Narrow-Spectrum Light. *J. Hosp. Infect.* **2010**, *76*, 247–251. [CrossRef]
194. Bache, S.E.; MacLean, M.; MacGregor, S.J.; Anderson, J.G.; Gettinby, G.; Coia, J.E.; Taggart, I. Clinical Studies of the High-Intensity Narrow-Spectrum Light Environmental Decontamination System (HINS-Light EDS), for Continuous Disinfection in the Burn Unit Inpatient and Outpatient Settings. *Burns* **2012**, *38*, 69–76. [CrossRef]
195. Maclean, M.; Booth, M.; Anderson, J.; MacGregor, S.; Woolsey, G.; Coia, J.; Hamilton, K.; Gettinby, G. Continuous Decontamination of an Intensive Care Isolation Room during Patient Occupancy Using 405 Nm Light Technology. *J. Infect. Prev.* **2013**, *14*, 176–181. [CrossRef]
196. MacLean, M.; Murdoch, L.E.; MacGregor, S.J.; Anderson, J.G. Sporicidal Effects of High-Intensity 405 nm Visible Light on Endospore-Forming Bacteria. *Photochem. Photobiol.* **2013**, *89*, 120–126. [CrossRef]
197. Dougall, L.; Anderson, J.G.; Timoshkin, I.V.; MacGregor, S.J.; Maclean, M. *Efficacy of Antimicrobial 405 nm Blue-Light for Inactivation of Airborne Bacteria*; Dai, T., Ed.; International Society for Optics and Photonics: San Fransisco, CA, USA, 2018; Volume 10479, p. 10479G.
198. Moreira, L.H.; de Souza, J.C.P.; de Lima, C.J.; Salgado, M.A.C.; Fernandes, A.B.; Andreani, D.I.K.; Villaverde, A.B.; Zângaro, R.A. Use of Photodynamic Therapy in the Treatment of Bovine Subclinical Mastitis. *Photodiagn. Photodyn. Ther.* **2018**, *21*, 246–251. [CrossRef]
199. Sellera, F.P.; Sabino, C.P.; Ribeiro, M.S.; Gargano, R.G.; Benites, N.R.; Melville, P.A.; Pogliani, F.C. In Vitro Photoinactivation of Bovine Mastitis Related Pathogens. *Photodiagn. Photodyn. Ther.* **2016**, *13*, 276–281. [CrossRef] [PubMed]
200. Sudheesh, P.S.; Al-Ghabshi, A.; Al-Mazrooei, N.; Al-Habsi, S. Comparative Pathogenomics of Bacteria Causing Infectious Diseases in Fish. *Int. J. Evol. Biol.* **2012**, *2012*, 16. [CrossRef] [PubMed]
201. Roh, H.J.; Kim, A.; Kang, G.S.; Kim, D.H. Photoinactivation of Major Bacterial Pathogens in Aquaculture. *Fish. Aquat. Sci.* **2016**, *19*, 1–7. [CrossRef]
202. Alves, E.; Faustino, M.A.F.; Tomé, J.P.C.; Neves, M.G.P.M.S.; Tomé, A.C.; Cavaleiro, J.A.S.; Cunha, Â.; Gomes, N.C.M.; Almeida, A. Photodynamic Antimicrobial Chemotherapy in Aquaculture: Photoinactivation Studies of Vibrio Fischeri. *PLoS ONE* **2011**, *6*, e20970. [CrossRef]

203. Arrojado, C.; Pereira, C.; Tomé, J.P.C.; Faustino, M.A.F.; Neves, M.G.P.M.S.; Tomé, A.C.; Cavaleiro, J.A.S.; Cunha, Â.; Calado, R.; Gomes, N.C.M.; et al. Applicability of Photodynamic Antimicrobial Chemotherapy as an Alternative to Inactivate Fish Pathogenic Bacteria in Aquaculture Systems. *Photochem. Photobiol. Sci.* **2011**, *10*, 1691–1700. [CrossRef]
204. Roh, H.J.; Kang, G.S.; Kim, A.; Kim, N.E.; Nguyen, T.L.; Kim, D.H. Blue Light-Emitting Diode Photoinactivation Inhibits Edwardsiellosis in Fancy Carp (Cyprinus Carpio). *Aquaculture* **2018**, *483*, 1–7. [CrossRef]
205. Villamizar, N.; Blanco-Vives, B.; Migaud, H.; Davie, A.; Carboni, S.; Sánchez-Vázquez, F.J. Effects of Light during Early Larval Development of Some Aquacultured Teleosts: A Review. *Aquaculture* **2011**, *315*, 86–94. [CrossRef]
206. Machado, S.G.; Baglinière, F.; Marchand, S.; Van Coillie, E.; Vanetti, M.C.D.; De Block, J.; Heyndrickx, M. The Biodiversity of the Microbiota Producing Heat-Resistant Enzymes Responsible for Spoilage in Processed Bovine Milk and Dairy Products. *Front. Microbiol.* **2017**, *8*, 302. [CrossRef]
207. Yuan, L.; Sadiq, F.A.; Burmølle, M.; Wang, N.; He, G. Insights into Psychrotrophic Bacteria in Raw Milk: A Review. *J. Food Prot.* **2019**, *82*, 1148–1159. [CrossRef]
208. Kaczmarek, M.; Avery, S.V.; Singleton, I. Microbes Associated with Fresh Produce: Sources, Types and Methods to Reduce Spoilage and Contamination. *Adv. Appl. Microbiol.* **2019**, *107*, 29–82.
209. Pellissery, A.J.; Vinayamohan, P.G.; Amalaradjou, M.A.R.; Venkitanarayanan, K. Spoilage Bacteria and Meat Quality. In *Meat Quality Analysis: Advanced Evaluation Methods, Techniques, and Technologies*; Biswas, A.K., Mandal, P.K., Eds.; Academic Press: London, UK, 2019; pp. 307–334.
210. Odeyemi, O.A.; Burke, C.M.; Bolch, C.C.J.; Stanley, R. Seafood Spoilage Microbiota and Associated Volatile Organic Compounds at Different Storage Temperatures and Packaging Conditions. *Int. J. Food Microbiol.* **2018**, *280*, 87–99. [CrossRef] [PubMed]
211. Lin, Y.; Hu, J.; Li, S.; Hamzah, S.S.; Jiang, H.; Zhou, A.; Zeng, S.; Lin, S. Curcumin-Based Photodynamic Sterilization for Preservation of Fresh-Cut Hami Melon. *Molecules* **2019**, *24*, 2374. [CrossRef] [PubMed]
212. Liu, F.; Li, Z.; Cao, B.; Wu, J.; Wang, Y.; Xue, Y.; Xu, J.; Xue, C.; Tang, Q.J. The Effect of a Novel Photodynamic Activation Method Mediated by Curcumin on Oyster Shelf Life and Quality. *Food Res. Int.* **2016**, *87*, 204–210. [CrossRef]
213. Gong, C.; Li, Y.; Gao, R.; Xiao, F.; Zhou, X.; Wang, H.; Xu, H.; Wang, R.; Huang, P.; Zhao, Y. Inactivation of Specific Spoilage Organism (Pseudomonas) of Sturgeon by Curcumin-Mediated Photodynamic Inactivation. *Photodiagn. Photodyn. Ther.* **2020**, *31*, 101827. [CrossRef] [PubMed]
214. Chui, C.; Aoki, A.; Takeuchi, Y.; Sasaki, Y.; Hiratsuka, K.; Abiko, Y.; Izumi, Y. Antimicrobial Effect of Photodynamic Therapy Using High-Power Blue Light-Emitting Diode and Red-Dye Agent on Porphyromonas Gingivalis. *J. Periodontal Res.* **2013**, *48*, 696–705. [PubMed]
215. Hamblin, M.R.; Abrahamse, H. Inorganic Salts and Antimicrobial Photodynamic Therapy: Mechanistic Conundrums? *Molecules* **2018**, *23*, 3190. [CrossRef]
216. Huang, L.; Szewczyk, G.; Sarna, T.; Hamblin, M.R. Potassium Iodide Potentiates Broad-Spectrum Antimicrobial Photodynamic Inactivation Using Photofrin. *ACS Infect. Dis.* **2017**, *3*, 320–328. [CrossRef]
217. Fotinos, N.; Convert, M.; Piffaretti, J.C.; Gurny, R.; Lange, N. Effects on Gram-Negative and Gram-Positive Bacteria Mediated by 5-Aminolevulinic Acid and 5-Aminolevulinic Acid Derivatives. *Antimicrob. Agents Chemother.* **2008**, *52*, 1366–1373. [CrossRef]
218. Nitzan, Y.; Salmon-Divon, M.; Shporen, E.; Malik, Z. ALA Induced Photodynamic Effects on Gram Positive and Negative Bacteria. *Photochem. Photobiol. Sci.* **2004**, *3*, 430–435. [CrossRef]
219. McKenzie, K.; Maclean, M.; Timoshkin, I.V.; MacGregor, S.J.; Anderson, J.G. Enhanced Inactivation of Escherichia Coli and Listeria Monocytogenes by Exposure to 405nm Light under Sub-Lethal Temperature, Salt and Acid Stress Conditions. *Int. J. Food Microbiol.* **2014**, *170*, 91–98. [CrossRef]
220. Ghate, V.; Kumar, A.; Zhou, W.; Yuk, H.G. Effect of Organic Acids on the Photodynamic Inactivation of Selected Foodborne Pathogens Using 461 nm LEDs. *Food Control* **2015**, *57*, 333–340. [CrossRef]
221. Kumar, A.; Ghate, V.; Kim, M.J.; Zhou, W.; Khoo, G.H.; Yuk, H.G. Antibacterial Efficacy of 405, 460 and 520 Nm Light Emitting Diodes on Lactobacillus Plantarum, Staphylococcus Aureus and Vibrio Parahaemolyticus. *J. Appl. Microbiol.* **2016**, *120*, 49–56. [CrossRef] [PubMed]
222. Franci, G.; Falanga, A.; Galdiero, S.; Palomba, L.; Rai, M.; Morelli, G.; Galdiero, M. Silver Nanoparticles as Potential Antibacterial Agents. *Molecules* **2015**, *20*, 8856–8874. [CrossRef] [PubMed]

223. Carbone, M.; Donia, D.T.; Sabbatella, G.; Antiochia, R. Silver Nanoparticles in Polymeric Matrices for Fresh Food Packaging. *J. King Saud Univ. Sci.* **2016**, *28*, 273–279. [CrossRef]
224. Akram, F.E.; El-Tayeb, T.; Abou-Aisha, K.; El-Azizi, M. A Combination of Silver Nanoparticles and Visible Blue Light Enhances the Antibacterial Efficacy of Ineffective Antibiotics against Methicillin-Resistant Staphylococcus Aureus (MRSA). *Ann. Clin. Microbiol. Antimicrob.* **2016**, *15*, 48. [CrossRef]
225. El Din, S.N.; El-Tayeb, T.A.; Abou-Aisha, K.; El-Azizi, M. In Vitro and in Vivo Antimicrobial Activity of Combined Therapy of Silver Nanoparticles and Visible Blue Light against Pseudomonas Aeruginosa. *Int. J. Nanomed.* **2016**, *11*, 1749–1758.
226. Yang, M.Y.; Chang, K.C.; Chen, L.Y.; Wang, P.C.; Chou, C.C.; Wu, Z.-B.; Hu, A. Blue Light Irradiation Triggers the Antimicrobial Potential of ZnO Nanoparticles on Drug-Resistant Acinetobacter Baumannii. *J. Photochem. Photobiol. B Biol.* **2018**, *180*, 235–242. [CrossRef]
227. Othman, L.; Sleiman, A.; Abdel-Massih, R.M. Antimicrobial Activity of Polyphenols and Alkaloids in Middle Eastern Plants. *Front. Microbiol.* **2019**, *10*, 911. [CrossRef]
228. Papuc, C.; Goran, G.V.; Predescu, C.N.; Nicorescu, V.; Stefan, G. Plant Polyphenols as Antioxidant and Antibacterial Agents for Shelf-Life Extension of Meat and Meat Products: Classification, Structures, Sources, and Action Mechanisms. *Compr. Rev. Food Sci. Food Saf.* **2017**, *16*, 1243–1268. [CrossRef]
229. Nakamura, K.; Yamada, Y.; Ikai, H.; Kanno, T.; Sasaki, K.; Niwano, Y. Bactericidal Action of Photoirradiated Gallic Acid via Reactive Oxygen Species Formation. *J. Agric. Food Chem.* **2012**, *60*, 10048–10054. [CrossRef]
230. Nakamura, K.; Ishiyama, K.; Sheng, H.; Ikai, H.; Kanno, T.; Niwano, Y. Bactericidal Activity and Mechanism of Photoirradiated Polyphenols against Gram-Positive and -Negative Bacteria. *J. Agric. Food Chem.* **2015**, *63*, 7707–7713. [CrossRef]
231. Nakamura, K.; Shirato, M.; Kanno, T.; Lingström, P.; Örtengren, U.; Niwano, Y. Photo-Irradiated Caffeic Acid Exhibits Antimicrobial Activity against Streptococcus Mutans Biofilms via Hydroxyl Radical Formation. *Sci. Rep.* **2017**, *7*. [CrossRef] [PubMed]
232. Tsukada, M.; Sheng, H.; Tada, M.; Mokudai, T.; Oizumi, S.; Kamachi, T.; Niwano, Y. Bactericidal Action of Photo-Irradiated Aqueous Extracts from the Residue of Crushed Grapes from Winemaking. *Biocontrol Sci.* **2016**, *21*, 113–121. [CrossRef] [PubMed]
233. Tsukada, M.; Sheng, H.; Kamachi, T.; Niwano, Y. Microbicidal Action of Photoirradiated Aqueous Extracts from Wine Lees. *J. Food Sci. Technol.* **2016**, *53*, 3020–3027. [CrossRef] [PubMed]
234. Hyldgaard, M.; Mygind, T.; Meyer, R.L. Essential Oils in Food Preservation: Mode of Action, Synergies, and Interactions with Food Matrix Components. *Front. Microbiol.* **2012**, *3*, 12. [CrossRef] [PubMed]
235. Chouhan, S.; Sharma, K.; Guleria, S. Antimicrobial Activity of Some Essential Oils—Present Status and Future Perspectives. *Medicines* **2017**, *4*, 58. [CrossRef] [PubMed]
236. Guerra-Rosas, M.I.; Morales-Castro, J.; Cubero-Márquez, M.A.; Salvia-Trujillo, L.; Martín-Belloso, O. Antimicrobial Activity of Nanoemulsions Containing Essential Oils and High Methoxyl Pectin during Long-Term Storage. *Food Control* **2017**, *77*, 131–138. [CrossRef]
237. Zhang, J.; Ye, K.P.; Zhang, X.; Pan, D.D.; Sun, Y.Y.; Cao, J.X. Antibacterial Activity and Mechanism of Action of Black Pepper Essential Oil on Meat-Borne Escherichia Coli. *Front. Microbiol.* **2017**, *7*, 2094. [CrossRef]
238. Marqués-Calvo, M.S.; Codony, F.; Agustí, G.; Lahera, C. Visible Light Enhances the Antimicrobial Effect of Some Essential Oils. *Photodiagn. Photodyn. Ther.* **2017**, *17*, 180–184. [CrossRef]
239. Verraes, C.; Van Boxstael, S.; Van Meervenne, E.; Van Coillie, E.; Butaye, P.; Catry, B.; De Schaetzen, M.-A.; Van Huffel, X.; Imberechts, H.; Dierick, K.; et al. Antimicrobial Resistance in the Food Chain: A Review. *Int. J. Environ. Res. Public Health* **2013**, *10*, 2643–2669. [CrossRef]
240. Zekar, F.M.; Granier, S.A.; Marault, M.; Yaici, L.; Gassilloud, B.; Manceau, C.; Touati, A.; Millemann, Y. From Farms to Markets: Gram-Negative Bacteria Resistant to Third-Generation Cephalosporins in Fruits and Vegetables in a Region of North Africa. *Front. Microbiol.* **2017**, *8*, 1569. [CrossRef] [PubMed]
241. Boss, R.; Overesch, G.; Baumgartner, A. Antimicrobial Resistance of Escherichia Coli, Enterococci, Pseudomonas Aeruginosa, and Staphylococcus Aureus from Raw Fish and Seafood Imported into Switzerland. *J. Food Prot.* **2016**, *79*, 1240–1246. [CrossRef] [PubMed]
242. Schrijver, R.; Stijntjes, M.; Rodríguez-Baño, J.; Tacconelli, E.; Babu Rajendran, N.; Voss, A. Review of Antimicrobial Resistance Surveillance Programmes in Livestock and Meat in EU with Focus on Humans. *Clin. Microbiol. Infect.* **2018**, *24*, 577–590. [CrossRef] [PubMed]

243. Reygaert, W. An Overview of the Antimicrobial Resistance Mechanisms of Bacteria. *AIMS Microbiol.* **2018**, *4*, 482–501. [CrossRef] [PubMed]
244. Rapacka-Zdonczyk, A.; Wozniak, A.; Pieranski, M.; Woziwodzka, A.; Bielawski, K.P.; Grinholc, M. Development of Staphylococcus Aureus Tolerance to Antimicrobial Photodynamic Inactivation and Antimicrobial Blue Light upon Sub-Lethal Treatment. *Sci. Rep.* **2019**, *9*, 9423. [CrossRef] [PubMed]
245. Fila, G.; Kawiak, A.; Grinholc, M.S. Blue Light Treatment of Pseudomonas Aeruginosa: Strong Bactericidal Activity, Synergism with Antibiotics and Inactivation of Virulence Factors. *Virulence* **2017**, *8*, 938–958. [CrossRef]
246. Tomb, R.M.; Maclean, M.; Coia, J.E.; MacGregor, S.J.; Anderson, J.G. Assessment of the Potential for Resistance to Antimicrobial Violet-Blue Light in Staphylococcus Aureus. *Antimicrob. Resist. Infect. Control* **2017**, *6*, 100. [CrossRef]
247. He, Y.; Huang, Y.Y.; Xi, L.; Gelfand, J.A.; Hamblin, M.R. Tetracyclines Function as Dual-Action Light-Activated Antibiotics. *PLoS ONE* **2018**, *13*, e0196485. [CrossRef]
248. Madsen, J.S.; Burmølle, M.; Hansen, L.H.; Sørensen, S.J. The Interconnection between Biofilm Formation and Horizontal Gene Transfer. *FEMS Immunol. Med. Microbiol.* **2012**, *65*, 183–195. [CrossRef]
249. Ferrer-Espada, R.; Liu, X.; Goh, X.S.; Dai, T. Antimicrobial Blue Light Inactivation of Polymicrobial Biofilms. *Front. Microbiol.* **2019**, *10*, 721. [CrossRef]
250. Ferrer-Espada, R.; Wang, Y.; Goh, X.S.; Dai, T. Antimicrobial Blue Light Inactivation of Microbial Isolates in Biofilms. *Lasers Surg. Med.* **2020**, *52*, 472–478. [CrossRef] [PubMed]
251. Leanse, L.G.; Harrington, O.D.; Fang, Y.; Ahmed, I.; Goh, X.S.; Dai, T. Evaluating the Potential for Resistance Development to Antimicrobial Blue Light (at 405 nm) in Gram-Negative Bacteria: In Vitro and in Vivo Studies. *Front. Microbiol.* **2018**, *9*, 2403. [CrossRef] [PubMed]
252. Adair, T.L.; Drum, B.E. RNA-Seq Reveals Changes in the Staphylococcus Aureus Transcriptome Following Blue Light Illumination. *Genom. Data* **2016**, *9*, 4–6. [CrossRef] [PubMed]
253. Tardu, M.; Bulut, S.; Kavakli, I.H. MerR and ChrR Mediate Blue Light Induced Photo-Oxidative Stress Response at the Transcriptional Level in Vibrio Cholerae. *Sci. Rep.* **2017**, *7*, 40817. [CrossRef]
254. Leanse, L.; Goh, X.; Cheng, J.; Hooper, D.; Dai, T. Dual-Wavelength Photo-Killing of Methicillin-Resistant Staphylococcus Aureus. *JCI Insight* **2020**, *5*, e134343. [CrossRef]
255. Cieplik, F.; Späth, A.; Leibl, C.; Gollmer, A.; Regensburger, J.; Tabenski, L.; Hiller, K.A.; Maisch, T.; Schmalz, G. Blue Light Kills Aggregatibacter Actinomycetemcomitans Due to Its Endogenous Photosensitizers. *Clin. Oral Investig.* **2014**, *18*, 1763–1769. [CrossRef]
256. Battisti, A.; Morici, P.; Ghetti, F.; Sgarbossa, A. Spectroscopic Characterization and Fluorescence Imaging of Helicobacter Pylori Endogenous Porphyrins. *Biophys. Chem.* **2017**, *229*, 19–24. [CrossRef]
257. Amin, R.M.; Bhayana, B.; Hamblin, M.R.; Dai, T. Antimicrobial Blue Light Inactivation of Pseudomonas Aeruginosa by Photo-Excitation of Endogenous Porphyrins: In Vitro and in Vivo Studies. *Lasers Surg. Med.* **2016**, *48*, 562–568. [CrossRef]
258. Wang, Y.; Wu, X.; Chen, J.; Amin, R.; Lu, M.; Bhayana, B.; Zhao, J.; Murray, C.K.; Hamblin, M.R.; Hooper, D.C.; et al. Antimicrobial Blue Light Inactivation of Gram-Negative Pathogens in Biofilms: In Vitro and in Vivo Studies. *J. Infect. Dis.* **2016**, *213*, 1380–1387. [CrossRef]
259. Hope, C.K.; Strother, M.; Creber, H.K.; Higham, S.M. Lethal Photosensitisation of Prevotellaceae under Anaerobic Conditions by Their Endogenous Porphyrins. *Photodiagn. Photodyn. Ther.* **2016**, *13*, 344–346. [CrossRef]
260. Hope, C.K.; Hindley, J.A.; Khan, Z.; de Josselin de Jong, E.; Higham, S.M. Lethal Photosensitization of Porphyromonas Gingivalis by Their Endogenous Porphyrins under Anaerobic Conditions: An in Vitro Study. *Photodiagn. Photodyn. Ther.* **2013**, *10*, 677–682. [CrossRef] [PubMed]
261. Bumah, V.V.; Aboualizadeh, E.; Masson-Meyers, D.S.; Eells, J.T.; Enwemeka, C.S.; Hirschmugl, C.J. Spectrally Resolved Infrared Microscopy and Chemometric Tools to Reveal the Interaction between Blue Light (470 Nm) and Methicillin-Resistant Staphylococcus Aureus. *J. Photochem. Photobiol. B Biol.* **2017**, *167*, 150–157. [CrossRef] [PubMed]
262. Kaidzu, S.; Sugihara, K.; Sasaki, M.; Nishiaki, A.; Igarashi, T.; Tanito, M. Evaluation of Acute Corneal Damage Induced by 222-nm and 254-nm Ultraviolet Light in Sprague–Dawley Rats. *Free Radic. Res.* **2019**, *53*, 611–617. [CrossRef] [PubMed]

263. Narita, K.; Asano, K.; Morimoto, Y.; Igarashi, T.; Nakane, A. Chronic Irradiation with 222-nm UVC Light Induces Neither DNA Damage nor Epidermal Lesions in Mouse Skin, Even at High Doses. *PLoS ONE* **2018**, *13*, e0201259. [CrossRef]
264. Buonanno, M.; Ponnaiya, B.; Welch, D.; Stanislauskas, M.; Randers-Pehrson, G.; Smilenov, L.; Lowy, F.D.; Owens, D.M.; Brenner, D.J. Germicidal Efficacy and Mammalian Skin Safety of 222-nm UV Light. *Radiat. Res.* **2017**, *187*, 493–501. [CrossRef]
265. Yamano, N.; Kunisada, M.; Kaidzu, S.; Sugihara, K.; Nishiaki-Sawada, A.; Ohashi, H.; Yoshioka, A.; Igarashi, T.; Ohira, A.; Tanito, M.; et al. Long-term Effects of 222-nm Ultraviolet Radiation C Sterilizing Lamps on Mice Susceptible to Ultraviolet Radiation. *Photochem. Photobiol.* **2020**, *96*, 853–862. [CrossRef]
266. Buonanno, M.; Randers-Pehrson, G.; Bigelow, A.W.; Trivedi, S.; Lowy, F.D.; Spotnitz, H.M.; Hammer, S.M.; Brenner, D.J. 207-nm UV Light—A Promising Tool for Safe Low-Cost Reduction of Surgical Site Infections. I: In Vitro Studies. *PLoS ONE* **2013**, *8*, e76968. [CrossRef]
267. Buonanno, M.; Stanislauskas, M.; Ponnaiya, B.; Bigelow, A.W.; Randers-Pehrson, G.; Xu, Y.; Shuryak, I.; Smilenov, L.; Owens, D.M.; Brenner, D.J. 207-Nm UV Light—A Promising Tool for Safe Low-Cost Reduction of Surgical Site Infections. II: In-Vivo Safety Studies. *PLoS ONE* **2016**, *11*, e0138418. [CrossRef]

foods

MDPI

Article

Growth Potential of *Listeria monocytogenes* on Refrigerated Spinach and Rocket Leaves in Modified Atmosphere Packaging

Paul Culliney and Achim Schmalenberger *

Department of Biological Sciences, University of Limerick, V94T9PX Limerick, Ireland; paul.culliney@ul.ie
* Correspondence: achim.schmalenberger@ul.ie; Tel.: +35-3612-3377-5

Received: 29 July 2020; Accepted: 28 August 2020; Published: 1 September 2020

Abstract: Minimally processed ready-to-eat (RTE) vegetables are increasingly consumed for their health benefits. However, they also pose a risk of being ingested with food-borne pathogens. The present study investigated the ability of RTE spinach and rocket to support the growth of *Listeria monocytogenes* as previous studies provided contradicting evidence. Findings were compared to growth on iceberg lettuce that has repeatedly been shown to support growth. Products were inoculated with a three-strain mix of *L. monocytogenes* at 10 and 100 cfu g^{-1} and stored in modified atmosphere (4 kPa O_2, 8 kPa CO_2) at 8 °C over 7–9 days. Spinach demonstrated the highest growth potential rate of 2 to 3 $\log_{10}$ cfu g^{-1} over a 9-day period with only marginal deterioration in its visual appearance. Growth potential on rocket was around 2 $\log_{10}$ cfu g^{-1} over 9 days with considerable deterioration in visual appearance. Growth potential of iceberg lettuce was similar to that of rocket over a 7-day period. Growth curves fitted closely to a linear growth model, indicating none to limited restrictions of growth over the duration of storage. The high growth potentials of *L. monocytogenes* on spinach alongside the limited visual deterioration highlight the potential risks of consuming this raw RTE food product when contaminated.

Keywords: *Listeria monocytogenes*; growth potential; ready-to-eat; iceberg lettuce; rocket; spinach; rucola; arugula

1. Introduction

The ready-to-eat (RTE) fruit and vegetable industry is a worldwide expanding sector. From 2000 to 2017, global production has increased by approximately 60% for vegetables [1]. Consumption of RTE vegetable salads has also increased within developing countries owing to a change in lifestyle patterns and growth of awareness regarding the positive relationship between human health and intake of RTE vegetables [2]. Indeed, leafy vegetables such as raw rocket and raw baby spinach contain many vitamins, minerals, antioxidants, and phytochemicals [3]. In the European Union, Ireland and Belgium have the highest rate of daily consumption of vegetables (84% of the population; [4]). The health benefits of RTE vegetables have driven consumer lifestyle towards increased consumption of this convenient and healthy type of food in RTE salads and smoothies [5,6]. Within the food industry, demand has increased for variation in terms of taste, color, and shape (in particular, baby sized leafy vegetables) for RTE green leafy vegetables [7].

As RTE vegetables are not at all or only minimally processed from farm to fork, further research is needed to study the risk of consumption of RTE vegetables in relation to foodborne illnesses including listeriosis [8]. Data assessing the occurrences of *Listeria monocytogenes*, the causative agent for the disease listeriosis, in RTE foods from investigations led by the European Union are compiled annually. In the case of fruits and vegetables, 1257 units were tested in 2018 (across 16 Member States) with an overall prevalence of *L. monocytogenes* of 1.8% (up from 0.6% in 2017 across 13 Member States). Additionally, for RTE salads, out of 2583 units, 1.5% of samples were confirmed positive in 2018 for *L. monocytogenes* [9].

Listeriosis can be life threatening, particularly for young, elderly, pregnant women and their unborn baby, and immuno-compromised individuals [10]. *L. monocytogenes* is ubiquitous in nature. It has exceptional physiological abilities to ensure its survival by adapting quickly and easily to harsh divergent physiological conditions [11]. Studies have shown that subjecting *L. monocytogenes* to food-related stresses including low storage (refrigeration) temperatures may induce increased expression levels of the organism's virulence genes, and thus increase the risk of listeriosis [12].

Storage period and temperature are important factors influencing the growth and survival of *L. monocytogenes* in foods such as RTE salads [13]. However, recent challenge studies also identified that inoculation densities for testing affect the outcomes of challenge studies as lower initial inoculation densities (100 cfu g^{-1}) may lead to greater growth potentials during shelf life [14]. Despite the possible underestimation of growth potential, assessing growth potentials at high inoculation densities (i.e., 10^5 cfu g^{-1}) in RTE food remains popular [15]. According to the guidance produced by the European Union Reference laboratory (EURL), if any food product shows growth potential (δ) greater than 0.50 $\log_{10}$ cfu g^{-1}, it is regarded as being permissive to the growth of *L. monocytogenes* [16,17]. Consequently, changes in the inoculation density or other environmental factors may affect the outcome of challenge studies with the potential to underestimate growth. This could lead to RTE products being falsely categorized as food unable to support the growth of *L. monocytogenes* according to Commission Regulation (EC No 2073/2005) [16]. Up to now, many studies have assessed the prevalence of *L. monocytogenes* on RTE leafy produce [18–20], while only few investigated the actual growth potential. In the latter, there have been contradictory findings of growth potential of *L. monocytogenes* on spinach and rocket, of which some reported growth [21], while others did not [22,23]. At the outset of the present study, our hypothesis was that the growth potential of *L. monocytogenes* on spinach and rocket is similar to that of iceberg lettuce. As the growth potential of *L. monocytogenes* on lettuce is well established [14], the consumption of raw spinach and rocket in salads could pose a potential risk for human infection if it is contaminated with low levels of *L. monocytogenes*.

The aim of this study was to determine the growth potential and survival of *L. monocytogenes* during a shelf life study at 8 °C, with initial inoculum densities of 10 and 100 cfu g^{-1} in spinach and rocket (arugula; rucola), and to compare these growth potentials to that of iceberg lettuce. Previously established protocols for testing growth of *L. monocytogenes* on iceberg lettuce were used in this study in order to minimize changes in environmental conditions that could potentially affect growth behavior.

2. Materials and Methods

2.1. Preparation of *L. monocytogenes* for Inoculation Experiments

Three different strains of *L. monocytogenes* from the Teagasc Food Research Centre strain collection (Moorepark, Ireland) were used, 959 (vegetable isolate), 1382 (EUR *Lm* reference strain), and 6179 (food processing plant isolate). For each of the three *L. monocytogenes* strains, 10 mL of tryptone soya broth (TSB, CM0129, Oxoid, Basingstoke, UK) was prepared and placed in 50 mL conical flasks. After autoclaving, single colonies from the previously streaked plates (*Listeria* selective agar, conforming to ALOA) of *L. monocytogenes* culture were transferred into each flask and incubated at 8 °C for 5 days. Spectrophotometry was used to verify the cell density (600 nm) [24]. Dilutions with phosphate buffered saline (PBS, pH 7.3, BR0014, Oxoid) were carried out to mix the three strains at equal cell densities to aim for inoculation at cell densities of either 10 or 100 cfu g^{-1}. This was confirmed by enumeration on *Listeria* selective agar, conforming to ALOA (Chromocult® LSA, Merck, Darmstadt, Germany).

2.2. Preparation of the Polypropylene Bags and RTE Leafy Vegetables and Subsequent Inoculation and Storage

Oriented polypropylene packaging film (35 μm thick) was used to create storage bags (18 cm × 10 cm) with a permeability to O_2 of 5.7 nmol m^{-2} s^{-1} kPa^{-1} and to CO_2 of 19 nmol m^{-2} s^{-1} kPa^{-1} (Amcor Flexibles, 120 Gloucester, UK). Twenty-eight bags were required for each product batch to allow for sampling at day 0, 2, 5, 7, and 9 in quadruplicates and to test for absence of *L. monocytogenes* at day 0 and day end in controls.

On the day of the experiment, the three vegetable products used in this experiment, Iceberg lettuce (Class 1 Spain), spinach (unwashed, origin Italy), and rocket (washed, origin Ireland), were all purchased from the local supplier (Supervalu, Castletroy, Ireland), where they were stored in a refrigerator. The shelf life of all products was at least 7 days at the time of purchase. The whole head of iceberg lettuce was prepared, using disinfected utensils (using 70% isopropanol), by removing the outer two to three layers of the head, and the core and stalk of the lettuce were also discarded. The remaining lettuce was chopped into strips of 1 cm by 3 cm. From this chopped lettuce, 20 g was weighed out and placed into the respective polypropene packaging. Similarly, 20 g of the uncut RTE spinach and rocket leaves was weighed and placed into the polypropene bags. No further processing of the rocket and spinach leaves was carried out (e.g., washing or chlorine dipping). This was repeated for the necessary number of bags prior to inoculation (quadruplicates for five sampling dates, eight bags as non-inoculated controls). Using the previously prepared *L. monocytogenes* dilutions, 100 μL of *L. monocytogenes* suspension (representing 10 or 100 cfu g^{-1} of food product) was distributed uniformly over the 20 g of leafy vegetable product within the polypropene bags (eight control bags were treated with 100 μL sterile PBS [14]). Each packaging containing 20 g sample of vegetable product was then atmospherically treated (8 kPa CO_2, 4 kPa O_2, 88 kPa N_2) using a vacuum packer (Multivac, Dublin, Ireland). The packages were then stored at 8 °C ± 0.5 °C (HR410, Foster Refrigerator, King's Lynn, UK) for 0–9 days. Storage temperatures were checked daily. Environmental conditions were identical to previous growth studies with iceberg lettuce [14,25].

2.3. Sampling of the Leafy RTE Vegetable Packs and Analysis

The specific sampling data points for these experiments were day 0, 2, 5, 7, and 9 (for iceberg lettuce, sampling of day 9 was abandoned owing to a high level of product deterioration). On each of these days, four bags of each product were removed from the storage area. Furthermore, control bags (without *L. monocytogenes* inoculation) were harvested at day 0 and day end, and the absence of *L. monocytogenes* was confirmed on *Listeria* selective agar (conforming to ALOA, see also below), with a detection limit of 1 cfu g^{-1} following methods described previously [25].

Before opening the packs, concentrations of oxygen were determined inside the packs, using a gas analyzer (PBI-Dansensor, PBI Development, Denmark, Model TIA-III LV) with an injection needle to penetrate the packs. Each bag was cut using disinfected utensils (70% iso-propanol), one at a time, directly underneath the heat seal for subsequent sample analysis. Visual appearance was determined on inoculated samples (spinach and rocket), by aseptically removing four leaves from one package for each product at each data point, using disinfected utensils. Images of these leaves and visual markers at each data point were captured using a digital camera. The consumer acceptability was visually assessed for gloss, freshness, and colour uniformity and (given an appearance score) by a sensory panel (postgraduate students, not specifically trained in grading visual appearance of food products [24]) consisting of 10 individuals scoring the products from 1 (mush/very poor condition) to 10 (pristine/excellent condition). A score of 6 was set as the lowest acceptable level for consumption [24]. Images of the samples were all taken in the same artificial light with a visual marker and the same angle, and were then coded and offered randomly to panelists.

Enumeration of *L. monocytogenes* counts was carried out at day 0, 2, 5, 7, and 9. The contents of each package were transferred into separate stomacher bags and homogenized using a stomacher (Seward 400, AGB Scientific, Dublin, Ireland), for 120 s at a high speed (260 rpm), in 20 mL of PBS. Following this, depending on anticipated low cell counts, samples were concentrated (via centrifugation at 4000 g for 240 s) by 10-fold resuspending in 100 μL PBS (10 cfu g^{-1}) or 5-fold using 200 μL PBS (100 cfu g^{-1}) (detection limit of 1 and 2 cfu g^{-1}, respectively, [24]). If necessary, samples were also diluted to achieve a countable number of colonies. Aliquots of 100 μL were then plated on Chromocult *Listeria* selective agar (ALOA) containing *Listeria* selective supplement (both Merck, Darmstadt, Germany). The plates were incubated at 37 °C for 24–48 h. Colony forming units (cfu) on days 0, 2, 5, 7, and 9 were transformed into $\log_{10}$ cfu g^{-1}, mean values and standard deviations were determined and plotted, areas under the curve were determined [14], and median values were used to calculate growth potentials. Maximum growth rates were calculated as outlined in Appendix A.

2.4. Total Bacteria Count

Total bacterial cell counts were repeated (as recommended by EURL) in quadruplicate for spinach and rocket at day 0 and day 9 and for iceberg lettuce at day 0 and day 7. The containments of each package were transferred into separate stomacher bags and homogenised as described above in 20 mL of PBS. Following this, a dilution series was aseptically carried out with PBS and plated on tryptone soy agar (TSA, CM0131, Oxoid). Total bacteria were enumerated after incubation at 37 °C for 48 h.

2.5. Product pH and Water Activity

Product pH and water activity were determined (as recommended by EURL) in quadruplicate at days 0, 2, 5, 7, and 9 for each product (day 9 was excluded for iceberg lettuce owing to advanced levels of product deterioration), and average values and standard deviations were reported. For pH measurements on homogenates of each product, a calibrated pH probe (Cole-Parmer, Saint Neots, UK) was used. In order to determine the water activity values, AQUALAB model Series 3TE water activity meter (LabCell Ltd., Four Marks, UK) was used (following the manufacturer's instructions).

2.6. Statistical Analysis

Populations were reported as the means of four replicates and (±) standard deviations and median values were used to calculate growth potentials. The experimental results were tested using SPSS (IBM, Armonk, NY, USA) for homoscedasticity (Leven's test) and normality (Shapiro–Wilk test). In situations of normality and homoscedasticity, pairwise comparisons (*t*-tests) and analysis of variance (ANOVA) with Tukey post hoc tests were carried out to determine significant differences. When homoscedasticity only was met, then ANOVA with Games–Howell post hoc was carried out. In the case where normality and homoscedasticity were not met, even after data transformation, a Kruskal–Wallis test and manual post hoc was applied in order to identify significant differences ($p \leq 0.05$ for all tests [14]).

For total bacteria count, the results from each food type were averaged from four replicates. In order to compare results from beginning and end of the experiment and other pairwise comparisons, an independent two-sample equal variance, two-tailed *t*-test was conducted. Significance was determined at $p \leq 0.05$ [26].

3. Results

3.1. Comparison of Growth of *L. monocytogenes* on RTE Leafy Vegetables Iceberg Lettuce, Spinach, and Rocket over 7 Days

The growth of *L. monocytogenes* was supported by all three vegetable products. The growth potentials in all cases exceeded 0.50 $\log_{10}$ cfu g^{-1}. As cfu for all iceberg lettuce samples was only determined until day 7, all three products were compared based on growth potentials calculated with day 7 being the end day of the shelf life study (Table 1). On the basis of the nine independent 100 cfu g^{-1} experiments carried out on all three products, spinach supported the growth and survival of *L. monocytogenes* on average with the largest growth potential of 2.40 $\log_{10}$ cfu g^{-1} (100 cfu g^{-1} $\delta = 2.16$, 2.46, and 2.58 $\log_{10}$ cfu g^{-1}). This was followed by the average growth potential on iceberg lettuce at 1.86 (100 cfu g^{-1} $\delta = 1.28$, 1.69, and 2.62 $\log_{10}$ cfu g^{-1}). The average growth potential on rocket was lower at 1.51 (100 cfu g^{-1} δ = 1.08, 1.70, and 1.76 $\log_{10}$ cfu g^{-1}). The highest growth potential in a single batch was demonstrated by iceberg lettuce (2.62 $\log_{10}$ cfu day^{-1}). A pairwise comparison of rocket and spinach identified a significant difference ($p = 0.024$, *t*-test). However, a comparison of all three products did not reach significance ($p > 0.05$, ANOVA). Established by the 10 cfu g^{-1} experiments, spinach supported the growth and survival of *L. monocytogenes* on average with the largest growth potential of 2.39 $\log_{10}$ cfu g^{-1}. However, in contrast to the 100 cfu g^{-1} experiments, this was followed by rocket (1.82 $\log_{10}$ cfu g^{-1}) and then iceberg lettuce (1.58 $\log_{10}$ cfu g^{-1}).

Table 1. Growth potentials (based on median values of results from day 0 and day 7) of *L. monocytogenes* in ready-to-eat leafy vegetables.

Product	Batch	Inoculation Density [cfu g^{-1}]	Day 0 Median Value (Log_{10} cfu g^{-1})	Day 7 Median Value (Log_{10} cfu g^{-1})	Growth Potential (δ) (Log_{10} cfu g^{-1})
Spinach	1	100	1.98	4.14	2.16
Spinach	2	100	1.73	4.19	2.46
Spinach	3	100	2.02	4.60	2.58
Spinach	4	10	0.88	3.27	2.39
Rocket	1	100	2.00	3.08	1.08
Rocket	2	100	1.61	3.37	1.76
Rocket	3	100	1.99	3.69	1.70
Rocket	4	10	0.92	2.74	1.82
Lettuce	1	100	1.79	3.07	1.28
Lettuce	2	100	1.96	3.65	1.69
Lettuce	3	100	1.34	3.96	2.62
Lettuce	4	10	0.97	2.55	1.58

In terms of inoculation density for rocket, the 10 cfu g^{-1} inoculum concentration produced a higher growth potential by day 7 than the higher inoculation 100 cfu g^{-1} (see above). For iceberg lettuce and spinach, such a trend was not detected as the growth potentials of 10 and 100 cfu g^{-1} inoculation density overlapped at day 7. Pairwise comparisons conducted on areas under the curve over 7 days identified a significant difference only between spinach and rocket ($p = 0.02$, *t*-test), while a comparison of all three areas under the curve did not reach significance ($p > 0.05$, ANOVA).

Spinach displayed the largest maximum growth rates on average (median) of 0.348 $\log_{10}$ cfu day^{-1}. Additionally, *L. monocytogenes'* maximum growth rates on lettuce were on average 0.255 $\log_{10}$ cfu day^{-1} and on rocket were 0.223 $\log_{10}$ cfu day^{-1} (Supplementary Materials Table S1).

3.2. Comparison of Growth of *L. monocytogenes* on RTE Leafy Vegetables Spinach and Rocket over 9 Days

Spinach continued to support the growth of *L. monocytogenes* over 9 days (7-day experiments extended by 2 days) with the average largest growth potential of 2.66 (100 cfu g^{-1} δ = 2.36, 2.78 and 2.83 $\log_{10}$ cfu g^{-1}; Table 2). Rocket supported the growth of *L. monocytogenes* the least, with an average growth potential of 1.83 (100 cfu g^{-1} δ = 1.67, 1.87, and 1.94 $\log_{10}$ cfu g^{-1}; Table 2). In terms of inoculation density for both spinach and rocket, the lower initial inoculum concentration of 10 cfu g^{-1} leads to greater growth potentials than all 100 cfu g^{-1} experiments by day end (Day 9). At 10 cfu g^{-1} (same conditions), the growth potential (2.90 $\log_{10}$ cfu g^{-1}) exceeded the highest growth potential in any 100 cfu g^{-1} batch in spinach (2.83 $\log_{10}$ cfu g^{-1}). (Figure 1A, Table 2). Likewise, when rocket was inoculated with 10 cfu g^{-1}, *L. monocytogenes* counts increased by 2.21 $\log_{10}$ cfu g^{-1}, which was 0.27 $\log_{10}$ cfu g^{-1} higher than the highest batch at 100 cfu g^{-1} (Figure 1B, Table 2). A pairwise comparison of the growth potential of rocket and spinach identified a significant difference at day 9

($p = 0.007$, *t*-test). However, a pairwise comparison conducted on areas under the curve for spinach and rocket, for the duration of 9 days, identified no significant differences (all *p*-values > 0.05).

Table 2. Growth potentials (based on median values of results from day 0 and day 9) of *L. monocytogenes* in ready-to-eat leafy vegetables.

Product	Batch	Inoculation Density [cfu g^{-1}]	Day 0 Median Value (Log_{10} cfu g^{-1})	Day 9 Median Value (Log_{10} cfu g^{-1})	Growth Potential (δ) (Log_{10} cfu g^{-1})
Spinach	1	100	1.98	4.34	2.36
Spinach	2	100	1.73	4.51	2.78
Spinach	3	100	2.02	4.85	2.83
Spinach	4	10	0.88	3.78	2.90
Rocket	1	100	2.00	3.67	1.67
Rocket	2	100	1.61	3.55	1.94
Rocket	3	100	1.99	3.86	1.87
Rocket	4	10	0.92	3.13	2.21

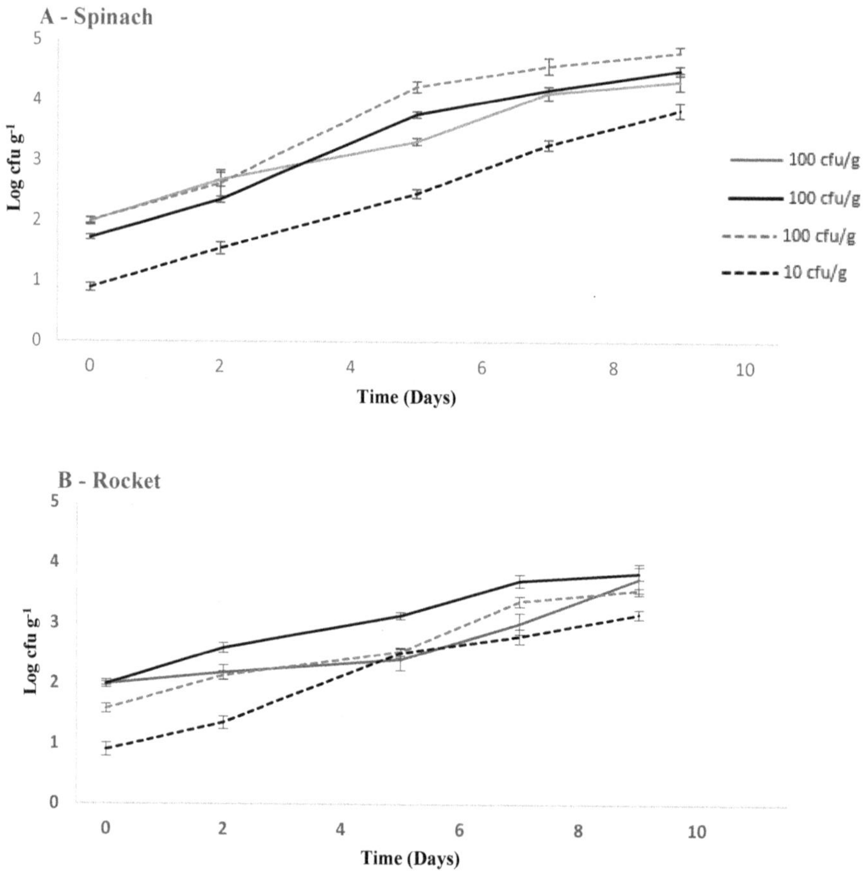

Figure 1. *Cont.*

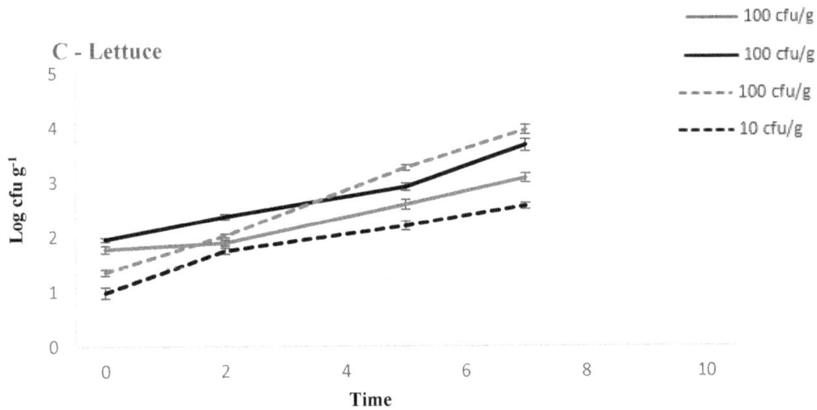

Figure 1. Growth and survival of *L. monocytogenes* in (**A**) spinach, (**B**) rocket, and (**C**) lettuce at 10–100 cfu g^{-1} inoculation densities at 8 °C. (±) error bars indicate standard deviation. Solid black, solid grey, and dashed grey lines represent experiments with 100 cfu g^{-1} starting inoculum density, and dashed black line represents experiment with 10 cfu g^{-1} starting inoculum density.

On the basis of day 9 data plotted to Baranyi and Roberts models (Supplementary Materials Table S2), spinach had the highest maximum growth rate on average 0.396 log10 cfu day^{-1}. Maximum growth rates for rocket were on average 0.282 log_{10} cfu day^{-1}. Day 9 data were also plotted to linear models (Supplementary Materials Table S2). There, spinach had the largest maximum growth rate on average at 0.314 log_{10} cfu day^{-1}. All maximum growth rates for spinach were higher compared with rocket maximum growth rates, which were on average 0.220 log_{10} cfu day^{-1}. Incubation experiments with lettuce were not extended to Day 9 owing to highly advanced levels of deterioration.

3.3. *Total Bacteria Count*

Bacteria counts from spinach, rocket, and iceberg lettuce revealed a log_{10} cfu g^{-1} at day 0 of 6.96, 5.94, and 7.11, respectively. Bacteria counts were also quantified at day 9 (for spinach and rocket) and were 8.86 and 7.97 log_{10} cfu g^{-1}, respectively (Supplementary Materials Table S3). Iceberg lettuce bacterial counts determined at day 7 were 8.69 log_{10} cfu g^{-1}. There were significant increases in the counts of total bacteria for spinach and rocket from day 0 to day 9 and for iceberg lettuce from day 0 to day 7 ($p < 0.05$).

3.4. *Product pH, Water Activity, and Atmosphere*

Spinach's pH values were highest during the present study. The pH values for spinach were 7.30 (day 0) and 7.25 (day 9), ranging from 6.93 to 7.30 with no trend over 9 days. Rocket's pH values were 6.55 (day 0) and 6.86 (day 9), ranging from 6.46 to 6.86 also with no trend observed over the course of the 9 days. Iceberg lettuce's pH values were lowest, at 6.34 (day 0) and 6.36 (day 7), ranging from 6.25 to 6.40 again with no trend demonstrated over 7 days. Water activity values for all three products ranged from 0.970 to 0.996 during this study (Supplementary Materials Table S4).

The oxygen concentration in the vegetable packs increased over the first seven days from the initial 4.0–4.2 kPa O_2 at day 0 to 9.12–11.7 by day 7 (lettuce, spinach, rocket) and, for spinach and rocket, the oxygen concentration stayed at 10.0–10.8 kPa (Supplementary Materials Table S5). No significant differences were observed between the three used vegetables.

3.5. Visual Appearance of Spinach and Rocket

For spinach (100 cfu g^{-1}), visual properties according to the untrained panelists showed a decrease from day 0 to day 9, from a score of above 9 to above 7 (Table 3, Supplementary Materials Figure S1). This decrease remained above the acceptable limit of 6. In comparison, rocket's (100 cfu g^{-1}) visual appearance analysis decreased from day 0 to day 7 to just above the acceptable range. At day 9, the untrained panelists deemed the visual appearance of rocket to be unacceptable (visual analysis score <6—lowest acceptable commercial score, Table 3).

Table 3. Visual (sensory) analysis of spinach and rocket leaves based on product appearance (i.e., gloss, freshness, and colour uniformity and intensity); average results ± standard deviations. 0 refers to ready-to-eat food products being in poor condition to 10 pristine/excellent condition, with 6 being the lowest acceptable commercial score. ‡ indicates product's sensory quality is unacceptable at that data point.

	Day 0	Day 2	Day 5	Day 7	Day 9
Spinach 100 cfu g^{-1}	9.2 ± 0.8	8.7 ± 0.8	8.2 ± 0.4	7.2 ± 0.6	7.1 ± 0.6
Rocket 100 cfu g^{-1}	8.8 ± 0.9	8.7 ± 1.1	6.3 ± 0.5	6.5 ± 0.6	5.6 ± 0.8

4. Discussion

Previous studies on the growth and survival of *L. monocytogenes* on spinach leaves provided apparently contradictory findings. Lokerse and colleagues [22] demonstrated that, by day 4 of storage/incubation, a relative increase of 0.70 $\log_{10}$ cfu g^{-1} *L. monocytogenes* was detectable on spinach at 7 °C. However, by day 5, a significant relative decrease to less than the starting inoculation density at day 0 was detected, which remained until then end of the experiment (day 10) [22]. The authors speculated that antimicrobial compounds present in spinach may cause bacteriostatic activity against *L. monocytogenes* growth. In contrast to the present study, Lokerse and colleagues [22] sealed their spinach in stomacher bags over the duration of the experiment, thus the atmosphere development was likely to be different from the present study, which had an oxygen concentration of around 10 kPa towards the end of the experiment. The same authors also tested the growth potential of *L. monocytogenes* on rocket leaves (rucola). Over the first 9 days of incubation, growth of *L. monocytogenes* on rucola was reported to vary within 0 to 0.9 $\log_{10}$ cfu g^{-1}, which was close to 0 again by day 9 [22]. Similarly, Söderqvist and colleagues [23] assessed *L. monocytogenes* growth with a starting cell density of 10^3 cfu g^{-1} at 8 °C on baby spinach (sealed within packages with water vapor and oxygen permeability). There, an increase by 0.30 $\log_{10}$ cfu g^{-1} was detected within the first three days, which was followed by a similar decrease by day 7 [23]. These findings are in contrast to the present study, where a continuous growth of *L. monocytogenes* was recorded, albeit at intervals that exceeded 24 h, hence smaller fluctuations between sampling events may have been missed. Nevertheless, Söderqvist and colleagues [23] found substantial growth of *L. monocytogenes* in a mixed-ingredient salad containing baby spinach and chicken, where growth continued to increase over a 7-day period that exceeded 1 $\log_{10}$ cfu g^{-1}.

Other studies identified growth potentials of *L. monocytogenes* on spinach, which were more similar to the findings from the present study. The validation results from predictive (Bayrani) models that investigated the effect of storage temperature on *L. monocytogenes* on fresh spinach leaves provided reliable estimates [27]. There, results showed growth potentials on spinach, where initial concentrations 2.28 ± 0.47 $\log_{10}$ cfu g^{-1} at 8 °C led to maximum population densities of 5.85 ± 0.67 $\log_{10}$ cfu g^{-1} over 16 days. With high initial inoculum densities of around 10^5 cfu g^{-1}, growth of *L. monocytogenes* was identified on freshly cut spinach leaves in ambient and modified atmosphere (low O_2, high CO_2) over 14 days at 10 °C storage [28]. Interestingly, under ambient atmosphere (filled with atmospheric air), cfu g^{-1} values dropped at day 7, but recovered subsequently to around 10^6 cfu g^{-1} by day 10. The high starting inoculation density may have played a major role in the more moderate increase in cfu g^{-1} when compared with the study from Omac and colleagues [27]. The findings from Ziegler and colleagues [15] seem to support this hypothesis. They investigated the growth potential of *L. monocytogenes* on rocket (Arugula) at initial inoculation densities of 5.4 $\log_{10}$ cfu g^{-1} under environmental conditions similar to the present study. While the authors reported some moderate growth to 5.9 $\log_{10}$ cfu g^{-1}, this was reported to be not significant.

The determination of growth potentials may have been systematically underestimated in the past and the use of high inoculation densities may have contributed to the contradicting findings of growth of *L. monocytogenes* on spinach and rocket. The present study followed the inoculation density recommendations in ANSES EURL *Lm* technical guidance document for conducting shelf-life studies on *L. monocytogenes* in RTE foods for determining growth potentials in challenge tests [29]. Recently, McManamon and colleagues [14] demonstrated the ability of lower *L. monocytogenes* contamination levels (100 cfu g^{-1}) to have higher growth potentials on iceberg lettuce when compared with higher initial densities (10^4 and 10^5 cfu g^{-1}) at 4 °C and 8 °C. They suggested that, when *L. monocytogenes* reaches higher cell densities (e.g., 10^6 cfu g^{-1}), intra-species competition plays a greater role at limiting growth. This finding could explain why the present study found higher growth potentials on spinach and rocket than the previous studies mentioned above.

Inoculation densities are not the only factor affecting growth of *L. monocytogenes.* According to Beaufort and colleagues [29], pH and water activity are important determinants of *L. monocytogenes* growth; they will not grow when food products have a pH $\leq$ 4.4 or a water activity value $\leq$ 0.920 or a combination of pH $\leq$ 5.0 and water activity $\leq$ 0.940. In this study, no growth inhibition was expected based on pH and water activity throughout the duration of the experiments for all leafy RTE vegetables tested.

In the present study, spinach had the highest average growth potential, while rocket had a considerably lower growth potential, similar to that of iceberg lettuce. Sant'Ana and colleagues [21] also tested spinach and rocket (among other RTE vegetables) for growth potential of *L. monocytogenes* at 7 and 15 °C. As in the present study, semi-permeable sealed bags were used with a comparable modified atmosphere and the inoculation density was 1000 cfu g^{-1}. In their study, rocket had a significantly higher growth potential (1.86 $\log_{10}$ cfu g^{-1}, 7 °C, day 6) compared with spinach (0.88 $\log_{10}$ cfu g^{-1}, same conditions). As the differences in results could not be explained by storage conditions, the difference in *L. monocytogenes* strains used and the origin of the produce may have played an important role. In both cases, the leafy vegetables were obtained from local supermarkets, which only revealed the country of origin, and variety or the environmental conditions during cultivation were not revealed. Nevertheless, spinach and rocket came from EU farms in the present study, while Sant'Ana and colleagues [21] received their produce from Brazil. Therefore, one can expect that variety and farming conditions were substantially different.

Growth of RTE vegetables in open fields has been linked to risks of microbial contamination [30]. Research has shown that handling procedures during the harvest greatly influence the presence of food-borne pathogens such as *L. monocytogenes* [31]; thus, environmental abiotic and biotic factors, as well agricultural pre-harvest practices may affect growth of pathogens on leafy vegetables during storage. For example, for products/plants grown in greenhouses or poly-tunnels, controlling the relative humidity (by limiting spells of prolonged high humidity) can serve as an intervention for decreasing *L. monocytogenes* incidences/populations [32]. While the present data clearly support the growth potential of *L. monocytogenes* on spinach and rocket, there is a need to take the pre-harvest environmental conditions as well as the harvest itself into consideration when it comes to identifying growth behavior of *L. monocytogenes* on leafy RTE vegetables in the future.

Natural background microbiota on baby spinach leaves have been reported to potentially affect the growth of *L. monocytogenes* and *Listeria innocua*, although any differences detected were not statistically significant [33]. A general feedback called the 'Jameson effect' may be responsible for the limitation of growth that includes *L. monocytogenes* owing to competition for resources when microbial populations are high in numbers. In the present study, total bacteria counts were already high, ranging from 5.94 logs to 7.11 logs cfu g^{-1} for all three tested products, and may have limited growth of *L. monocytogenes* towards day 9. Fitting the growth curve of *L. monocytogenes* on spinach and rocket over 9 days to a linear and sigmoidal function revealed similar relative and absolute measures of fit. This suggests that a slowdown of growth may have just started by day 9. Future experiments with extended storage beyond 9 days might be able to demonstrate this Jameson effect. Bacteria counts of 10^7 cfu g^{-1} are not uncommon on leafy vegetables. Valentin-Bon and colleagues [34] carried out microbiological counts on both conventional and organic types of spinach (7.7 and 7.2 log_{10} cfu g^{-1}) and iceberg lettuce (7.0, 7.3 log_{10} cfu g^{-1}), respectively. Likewise, Allende and colleagues [35] found an initial microbial load on baby spinach leaves at 7.2 ± 0.1 log_{10} cfu g^{-1} that increased to 9 logs within 12 days.

Recent work on iceberg lettuce identified a visual degradation of the vegetable that was considered unsuitable for consumption within 5–7 days of storage [24]. Inversely, when the current study identified counts of *L. monocytogenes* on spinach that were higher than on iceberg lettuce, the visual appearance of the spinach decreased less and was still considered to be acceptable by day 9. This is potentially dangerous as, judging from appearance, consumers would likely underestimate the potential risk of high-level contamination with *L. monocytogenes*. In a related study, five panelists (trained in scoring quality attributes) assessed the quality deterioration of commercially packaged baby spinach stored at 8 °C (without contamination) and deemed the product acceptable until day 8 [36]. Fortunately for rocket, its visual quality decreased further than that of spinach in the present study, while growth of *L. monocytogenes* was at unacceptable levels, thus appearance may potentially deter consumers from eating contaminated rocket. Chlorophyll degradation has been identified as the reason for the limited shelf life of rocket [37].

5. Conclusions

In conclusion, the present study has confirmed that rocket and especially spinach support the growth of *L. monocytogenes*, with the latter showing very little visual deterioration; therefore, contaminated spinach may pose a serious health risk to consumers. Furthermore, this study identified a range of environmental factors that could explain why many other studies found contradicting evidence of growth of *L. monocytogenes* on rocket and spinach. Indeed, preliminary tests by the authors suggest that rocket cultivated in tunnel or open field influences the natural microbiome of the vegetable, which in turn putatively affects the growth rate of *L. monocytogenes* (data not shown). Therefore, the influence of different varieties of spinach and rocket, soil and climatic conditions, the development of the natural microbiome, and product washing has to be considered in future studies to evaluate the growth potential of *L. monocytogenes* on leafy RTE vegetables in greater detail.

Supplementary Materials: The following are available online at http://www.mdpi.com/2304-8158/9/9/1211/s1, Figure S1: Images of rocket (day 0, 100 cfu g^{-1}, top left), rocket (day 9, 100 cfu g^{-1}, top right), spinach (day 0, 100 cfu g^{-1} bottom left) and spinach (day 9, 100 cfu g^{-1}, bottom right) for visual appearance analysis. Table S1: ComBase Data based on data over 7 days, Table S2: ComBase Data based on data over 9 days, Table S3: Total heterotrophic bacteria in ready-to-eat salad vegetables [log CFU g^{-1}] ± standard deviations, Table S4: Water Activity and pH values ± standard deviations, Table S5: Oxygen concentrations (%).

Author Contributions: A.S. was responsible for securing funding for the study and designed the study; A.S. shared the writing of the manuscript with P.C. and led the review and editing process; P.C. carried out the experiments, analyzed the data, and was instrumental in writing the first draft and contributed to editing and review. All authors have read and agreed to the published version of the manuscript.

Funding: This study was funded by the Department of Agriculture, Food, and the Marine (Ireland), grant number 17F/244.

Acknowledgments: We are grateful to Lisa O'Connor and Mary Lenahan (Food Safety Authority Ireland) for providing feedback.

Conflicts of Interest: The authors have no conflicts of interest to declare. The funders had no role in the design of the study; in the collection, analyses, or interpretation of data; in the writing of the manuscript; or in the decision to publish the results.

Appendix A

A predictive microbiological software called ComBase was used to predict the maximum growth rate of *L. monocytogenes* in each experiment by plotting and fitting *L. monocytogenes* data to linear models (Tables S1 and S2) and the Baranyi and Roberts model (Table S2). The model of Baranyi and Roberts (1994) [38] describes a sigmoid bacterial curve (lag phase, acceleration phase, exponential phase, deceleration phase), whereas the linear model describes where the bacterial counts describe only the growth/death phase [38]. The potential of these models to accurately predict the growth of *L. monocytogenes* was evaluated using coefficient of determination (R^2) and RMSE (root mean squared error). R^2 statistic (also referred to as coefficient of determination) in predictive microbiology is a measure of the goodness-of-the-fit (value of 1 equates the best fit) [39]. RMSE is the difference between observed and the predicted data. Therefore, an RMSE close to 0 is the ideal value as it implies that predicted and observed data were very close. These factors are calculated as follows [40]:

$$R^2 = 1 - \sqrt{\frac{sum\ of\ squares\ of\ residuals}{total\ sum\ of\ squares}}$$

$$RMSE = \sqrt{\frac{sum\ of\ square\ of\ errors}{number\ of\ observations - number\ of\ model\ parameters}}$$

References

1. FAO (FAOSTAT—Food and Agriculture Organization of the United Nations), 2017. Available online: http://www.fao.org/faostat/en/#data/QC (accessed on 19 August 2019).
2. Mir, S.; Shah, M.; Mir, M.; Dar, B.; Greiner, R.; Roohinejad, S. Microbiological contamination of ready-to-eat vegetable salads in developing countries and potential solutions in the supply chain to control microbial pathogens. *Food Control* **2018**, *85*, 235–244. [CrossRef]
3. Santos, J.; Herrero, M.; Mendiola, J.; Oliva-Teles, M.; Ibáñez, E.; Delerue-Matos, C.; Oliveira, M. Assessment of nutritional and metabolic profiles of pea shoots: The new ready-to-eat baby-leaf vegetable. *Food Res. Int.* **2014**, *58*, 105–111. [CrossRef]
4. Eurostat 2017. Available online: https://ec.europa.eu/eurostat/web/agriculture/data/database (accessed on 21 August 2019).
5. González-Tejedor, G.; Garre, A.; Esnoz, A.; Artés-Hernández, F.; Fernández, P. Effect of storage conditions in the response of *Listeria monocytogenes* in a fresh purple vegetable smoothie compared with an acidified TSB medium. *Food Microbiol.* **2018**, *72*, 98–105. [CrossRef] [PubMed]
6. Thienhirun, S.; Chung, S. Consumer attitudes and preferences toward cross-cultural ready-to-eat (RTE) food. *J. Food Prod. Mark.* **2017**, *24*, 56–79. [CrossRef]
7. Martínez-Sánchez, A.; Luna, M.; Selma, M.; Tudela, J.; Abad, J.; Gil, M. Baby-leaf and multi-leaf of green and red lettuces are suitable raw materials for the fresh-cut industry. *Postharvest Biol. Technol.* **2012**, *63*, 1–10. [CrossRef]
8. Critzer, F.; Doyle, M. Microbial ecology of foodborne pathogens associated with produce. *Curr. Opin. Biotechnol.* **2010**, *21*, 125–130. [CrossRef]
9. EFSA and ECDC—European Food Safety Authority, and European Centre for Disease Prevention and Control. The European Union One Health 2018 Zoonoses Report. *EFSA J.* **2019**, *17*. [CrossRef]
10. Montero, D.; Bodero, M.; Riveros, G.; Lapierre, L.; Gaggero, A.; Vidal, R.; Vidal, M. Molecular epidemiology and genetic diversity of *Listeria monocytogenes* isolates from a wide variety of ready-to-eat foods and their relationship to clinical strains from listeriosis outbreaks in Chile. *Front. Microbiol.* **2015**, *6*, 384. [CrossRef]
11. Radoshevich, L.; Cossart, P. *Listeria monocytogenes*: Towards a complete picture of its physiology and pathogenesis. *Nat. Rev. Microbiol.* **2017**, *16*, 32–46. [CrossRef]
12. Pereira, S.; Alves, Â.; Ferreira, V.; Teixeira, P. The Impact of Environmental Stresses in the Virulence Traits of *Listeria monocytogenes* Relevant to Food Safety. *List. Monocytogenes* **2018**. [CrossRef]

13. Francis, G.; O'Beirne, D. Effects of vegetable type, package atmosphere and storage temperature on growth and survival of *Escherichia coli O157:H7* and *Listeria monocytogenes*. *J. Ind. Microbiol. Biotechnol.* **2001**, *27*, 111–116. [CrossRef] [PubMed]
14. McManamon, O.; Scollard, J.; Schmalenberger, A. Inoculation density is affecting growth conditions of *Listeria monocytogenes* on fresh cut lettuce. *World J. Microbiol. Biotechnol.* **2017**, *33*, 217. [CrossRef] [PubMed]
15. Ziegler, M.; Kent, D.; Stephan, R.; Guldimann, C. Growth potential of *Listeria monocytogenes* in twelve different types of RTE salads: Impact of food matrix, storage temperature and storage time. *Int. J. Food Microbiol.* **2019**, *296*, 83–92. [CrossRef] [PubMed]
16. EURL—European Union Reference Laboratory 2019. *Lm* Technical Guidance Document for Conducting Shelf-Life Studies on *Listeria monocytogenes* in Ready-to-Eat Foods. Available online: https://ec.europa.eu/food/sites/food/files/safety/docs/biosafety_fh_mc_tech-guide-doc_listeria-in-rte-foods_en.pdf (accessed on 15 January 2020).
17. Ziegler, M.; Rüegg, S.; Stephan, R.; Guldimann, C. Growth potential of *Listeria monocytogenes* in six different RTE fruit products: Impact of food matrix, storage temperature and shelf life. *Ital. J. Food Saf.* **2018**, *7*, 7581. [CrossRef]
18. Denis, N.; Zhang, H.; Leroux, A.; Trudel, R.; Bietlot, H. Prevalence and trends of bacterial contamination in fresh fruits and vegetables sold at retail in Canada. *Food Control* **2016**, *67*, 225–234. [CrossRef]
19. Cordano, A.; Jacquet, C. *Listeria monocytogenes* isolated from vegetable salads sold at supermarkets in Santiago, Chile: Prevalence and strain characterization. *Int. J. Food Microbiol.* **2009**, *132*, 176–179. [CrossRef]
20. Kramarenko, T.; Roasto, M.; Meremäe, K.; Kuningas, M.; Põltsama, P.; Elias, T. *Listeria monocytogenes* prevalence and serotype diversity in various foods. *Food Control* **2013**, *30*, 24–29. [CrossRef]
21. Sant'Ana, A.; Barbosa, M.; Destro, M.; Landgraf, M.; Franco, B. Growth potential of *Salmonella* spp. and *Listeria monocytogenes* in nine types of ready-to-eat vegetables stored at variable temperature conditions during shelf-life. *Int. J. Food Microbiol.* **2012**, *157*, 52–58. [CrossRef]
22. Lokerse, R.; Maslowska-Corker, K.; van de Wardt, L.; Wijtzes, T. Growth capacity of *Listeria monocytogenes* in ingredients of ready-to-eat salads. *Food Control* **2016**, *60*, 338–345. [CrossRef]
23. Söderqvist, K.; Lambertz, S.; Vågsholm, I.; Fernström, L.; Alsanius, B.; Mogren, L.; Boqvist, S. Fate of *Listeria monocytogenes*, pathogenic *Yersinia enterocolitica*, and *Escherichia coli O157:H7*gfp+in ready-to-eat salad during cold storage: What is the risk to consumers? *J. Food Prot.* **2017**, *80*, 204–212. [CrossRef]
24. McManamon, O.; Kaupper, T.; Scollard, J.; Schmalenberger, A. Nisin application delays growth of *Listeria monocytogenes* on fresh-cut iceberg lettuce in modified atmosphere packaging, while the bacterial community structure changes within one week of storage. *Postharvest Biol. Technol.* **2019**, *147*, 185–195. [CrossRef]
25. Scollard, J.; McManamon, O.; Schmalenberger, A. Inhibition of *Listeria monocytogenes* growth on fresh-cut produce with thyme essential oil and essential oil compound verbenone. *Postharvest Biol. Technol.* **2016**, *120*, 61–68. [CrossRef]
26. Hunt, K.; Blanc, M.; Álvarez-Ordóñez, A.; Jordan, K. Challenge studies to determine the ability of foods to support the growth of *Listeria monocytogenes*. *Pathogens* **2018**, *7*, 80. [CrossRef]
27. Omac, B.; Moreira, R.; Castell-Perez, E. Quantifying growth of cold-adapted *Listeria monocytogenes* and *Listeria innocua* on fresh spinach leaves at refrigeration temperatures. *J. Food Eng.* **2018**, *224*, 17–26. [CrossRef]
28. Boyacioglu, O.; Sulakvelidze, A.; Sharma, M.; Goktepe, I. Effect of a bacteriophage cocktail in combination with modified atmosphere packaging in controlling *Listeria monocytogenes* on fresh-cut spinach. *Ir. J. Agric. Food Res.* **2016**, *55*, 74–79. [CrossRef]
29. Beaufort, A.; Bergis, H.; Lardeux, A.-L.; Lombard, B.; Polet, M.; Botteldoorn, N.; Papageorgiou, G.; Andersen, J.K.; Boel, J.; Hickey, B.; et al. *Technical Guidance Document for Conducting Shelf-Life Studies on Listeria monocytogenes in Ready-to-Eat Foods*; EURL *Lm* Technical Guidance Document; European Union Reference Laboratory for Listeria Monocytogenes. 2019; pp. 1–47. Available online: https://eurl-listeria.anses.fr/en/system/files/LIS-Cr-201909D2.pdf (accessed on 10 January 2020).
30. Castro-Ibáñez, I.; Gil, M.; Allende, A. Ready-to-eat vegetables: Current problems and potential solutions to reduce microbial risk in the production chain. *LWT Food Sci Technol* **2017**, *85*, 284–292. [CrossRef]
31. Miceli, A.; Settanni, L. Influence of agronomic practices and pre-harvest conditions on the attachment and development of *Listeria monocytogenes* in vegetables. *Ann. Microbiol.* **2019**, *69*, 185–199. [CrossRef]

32. Bardsley, C.; Boyer, R.; Rideout, S.; Strawn, L. Survival of *Listeria monocytogenes* on the surface of basil, cilantro, dill, and parsley plants. *Food Control* **2019**, *95*, 90–94. [CrossRef]
33. Omac, B.; Moreira, R.; Castillo, A.; Castell-Perez, E. Growth of *Listeria monocytogenes* and *Listeria innocua* on fresh baby spinach leaves: Effect of storage temperature and natural microflora. *Postharvest Biol. Technol.* **2015**, *100*, 41–51. [CrossRef]
34. Valentin-Bon, I.; Jacobson, A.; Monday, S.; Feng, P. Microbiological quality of bagged cut spinach and lettuce mixes. *Appl. Environ. Microbiol.* **2007**, *74*, 1240–1242. [CrossRef]
35. Allende, A.; Luo, Y.; McEvoy, J.; Artés, F.; Wang, C. Microbial and quality changes in minimally processed baby spinach leaves stored under super atmospheric oxygen and modified atmosphere conditions. *Postharvest Biol. Technol.* **2004**, *33*, 51–59. [CrossRef]
36. Kou, L.; Luo, Y.; Park, E.; Turner, E.; Barczak, A.; Jurick, W. Temperature abuse timing affects the rate of quality deterioration of commercially packaged ready-to-eat baby spinach. Part I: Sensory analysis and selected quality attributes. *Postharvest Biol. Technol.* **2014**, *91*, 96–103. [CrossRef]
37. Koukounaras, A.; Siomos, A.; Sfakiotakis, E. Postharvest CO2 and ethylene production and quality of rocket (*Eruca sativa* Mill.) leaves as affected by leaf age and storage temperature. *Postharvest Biol. Technol.* **2007**, *46*, 167–173. [CrossRef]
38. Baranyi, J.; Roberts, T. A dynamic approach to predicting bacterial growth in food. *Int. J. Food Microbiol.* **1994**, *23*, 277–294. [CrossRef]
39. Pla, M.; Oltra, S.; Esteban, M.; Andreu, S.; Palop, A. Comparison of Primary Models to Predict Microbial Growth by the Plate Count and Absorbance Methods. *BioMed Res. Int.* **2015**, *2015*, 365025. [CrossRef] [PubMed]
40. Guzel, M.; Moreira, R.; Omac, B.; Castell-Perez, M. Quantifying the effectiveness of washing treatments on the microbial quality of fresh-cut romaine lettuce and cantaloupe. *LWT* **2017**, *86*, 270–276. [CrossRef]

MDPI
St. Alban-Anlage 66
4052 Basel
Switzerland
Tel. +41 61 683 77 34
Fax +41 61 302 89 18
www.mdpi.com

Foods Editorial Office
E-mail: foods@mdpi.com
www.mdpi.com/journal/foods

www.ingramcontent.com/pod-product-compliance
Lightning Source LLC
Chambersburg PA
CBHW040803250926
51850CB00007BA/1068
* 9 7 8 3 0 3 6 5 4 4 3 2 8 *